Springer Series in Statistics

Springer Series in Statistics

Alho/Spencer: Statistical Demography and Forecasting
Andersen/Borgan/Gill/Keiding: Statistical Models Based on Counting Processes
Atkinson/Riani: Robust Diagnostic Regression Analysis
Atkinson/Riani/Ceriloi: Exploring Multivariate Data with the Forward Search
Berger: Statistical Decision Theory and Bayesian Analysis, 2nd edition
Borg/Groenen: Modern Multidimensional Scaling: Theory and Applications, 2nd edition
Brockwell/Davis: Time Series: Theory and Methods, 2nd edition
Bucklew: Introduction to Rare Event Simulation
Cappé/Moulines/Rydén: Inference in Hidden Markov Models
Chan/Tong: Chaos: A Statistical Perspective
Chen/Shao/Ibrahim: Monte Carlo Methods in Bayesian Computation
Coles: An Introduction to Statistical Modeling of Extreme Values
Devroye/Lugosi: Combinatorial Methods in Density Estimation
Diggle/Ribeiro: Model-based Geostatistics
Dudoit/ Van der Laan: Multiple Testing Procedures with Applications to Genomics
Efromovich: Nonparametric Curve Estimation: Methods, Theory, and Applications
Eggermont/LaRiccia: Maximum Penalized Likelihood Estimation, Volume I: Density Estimation
Fahrmeir/Tutz: Multivariate Statistical Modeling Based on Generalized Linear Models, 2nd edition
Fan/Yao: Nonlinear Time Series: Nonparametric and Parametric Methods
Ferraty/Vieu: Nonparametric Functional Data Analysis: Theory and Practice
Ferreira/Lee: Multiscale Modeling: A Bayesian Perspective
Fienberg/Hoaglin: Selected Papers of Frederick Mosteller
Frühwirth-Schnatter: Finite Mixture and Markov Switching Models
Ghosh/Ramamoorthi: Bayesian Nonparametrics
Glaz/Naus/Wallenstein: Scan Statistics
Good: Permutation Tests: Parametric and Bootstrap Tests of Hypotheses, 3rd edition
Gouriéroux: ARCH Models and Financial Applications
Gu: Smoothing Spline ANOVA Models
Györfi/Kohler/Krzyżak/Walk: A Distribution-Free Theory of Nonparametric Regression
Haberman: Advanced Statistics, Volume I: Description of Populations
Hall: The Bootstrap and Edgeworth Expansion
Härdle: Smoothing Techniques: With Implementation in S
Harrell: Regression Modeling Strategies: With Applications to Linear Models, Logistic Regression, and Survival Analysis
Hart: Nonparametric Smoothing and Lack-of-Fit Tests
Hastie/Tibshirani/Friedman: The Elements of Statistical Learning: Data Mining, Inference, and Prediction
Hedayat/Sloane/Stufken: Orthogonal Arrays: Theory and Applications
Heyde: Quasi-Likelihood and its Application: A General Approach to Optimal Parameter Estimation
Huet/Bouvier/Poursat/Jolivet: Statistical Tools for Nonlinear Regression: A Practical Guide with S-PLUS and R Examples, 2nd edition
Ibrahim/Chen/Sinha: Bayesian Survival Analysis
Jiang: Linear and Generalized Linear Mixed Models and Their Applications
Jolliffe: Principal Component Analysis, 2nd edition
Knottnerus: Sample Survey Theory: Some Pythagorean Perspectives

(continued after index)

Peter X.-K. Song

Correlated Data Analysis: Modeling, Analytics, and Applications

Springer

Peter X.-K. Song
Department of Statistics and Actuarial Science
University of Waterloo
200 University Avenue West
Waterloo, Ontario, Canada N2L 3G1
song@uwaterloo.ca

Library of Congress Control Number: 2007929730

ISBN 978-0-387-71392-2 e-ISBN 978-0-387-71393-9

Printed on acid-free paper.

9 8 7 6 5 4 3 2 1

springer.com

To Ru and Angela

Preface

This book, like many other books, was delivered under tremendous inspiration and encouragement from my teachers, research collaborators, and students. My interest in longitudinal data analysis began with a short course taught jointly by K.Y. Liang and S.L. Zeger at the Statistical Society of Canada Conference in Acadia University, Nova Scotia, in the spring of 1993. At that time, I was a first-year PhD student in the Department of Statistics at the University of British Columbia, and was eagerly seeking potential topics for my PhD dissertation. It was my curiosity (driven largely by my terrible confusion) with the generalized estimating equations (GEEs) introduced in the short course that attracted me to the field of correlated data analysis. I hope that my experience in learning about it has enabled me to make this book an enjoyable intellectual journey for new researchers entering the field. Thus, the book aims at graduate students and methodology researchers in statistics or biostatistics who are interested in learning the theory and methods of correlated data analysis.

I have attempted to give a systematic account of regression models and their applications to the modeling and analysis of correlated data. Longitudinal data, as an important type of correlated data, has been used as a main venue for motivation, methodological development, and illustration throughout the book. Given the many applied books on longitudinal data analysis already available, this book is inclined more towards technical details regarding the underlying theory and methodology used in software-based applications. I hope the book will serve as a useful reference for those who want theoretical explanations to puzzles arising from data analyses or deeper understanding of underlying theory related to analyses. This book has evolved from lecture notes on longitudinal data analysis, and may be considered suitable as a textbook for a graduate course on correlated data analysis.

This book emphasizes some recent developments in correlated data analysis.

First, it takes the perspective of Jørgensen's theory of dispersion models for the discussion of generalized linear models (GLMs) in Chapter 2. It is known that the class of generalized linear models plays a central role in the regression analysis of nonnormal data. In the context of correlated data analysis, these models constitute marginal components in a joint model formulation. One benefit from such a treatment is that it enables this book to cover a broader range of data types than the traditional GLMs. Two types that are of particular interest and discussed in detail in the book are compositional (or continuous proportional) data and directional (or circular) data.

Second, it gives a systematic treatment for the theory of inference functions (or estimating functions) in Chapter 3. The popular GEE methods presented in Chapter 5 are then easily introduced and studied as a special class of inference functions. Building upon Chapter 3, some alternative estimating function methods can be readily discussed. Recent work on quadratic inference functions (QIF) is an example that benefits from Chapter 3.

Third, it presents a joint modeling approach to regression analysis of correlated data via the technique of parametric copulas. Copulas are becoming increasingly popular in the analysis of correlated data, and Chapter 6 focuses on Gaussian copulas, for which both theory and numerical examples are illustrated.

Fourth, it deals with state space models for longitudinal data from long time series. In contrast to longitudinal data from short time series, modeling stochastic patterns or transitional behaviors becomes a primary task. In such a setting, asymptotics may be established by letting the length of the time series tend to ∞, as opposed to letting the number of subjects tend to ∞, as in the case of data consisting of many short time series. Chapters 10, 11, and 12 are devoted to this topic.

Fifth, this book covers two kinds of statistical inferences in generalized linear mixed effects models (GLMMs): maximum likelihood inference in Chapter 7 and Bayesian inference based on Markov Chain Monte Carlo (MCMC) in Chapter 8. In Chapter 8, the analysis of multi-level data is also discussed in the framework of hierarchical models. Inference can be dealt with easily by the MCMC method, as an extension from the GLMMs with little extra technical difficulty.

The book contains some other topics that are highly relevant to the analysis of correlated data. For example, Chapter 13 concerns missing data problems arising particularly from longitudinal data.

The presentation of some material in the book is a little technical in order to achieve rigor of exposition. Readers' backgrounds should include mathematical statistics, generalized linear models, and some knowledge of statistical computing, such as represented R and SAS software. The following chart displays the relationship among the thirteen chapters, and readers can follow a particular path to reach a topic of interest.

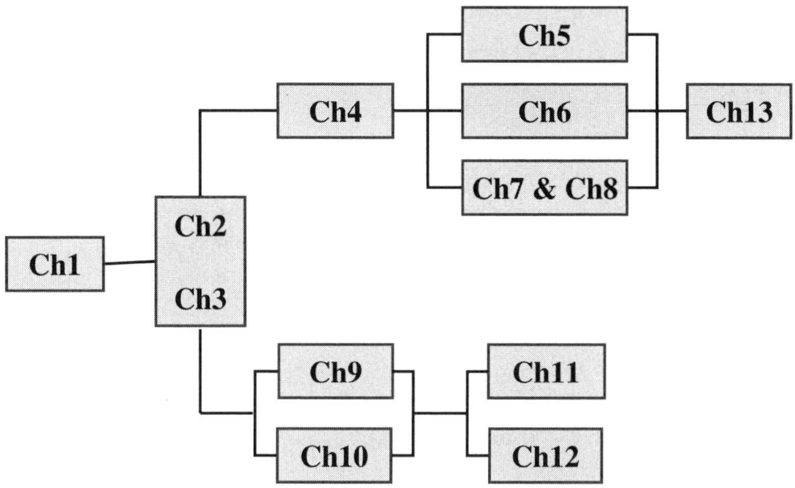

A webpage has been created to provide some supplementary material for the book. The URL address is

`http://www.stats.uwaterloo.ca/~song/BOOKLDA.html`

All data sets used in the book are available. A SAS Macro QIF is available for a secured download; that is, an interested user needs to submit an online request for permission in order to download this software package. In addition, some figures that are printed in reduced size in the book are supplied in their full sizes. Exercise problems for some of the thirteen chapters are posted, which may be useful when the book is used as a text for a course.

I would like to acknowledge my debt to many people who have helped me to prepare the book. I was fortunate to begin my research in this field under the supervision of Bent Jørgensen, who taught me his beautiful theory of dispersion models. At UBC, I learned the theory of copulas from Harry Joe. This book has benefited from some of the PhD theses that I supervised in the past ten years or so, including Zhenguo (Winston) Qiu, Dingan Feng, Baifang Xing, and Peng Zhang, as well as from a few data analysis projects that graduate students did in my longitudinal data analysis course; thanks go to Eric Bingshu Chen, Wenyu Jiang, David Tolusso, and Wanhua Su. Many graduate students in my course pointed out errors in an early draft of the book. Qian Zhou helped me to draw some figures in the book, and Zichang Jiang worked with me to develop SAS MACRO QIF, which is a software package to fit marginal models for correlated data.

I am very grateful to my research collaborators for their constant inspiration and valuable discussions on almost every topic presented in the book. My great appreciation goes to Annie Qu, Jack Kalbfleisch, Ming Tan, Claudia Czado, Søren Lundbye-Christensen, Jianguo (Tony) Sun, and Mingyao Li. I would also like to express my sincere gratitude to people who generously provided and allowed me to analyze their datasets in the book, including John

Petkau and Angela D'Elia. Zhenguo Qiu, Grace Yi, and Jerry Lawless provided with me their valuable comments on drafts of the book. My research in the field of correlated data analysis has been constantly supported by grants from the Natural Sciences and Engineering Research Council of Canada. I thank John Kimmel and Frank Ganz from Springer for their patience and editorial assistance.

I take full responsibility for all errors and omissions in the book. Finally, I would like to say that given the vast amount of published material in the field of correlated data analysis, the criterion that I adopted for the selection of topics for the book was really my own familiarity. Because of this and space limitations, some worthwhile topics have no doubt been excluded. Research in this field remains very active with many new developments. I would be grateful to readers for their critical comments and suggestions for improvement, as well as corrections.

Waterloo, Ontario, Canada *P.X.-K. Song*
December 2006

Contents

1

Introduction and Examples

1.1 Correlated Data

Regression analysis of correlated data is undertaken in many practical areas. In this book, *correlated data* refers to a collection of multi-dimensional measurements with correlated response variables. Depending on the setting from which the data is collected, the nature of correlation among multiple outcomes can differ from one case to another. Thus, in the literature correlated data are classified into different types, such as longitudinal data, clustered data, spatial data, and multi-level data. In spite of certain specific features attached with each data type, in general correlated data share many commonalities, which is the rationale that it is possible to develop statistical modeling and inference within one framework. This book will utilize longitudinal data as a main venue to illustrate the theory and methods in the analysis of correlated data, with supplementary discussions about analyzing other data types whenever applicable.

Longitudinal data is a data type frequently encountered in many subject-matter areas such as biology, medical and public health sciences, and social sciences. Sequentially observed over time, longitudinal data may be collected either from an observational study or a designed experiment, in which response variables pertain to a sequence of events or outcomes recorded at certain time points during a study period. In essence, longitudinal data may be regarded as a collection of many time series, each for one subject.

Clustered data refers to a set of measurements collected from subjects that are structured in clusters, where a group of related subjects constitutes a cluster, such as a group of genetically related members from a familial pedigree. Obviously, settings where clustered data arise can be independent of time. It is interesting to note that sometimes longitudinal data may be thought of as a special kind of clustered data by treating a subject as a cluster, so each subject's time series forms a set of correlated observations. This perspective is mainly for technical convenience, because, technically, similar tools can be applied to analyze longitudinal data or clustered data with, however, possibly different modeling of dependence structures. In longitudinal data analysis, serial correlation is commonly assumed, while in clustered data analysis equal pairwise within-cluster correlation is popular. On this line, when clusters are represented by spatial groups, such as geographic regions, *spatial data* may also be treated as a special case of clustered data. Consequently, in such a case modeling spatial correlation becomes an essential task.

In many biomedical studies, design or sampling protocols play crucial roles in the data collection. As far as design or sampling protocol concerns, longitudinal and clustered data collection procedures are fundamentally different, and therefore it is important to distinguish whether data are collected from a longitudinal study or from a clustered study. Factors that are administrated and investigated in a longitudinal study design can be very different from those considered in a clustered design. In contrast, because of the similarity in the methodological development, it seems convenient to include the two study designs in one framework. Given this circumstance, the term of *repeated measurements* becomes useful to denote either longitudinal data or clustered data. A study protocol that combines both clustered and time-course features (e.g., familial data measured over time) gives rise to a more complex data structure. Data collected from such multi-dimensional hierarchies are referred to as *multi-level data*. An interesting type of multi-level data is *spatio-temporal* data that comprise of repeated measurements recorded jointly over time and across spatial locations.

Data with multiple outcomes or simply *vector data* refers to a dataset in that a vector of response variables is measured for each of many subjects. Comparing to longitudinal or clustered data, vector data is constrained with the equal dimension of the response vector, but it is more flexible to allow the components of the response vector to follow different marginal distributions. Examples of vector data include clustered data with an equal cluster size, longitudinal data of time series of equal length, and spatial data collected from an equal number of spatial locations. Moreover, multi-dimensional data of mixed types is another example of the vector data.

1.2 Longitudinal Data Analysis

The primary interest of longitudinal data analysis lies in the mechanism of change over time, including growth, aging, time profiles or effects of covariates. Some main advantages of a longitudinal study are listed as follows.

(1) It allows researchers to investigate how the variability of the response varies in time with covariates. For the instance of a clinical trial that presumably aims to investigate the effectiveness of a new drug treating a disease, it is often of interest to examine the pharmacokinetic behavior of the drug when it is applied to experimental animals or patients. Most drugs do not have constant efficacy over time, possibly due to drug resistance. Such time-varying treatment effectiveness can be examined through a longitudinal

study in which responses to the drug treatment are monitored over time. Obviously, it is hard or impossible to study such a time-dependent behavior via a cross-sectional study.

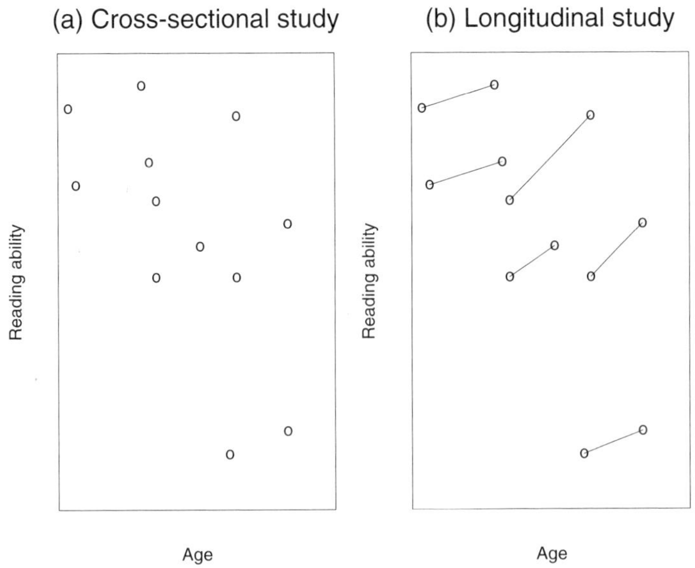

Fig. 1.1. Scatterplots of hypothetical data for reading ability and age.

(2) It enables researchers to separate the so-called *cohort* and *age* (or time) effects; Diggle et al. (2002) presents a hypothetical example of reading ability versus age, which clearly illustrates this point. Figure 1.1 displays two scatterplots of hypothetical measurements of reading ability against age. Panel (a) considers a cross-sectional analysis where all data points are treated as drawn independently by different individuals. It is easy to see from this plot a trend of deterioration of reading ability in age. In contrast, when a pair of reading ability measurements from each individual is linked by a line (namely, the data being plotted in the form of time series) in order to reflect the longitudinal nature, panel (b) tells an opposite story to that of the cross-sectional study. That is, each individual improves his reading ability when he grows older. From this example, we learn that: *(a)* this contrary indicates the importance of analyzing longitudinal data based on individual time series trajectories, and it could produce misleading results if longitudinal data were modeled and analyzed as of cross-sectional data; *(b)* a longitudinal study can characterize changes over time within individuals (e.g., age effect) from differences among people in the reference

to their baseline status (or cohort effects), but a cross-sectional study cannot.

(3) It helps the recruitment of subjects–collecting repeated outcomes from one subject may help to reduce the burden of recruiting a sizable number of subjects required for a cross-sectional study. This is sometimes prohibited. For instance, in studies of rare diseases, the number of available patients in the population is typically insufficient for simple randomized trials. One solution to this difficulty is the so-called *cross-over clinical trial* where subjects serve as their own controls. In effect, a cross-over trial administrates each patient with active drug and placebo at two separate time periods.

On the other hand, comparing to cross-sectional studies, some challenges of a longitudinal study include:

(1) Analyzing longitudinal data becomes technically more demanding, due to the complexity of underlying probability mechanisms of data generation. In most cases, the maximum likelihood inference is either unavailable or numerically too intricate to be implemented. One of the popular compromises towards this difficulty is the generalized estimating equations (GEE) approach proposed by Liang and Zeger (1986), which does not require to specify a complete probability model for data analysis. In fact, GEE method is a quasi-likelihood inference, which only requires to correctly specify the first two moments of the underlying distribution of data and treats the correlation as nuisance parameters (not modeled) in the data analysis.

(2) It is more difficult to deal with missing data in longitudinal studies. This is because missing data patterns appear much more sophisticated than those in cross-sectional studies. For instance, in cross-sectional studies, each individual contributes one data point, and when a data point is missing the corresponding individual might be deleted from the analysis. This is not the case in longitudinal studies; a data point missing at a time point does not imply that the corresponding individual is completely noninformative, because it possibly has measurements recorded at some other time points. It is not a trivial issue to properly analyze repeated measurements in the presence of missing values, with the constraint of preserving the same correlation structure as that being completely observed.

(3) When the length of time series is not short, modeling stochastic patterns or transitional behaviors of longitudinal data becomes a primary task. In this case, the development of statistical inference is more challenging since the correlation structure is no longer a nuisance.

To facilitate further discussions, let us first introduce some necessary notations. Denote a longitudinal data by

$$(y_{ij}, \mathbf{x}_{ij}, t_{ij}), \ j = 1, \ldots, n_i, i = 1, \ldots, K,$$

where the response variable y_{ij} is observed at time t_{ij}. When data are observed at equally spaced time points, t_{ij} may be simplified as t. This book always assumes that time series across different subjects are statistically independent; that is, vectors of repeated outcomes

$$\mathbf{y}_i = (y_{i1}, \ldots, y_{in_i})^T, i = 1, \ldots, K$$

are independent. However, individual measurements, $y_{i1}, \ldots, y_{in_i}$, from the same subject are not independent.

In particular, when $n_i \equiv 1, i = 1, \ldots, K$, i.e., only one observation recorded for each subject, longitudinal data reduces to a *cross-sectional data* which is the data type considered in the theory of generalized linear models.

Associated with time series (y_{ij}, t_{ij}) are p-element vectors of covariates (or explanatory variables) $\mathbf{x}_{ij}$, either time-dependent (e.g., age and blood pressure reading) or time-independent (e.g., sex and ethnicity) during a study period.

Some main features of longitudinal data are given as follows:

(1) The presence of repeated measurements for each subject implies that the observations from the same subject are *autocorrelated* or serially correlated. This requires us to develop statistical methodology that takes the serial correlation into account.

(2) Longitudinal data can be roughly divided into two categories in terms of the length of time series: longitudinal data from short time series (n_i is small) or from long time series (n_i is large). For the case of short time series, the primary interest is to model a population-averaged pattern, and the dynamic or stochastic pattern is not of interest because of little information available over time. In contrast, longitudinal data of long time series provide a wealth of information over time, which enables investigators to make statistical inference over time. In this case, an objective would be the modeling of dynamic evolution or transitional behavior among the states of response variables.
    Although the cutoff for the length of time series is somewhat subjective, it is important to make such a distinction in light of the primary interests of modeling and inference. Moreover, this distinction is closely related to model formulation for longitudinal data, in which different strategies would be invoked to handle serial correlation. As a matter of fact, in the case of many short time series, modeling will focus on the cause-and-effect relationship between the response and covariates at the mean level (i.e., the first moment), where the correlation is treated as a nuisance, as opposed to the case of long time series in that the correlation will be modeled explicitly via a certain stochastic process.

(3) Many longitudinal datasets used in this book are from biomedical studies, where the response variables are often in categorical scales. Hence, this book is devoted largely to the regression analysis of correlated nonnormal data. Borrowing the strength from the theory of generalized linear models is crucial to build up marginal components for a joint model formulation in the analysis of correlated nonnormal data.

(4) It is often true in practice that at a given time t_{ij}, a multi-dimensional measurement is recorded, giving rise to data of repeated response vectors. The complication associated with such data arises from the fact that there exist two levels of correlation to be accounted for, namely the serial correlation and the correlation across the components of the response vector.

(5) Most longitudinal data from practical studies contain missing data. Dealing with missing data, when the missing data mechanism is informative, is generally nontrivial. To make a proper statistical inference, one has to rely on the information that is supposed to be, but actually not, observed. The degree of complication depends on the amount and patterns of missingness. The most difficult case is the non-ignorable missing pattern or informative missingness, which refers to a missing data process under which the probability of missingness is related to unobserved outcomes of the response variable. The other two types of missing patterns are missing completely at random (MCAR) and missing at random (MAR). See more discussions in Chapter 13.

1.3 Data Examples

To motivate both theoretical and methodological developments given in the subsequent chapters, a few real world datasets will be used for illustration throughout the book. Also, these examples help readers to grasp the data features discussed in the previous section. It begins with examples of short time series, and then examples of long time series.

1.3.1 Indonesian Children's Health Study

Table 1.1. A summary of the data involving 275 subjects.

		Age						
Xerophthalmia	RI	1	2	3	4	5	6	7
No	No	90	236	330	176	143	65	5
	Yes	8	36	39	9	7	1	0
Yes	No	0	2	18	15	8	4	1
	Yes	0	0	7	0	0	0	0

The description of the data is adapted from Diggle et al. (2002). Sommer et al. (1984) reported a study in West Java, Indonesia to determine the causes and effects of vitamin A deficiency in preschool children. Over 3000 children were medically examined quarterly for up to six visits to assess whether they suffered from respiratory or diarrheal infection (RI) and xerophthalmia, an ocular manifestation of vitamin A deficiency. Weight and height variables were also measured. Table 1.1 contains statistics on only 275 children whose measurements are summarized by a three-way table.

This longitudinal data is recorded at equally spaced time points and can be denoted by

$$(y_{it}, \mathbf{x}_{it}, t), t = 1, \ldots, n_i, i = 1, \ldots, 275,$$

where binary response variable $y_{ij} = 1$ if child i had RI at visit t and 0, otherwise. A key covariate of interest is xerophthalmia, which is an indicator of xerophthalmia symptom (1 for the presence and 0 for the absence of xerophthalmia), as well as other baseline covariates such as age, weight, and height. Here, the visit time is equally spaced at $t_{ij} = t = 1, 2, 3, 4, 5, 6$, set apart by a three-month intervals. The length of time series is $n_i \leq 6$, and the total number of subjects is $K = 275$.

The primary objective of this study was to assess the increase in risk of RI for kids who were vitamin A deficient, which was measured indirectly via xerophthalmia. It was also of interest to evaluate the degree of heterogeneity in the risk of disease among the kids.

In summary, this longitudinal data is a collection of many quarterly binary short time series. Since the data contains a large number of subjects ($K = 275$) in comparison to the length of time series, $n_i \leq 6$, it seems natural to draw statistical inference by gathering rich information across subjects. This means that, technically, one should derive asymptotics by letting the number of subjects $K \to \infty$, rather than letting the length $n_i \to \infty$.

This longitudinal data has been analyzed in many published books and articles; for example, see Diggle et al. (2002). However, in this book it is treated as an exercise dataset, and interested readers are encouraged to apply models and inferential methods learned from the book to their analyses of the data. Of course, solutions can be easily found in published works.

1.3.2 Epileptic Seizures Data

Reported by Thall and Vail (1990), Table 1.2 comprises data from a clinical trial of 59 epileptics, which aimed to examine the effectiveness of the drug progabide in treating epileptic seizures. For each patient, the number of epileptic seizures was recorded during a baseline period of eight weeks. Patients were then randomized to two treatment arms, one with the anti-epileptic drug (i.e., progabide) and the other with a placebo, in addition to a standard chemotherapy. The number of seizures was recorded in four consecutive two-week periods after the randomization. The scientific question was whether the drug progabide helps to reduce the rate of epileptic seizures. Given that there are a few outliers in the data, e.g., ID 207, Table 1.3 lists the sample medians and IQRs (interquartile ranges) at the baseline period and each of two-week time intervals across the two treatment arms.

Table 1.2. Bi-weekly epileptic seizure counts and covariates over eight weeks of 59 patients.

ID	Y_1	Y_2	Y_3	Y_4	Trt.	Base	Age	ID	Y_1	Y_2	Y_3	Y_4	Trt.	Base	Age
104	5	3	3	3	0	11	31	103	0	4	3	0	1	19	20
106	3	5	3	3	0	11	30	108	3	6	1	3	1	10	30
107	2	4	0	5	0	6	25	110	2	6	7	4	1	19	18
114	4	4	1	4	0	8	36	111	4	3	1	3	1	24	24
116	7	18	9	21	0	66	22	112	22	17	19	16	1	31	30
118	5	2	8	7	0	27	29	113	5	4	7	4	1	14	35
123	6	4	0	2	0	12	31	117	2	4	0	4	1	11	27
126	40	20	23	12	0	52	42	121	3	7	7	7	1	67	20
130	5	6	6	5	0	23	37	122	4	18	2	5	1	41	22
135	14	13	6	0	0	10	28	124	2	1	1	0	1	7	28
141	26	12	6	22	0	52	36	128	0	2	4	0	1	22	23
145	12	6	8	4	0	33	24	129	5	4	0	3	1	13	40
201	4	4	6	2	0	18	23	137	11	14	25	15	1	46	33
202	7	9	12	14	0	42	36	139	10	5	3	8	1	36	21
205	16	24	10	9	0	87	26	143	19	7	6	7	1	38	35
206	11	0	0	5	0	50	26	147	1	1	2	3	1	7	25
210	0	0	3	3	0	18	28	203	6	10	8	8	1	36	26
213	37	29	28	29	0	111	31	204	2	1	0	0	1	11	25
215	3	5	2	5	0	18	32	207	102	65	72	63	1	151	22
217	3	0	6	7	0	20	21	208	4	3	2	4	1	22	32
219	3	4	3	4	0	12	29	209	8	6	5	7	1	41	25
220	3	4	3	4	0	9	21	211	1	3	1	5	1	32	35
222	2	3	3	5	0	17	32	214	18	11	28	13	1	56	21
226	8	12	2	8	0	28	25	218	6	3	4	0	1	24	41
227	18	24	76	25	0	55	30	221	3	5	4	3	1	16	32
230	2	1	2	1	0	9	40	225	1	25	19	8	1	22	26
234	3	1	4	2	0	10	19	228	2	3	0	1	1	25	21
238	13	15	13	12	0	47	22	232	0	0	0	0	1	13	36
101	11	14	9	8	1	76	18	236	1	4	3	2	1	12	37
102	8	7	9	4	1	38	32								

Table 1.3 indicates that over time, the bi-weekly median count appears to slightly decrease for the progabide group, whereas the median count remains nearly constant for the placebo group. In contrast to the steady pattern of the medians, the IQR appears to vary largely over time and across the treatment groups. A regression model invoked to analyze this data needs to address such strong variation.

To echo the notation introduced earlier in this chapter, let y_{it} be the bi-weekly number of seizures for patient i at equally spaced time $t = 1, 2, 3, 4$, and let $\mathbf{x}_{it}$ be the vector of covariates, including baseline seizure count, treatment,

Table 1.3. Sample medians and interquartile ranges of raw seizure counts per two weeks at the baseline and four successive two-week intervals.

	Progabide					Placebo				
Stats	Base	T_1	T_2	T_3	T_4	Base	T_1	T_2	T_3	T_4
Median	6.00	4.00	5.00	4.00	4.00	4.75	5.00	4.50	5.00	5.00
IQR	4.53	6.58	5.48	7.13	4.21	4.95	6.36	5.29	6.04	4.96

age, and possibly the interaction between treatment and age. It is interesting to note that all the covariates in this study are time independent, i.e., $\mathbf{x}_{it} = \mathbf{x}_i$.

In summary, this data is a collection of $K = 59$ bi-weekly short series of seizure counts, each having the same length of $n_i = 4$, with time-independent covariates and with no missing data.

1.3.3 Retinal Surgery Data

Meyers et al. (1992) reported the data from a prospective study in ophthalmology where intraocular gas was used in complex retinal surgeries to provide internal tamponade of retinal breaks in the eye. Three gas concentration levels were randomly administrated to 31 patients, who were then visited three to fifteen times over a three-month period after gas injection. The volume of the gas in their eyes at each follow-up was recorded as a *percentage* to the initial gas volume. Figure 1.2 displays a longitudinal (or spaghetti) plot of the data, where each trajectory represents a time series of a patient. Overall, a clear decreasing trend in time is shown in the plot. The primary objective of this study was to estimate the kinetics such as the decay rate of gas disappearance across three gas concentration levels.

Let y_{ij} be the percentage of gas volume for patient i at time t_{ij}, which is measured as a ratio of the gas volume V_{ij} at time t_{ij} over the initial gas volume V_{i0}, namely,

$$y_{ij} = \frac{V_{ij}}{V_{i0}}, j = 1, \ldots, n_i, i = 1, \ldots, 31,$$

where both V_{ij} and V_{i0} were not recorded in the study. Also, the visits for each patient did not take place regularly, so that the repeated measurements were collected at unequally spaced time points. Obviously, the response y_{ij} is confined in the unitary interval $(0, 1)$, with zero probability of taking a value beyond this interval. According to Song and Tan (2000), the data appears to be highly overdispersed and marginally skewed to the left. Based on these features, a regression analysis using a normal distribution is doubtful. This

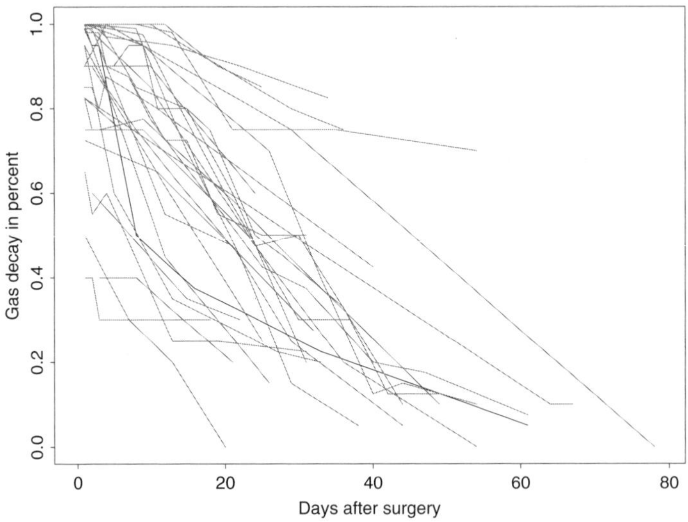

Fig. 1.2. Longitudinal plot of ophthalmological surgery data.

book will demonstrate the utility of the simplex distribution to carry out data analysis.

In summary, the data is a collection of $K = 31$ unequally spaced short time series with responses of continuous proportions confined between 0 and 1. The covariate vector $\mathbf{x}_{ij}$ consists of time after surgery (in days) and gas concentration level, possibly as well as their interaction.

1.3.4 Orientation of Sandhoppers

Borgioli et al. (1999) reported a longitudinal study to understand the mechanism regarding the orientation of sandhoppers (*talitrus saltators*) escaping towards the sea in oder to avoid the risk of high dehydration. It is believed that sandhoppers will take a course perpendicular to the shoreline, known as the *theoretical escape direction (TED)*, which was 201° at the site of Castiglione della Pescaia beach in Italy, where the experiment was performed. Sixty-five (K) sandhoppers were sequentially released five times, and their escape direction was recorded after each release, along with measurements of covariates including wind speed, sun azimuth, and eye asymmetry. The primary objective was to examine which covariates would significantly affect the escape direction of sandhoppers.

As shown in Figure 1.3, this dataset contains a collection of 65 short time series with angular responses, each having the same length of $n = 5$ repeated measurements. In this longitudinal plot, 0° is set for the north.

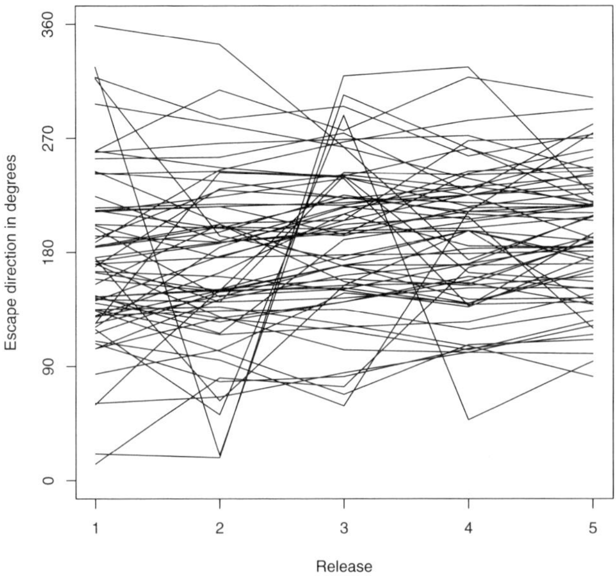

Fig. 1.3. Longitudinal plot of escape orientations for sandhoppers over five releases.

1.3.5 Schizophrenia Clinical Trial

Now let us look at a dataset with missing values. The schizophrenia data is collected from a double-blind randomized clinical trial on patients with schizophrenia. In the study, the patients were scheduled to be examined six times during the six-week period of the study. The response variable is the Brief Psychiatric Rating Scale (BPRS) of the patients, ranging from 0 to 108 with higher scores indicating more severe symptoms, which was scheduled to be measured at each examination time and used to assess schizophrenia status. The main objective of the trial was to evaluate a new treatment (NT) against a standard treatment (ST) (anti-psychotic medication), of which three doses (low, medium, and high) were administered in the trial. Figure 1.4 displays individual BPRS trajectories over six visits across two treatment arms, respectively. It is easy to see in the figure that many patients did not complete the six visits and dropped out of the study. For more details about the trial, refer to Shih and Quan (1997) or Hogan and Laird (1997).

Dropouts frequently occurred in this type of clinical trial due to the nature of the disease. From Table 1.4, it is seen that about 35% of patients in NT cohort and approximately 50% patients in ST cohort dropped out from the trial. In fact, a large proportion of patients dropped out of the study before

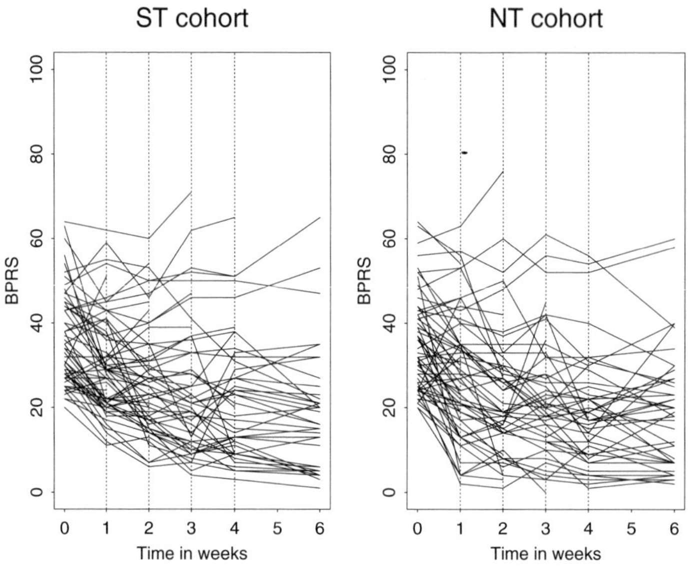

Fig. 1.4. Longitudinal plot of individual BPRS trajectories across the new and standard treatment arms.

Table 1.4. A summary of withdrawal patters by treatment group.

Treatment	Completers	Withdrawals by reason			
		Adverse experience	Lack of efficacy	Other reason	Total
New	40 (65.6%)	1 (1.6%)	7 (11.5%)	13 (21.3%)	61
Standard	34 (53.9%)	12 (19.1%)	11 (17.5%)	6 (9.5%)	63

week six for various reasons, which have been documented by the clinician of the trial. Essentially, there were three reasons leading to patients' dropouts: lack of treatment effect, adverse experience, and other reasons. These documented reasons of withdrawal from the trial provide additional information for understanding and modeling missing data mechanisms. This information would be particularly valuable in the situation where missing values are not missing at random. More discussions are available in Chapter 13.

1.3.6 Multiple Sclerosis Trial

Two examples of multi-level longitudinal data will be presented in this and the next sections, respectively. The first example concerns a longitudinal clinical trial to assess the effects of neutralizing antibodies on interferon beta-1b (IFNB) in relapsing-remitting multiple sclerosis (MS), in which multi-dimensional time series was recorded for each patient (Petkau et al. (2004); Petkau and White (2003)).

Multiple sclerosis is a disease that destroys the myelin sheath that surrounds the nerves. The data are from six-weekly frequent Magnetic Resonance Imaging (MRI) sub-study of the Betaseron clinical trial conducted at University of British Columbia in relapsing-remitting multiple sclerosis involving 52 patients. At each of 17 scheduled visits, three response variables measured on each patient include *active scan*, a binary response recorded for each scan subsequent to the baseline scan; *exacerbation*, a binary response recorded at the time of each scan according to whether an exacerbation began since the previous scan; and *burden of disease*, a positive continuous response recorded as the total area (in units of mm^2) of MS lesions on all slices of each scan. The objective of this trial was to examine the effects of the drug treatment in reducing the disease symptoms.

The patients were randomized into three treatment groups, with the allocation of 17 patients being treated by placebo, 17 by low dose, and 16 by high dose. Baseline covariates include age, duration of disease (in years), gender, and initial EDSS (Expanded Disability Status Scale) scores.

In summary, the data is a collection of 52 equally spaced short multi-dimensional time series of mixed types, where the response vector comprises of two binary and one positive continuous variables.

1.3.7 Tretinoin Emollient Cream Trial

Y. Qu and M. Tan (1998) reports a multi-level longitudinal dataset that was collected from a controlled clinical trial that was conducted to assess the effectiveness of tretinoin emollient cream (TEC) in treating photo-aged skin. A total of 32 patients were randomly assigned to the TEC group and a placebo cream group for facial application of TEC over a period of 24 weeks. In the meantime, one arm of each patient was randomly selected to receive TEC and the other arm to receive placebo for a longer period of 48 weeks. At both the 24th and 48th weeks, each patient was examined for the patient's overall improvement in photoaging, which was measured by a score variable $y_{it} \in \{1, 2, 3, 4\}$, with 1 suggesting no change or worsening from baseline, 2 suggesting slight improvement, 3 suggesting improvement, and 4 suggesting great improvement. Thus, for each patient five repeated ordinal outcomes were recorded, four of which were taken on the two arms at the 24th and 48th weeks and the fifth one was recorded from the face at the 24th week. Since for each subject measurements were recorded in time and across multiple locations (or a cluster), two-level correlation, namely the serial correlation and the within-cluster correlation, has to be accounted for in the data analysis.

1.3.8 Polio Incidences in USA

Now let us turn to longitudinal data of long time series. The first example is based on a dataset from Zeger (1988) that reports a single time series of monthly polio incidences in the USA from 1970 to 1983. The data is plotted in Figure 1.5. The primary objective is to assess whether the data provide evidence of a decreasing trend in the rate of US polio infections over time, after the country implemented a nationwide anti-polio vaccination policy in early 1970s. According to this plot, the time series seems to display certain seasonal patterns in addition to a decreasing trend over time. This suggests that the data is clearly a non-stationary time series. Also, because of low counts, the data cannot be properly analyzed by using the conventional Box and Jenkins (1976) ARIMA models with normally distributed errors.

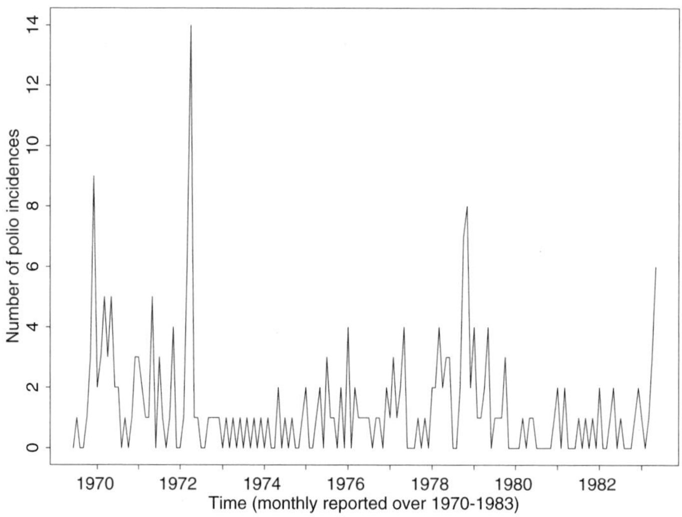

Fig. 1.5. Monthly counts of polio incidences.

Denote the data by

$$(y_t, \mathbf{x}_t),\ t = 1, \ldots, 168,$$

where y_t is the count of polio cases in a given month, and covariates include time t and seasonality patterns described in the forms of *sin* and *cos* functions with bi-yearly or quarterly periods; that is,

$$\mathbf{x}_t = [1, t, \cos(2\pi t/6), \sin(2\pi t/6), \cos(2\pi t/12), \sin(2\pi t/12)].$$

When a regression model is invoked to analyze this data, the related statistical inference has to take the serial correlation into account. In addition, in such a case, the theory of asymptotics in statistical inference has to be established on the basis of the series length n tending to ∞.

1.3.9 Tokyo Rainfall Data

The rainfall data, reported in Kitagawa (1987), consists of daily numbers of occurrences of rainfall in Tokyo area during years 1983 and 1984. Figure 1.6 plots the aggregated series over the two years, so at a given day t, it is possible to have two, one or zero occurrence of rainfall. Let y_t be the number of occurrences of rainfall at calendar day t of a year. Then y_t follows marginally a binomial distribution Binomial$(2, p_t)$, except for February 29 that only exists in 1984, where p_t is the probability of rainfall. Therefore, the data forms a single time series of binomial observations, and the objective is to estimate the probability of rainfall, p_t, over the period of a year, so different seasons such as dry or wet periods in Tokyo can be identified through the series of estimated probabilities.

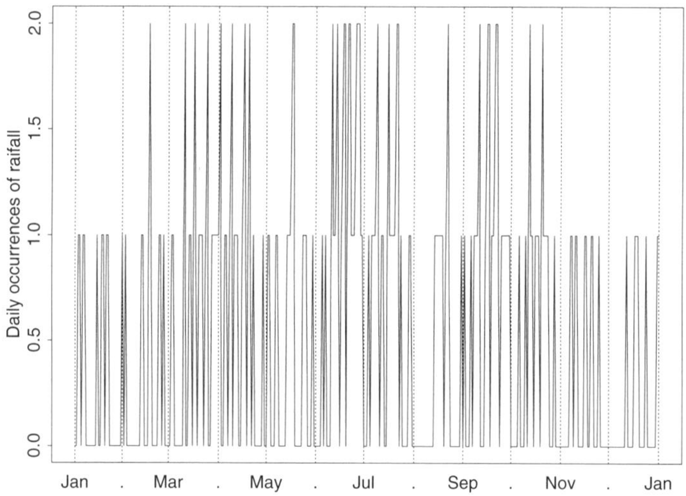

Fig. 1.6. Time series plot of the aggregated daily occurrences of rainfall during 1983-1984 in Tokyo area.

Estimating the rainfall probabilities is challenging, as the number of such probabilities is equal to the length of the time series. In other words, one faces a problem involving a considerably large number of parameters. One way to reduce the dimensionality is to utilize the serial correlation by assuming that observations at past times may predict the state of rainfall at the current day. For example, one may consider the state of rainfall is mostly driven by an underlying meteorological variable, say *moisture*, θ_t, which is a latent variable determining the rainfall probability through, say, a logistic model,

$$p_t = \frac{e^{\theta_t}}{1 + e^{\theta_t}}. \tag{1.1}$$

To address the serial correlation of the moisture process, one may consider a Markov model for the underlying latent continuum θ_t; for instance, an autoregressive model of order 1, $\theta_t = \alpha\theta_{t-1} + \epsilon_t$. Under this setup, one needs to estimate the latent process θ_t in order to estimate p_t and other model parameters such as the autocorrelation parameter α and the variance parameter σ_ϵ^2 of the white noise ϵ_t. Kalman filter and smoothing techniques can be developed in this context to carry out the estimation. See more details in Chapter 11.

1.3.10 Prince George Air Pollution Study

Assessing the impact of air pollution on the public health is of great importance in health sciences. The monitoring of air pollution in Prince George, British Columbia, Canada (e.g., Lambert et al. 1987) shows that there are frequent excursions above the provincial air quality standards, and there has long been public concern in Prince George that the air quality in the city may be adversely affecting the health of the residents.

This data was collected from Prince George, consisting of daily counts of emergency room (ER) visits for respiratory diseases, classified into four categories (asthma, bronchitis, ear infections, and others) for the period of 1984 to 1986, along with daily measurements of air pollution and meteorological variables. Figure 1.7 displays four time series of daily ER visits, each for one disease category. The air pollution variables were sulphur (total reduced sulphur compounds) and particulates (total suspended particulates), and the meteorological variables include average daily temperature and daily minimum and maximum humidity readings, all of which are plotted in Figure 1.8. Refer to Jørgensen et al. (1996b) for more details of the data description.

The main objective of the investigation was to examine the relationship between air pollution and respiratory morbidity. Essentially, the data are a collection of a four-dimensional time series of daily RE visits over a period of 730 days. Denote the data by

$$(\mathbf{y}_t, \mathbf{x}_t), \ t = 1, \ldots, 730$$

where $\mathbf{y}_t = (y_{1t}, y_{2t}, y_{3t}, y_{4t})^T$ with

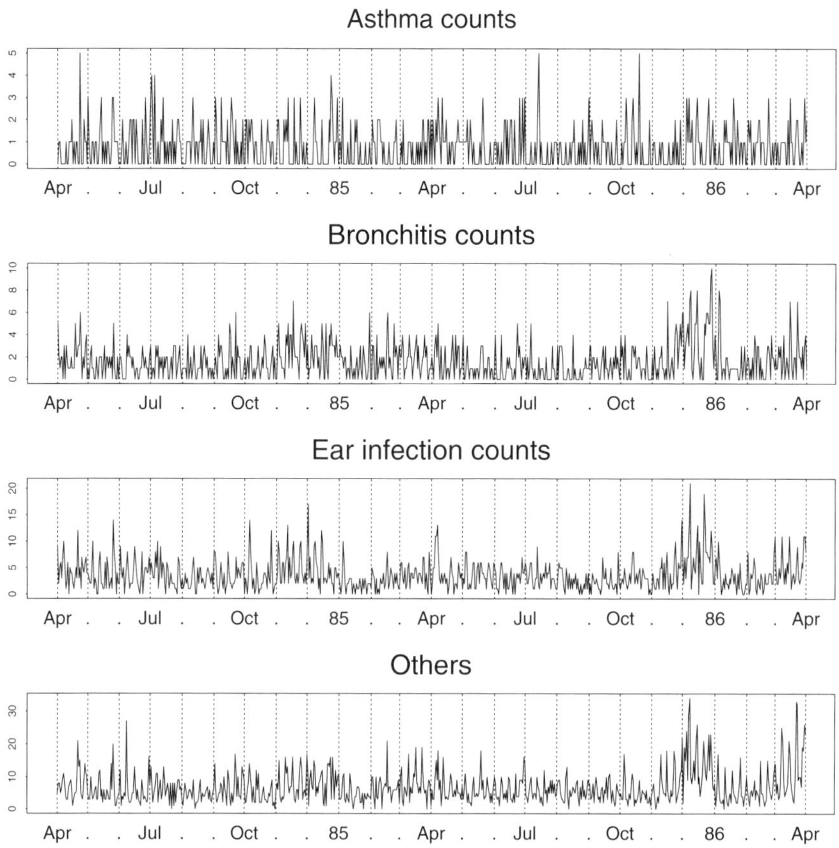

Fig. 1.7. Time series plots of daily emergency room visits during April, 1984 and March, 1986 in Prince George, British Columbia, Canada.

$$y_{1t} = \text{the daily number of ER visits due to asthma,}$$
$$y_{2t} = \text{the daily number of ER visits due to bronchitis,}$$
$$y_{3t} = \text{the daily number of ER visits due to ear infections,}$$
$$y_{4t} = \text{the daily number of ER visits due to other symptoms,}$$

and $\mathbf{x}_t = (x_{1t}, x_{2t}, x_{3t}, x_{4t})^T$ with daily average (or maximum) measurements of covariates, $x_{1t} = \text{temperature}, x_{2t} = \text{humidity}, x_{3t} = \text{sulphur}, x_{4t} =$ particulates.

Two main features of the data need to be addressed in the data analysis. First, it is necessary to account for both serial correlation and rich information over time in the development of a statistical model for the data. Second, the modeling needs to distinguish the different ways that the meteorological variables and the air pollution variables affect the respiratory morbidity. The

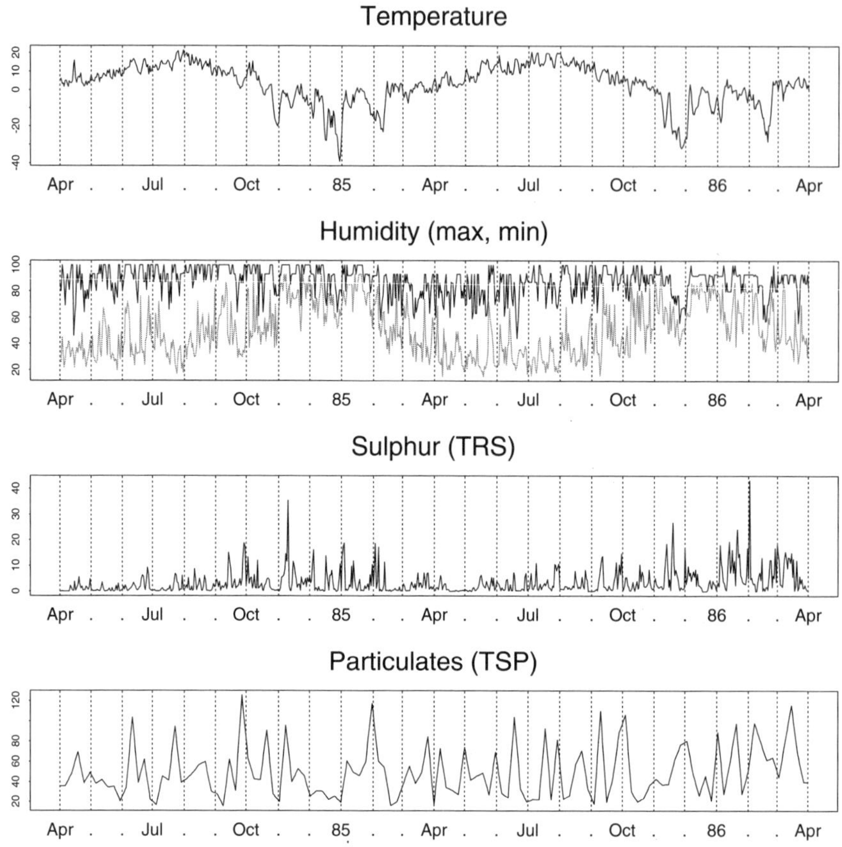

Fig. 1.8. Time series plots of daily air pollution and meteorological readings during April, 1984 and March, 1986 in Prince George, British Columbia, Canada.

effect of the meteorological variables on respiratory morbidity tends to be more acute than that of the air pollution variables, which usually appears lagged in time. For example, a comparison between Figures 1.7 and 1.8 unveils that the sulphur level was high in November, 1985, which seems responsible for the peaks that occurred in December of 1985 for the number of ER visits by patients of bronchitis, ear infections, and others. This indicates a delayed effect of sulphur by one month or so. Similarly, another occurrence of high sulphur level in January, 1986 could be linked with the increasing number of ER visits due to ear infections and others in March, 1986. This case gives a lagged effect of sulphur by approximately two months. Models used to analyze the data should have the flexibility in addressing these observed features learned from the preliminary analysis of the data.

1.4 Remarks

In Section 1.3, we have seen various types of correlated data arising in practice. Some are collections of many short time series and others are collections of several long time series. It is necessary to make such a distinction because statistical inference will be established differently, either by letting the number of subjects tend to infinity or by letting the length of time series go to infinity.

Since many books (e.g., Diggle et al. 2002; Fitzmaurice et al. 2004; McCulloch and Searle 2001; Lindsey 1999; Davis 2002) have extensively covered the analysis of longitudinal data with normal responses, this book is inclined to focus more on nonnormal longitudinal data. This is why Section 1.3 did not illustrate examples of normal longitudinal data.

For the regression analysis of univariate nonnormal data, the class of generalized linear models has been widely used. In particular, Chapter 2 of this book will present generalized linear models from the perspective of Jørgensen's (1997) theory of dispersion models. An advantage of this perspective is that a broader class of regression models can be covered under a unified framework, including regression models for both compositional (or continuous proportional) data and directional (or circular or angular) data. These two data types are not treated in the classical theory of generalized linear models presented by, for example, by McCullagh and Nelder (1989).

To deal with multi-level data, a joint modeling approach may be taken to carry out a simultaneous statistical inference. In the case of the multiple sclerosis data in section 1.3.6, statistical inference needs to take two-level correlation into account. One is the serial correlation and the other is the correlation across the components of multiple outcomes at a given time. Another complicating factor in this data is the mixed types of responses, where the components follow different marginal distributions.

When faced with long time series where modeling the stochastic pattern becomes essential, one has to assume either stationarity or non-stationarity in a certain aspect of a model, such as the θ_t in the model of Tokyo rainfall probability (1.1) where θ_t is assumed to be a stationary AR(1) process. However, this process may be assumed to be nonstationary, such as a random walk process. The assumption of stationarity or non-stationarity, if relevant in the modeling, represents a fundamentally different stochastic mechanism governing the structure and behavior of transition over time.

Missing data, especially those of informative or non-ignorable type, presents great difficulty to the development of proper statistical inference. This is the difficulty that one is not willing to, but must, face in data analysis. This difficulty may be alleviated, to some extent, if additional information such as reasons leading to missing values is documented in a study.

Tools of analyzing correlated data are largely data-driven not only by specific features associated with the data, but also by particular scientific questions in data analysis. Correlated data could be very complex in practical studies. A strategy would be dividing a large study into a few manageable sub-studies, each answering one specific question. Also, it is always recommended to try different models in the data analysis, so the robustness of results may be evaluated.

1.5 Outline of Subsequent Chapters

This book is roughly partitioned into three parts: the first part (Chapters 2–3) introduces the dispersion models and inference functions, which are the basis of the theory and methods developed in the subsequent chapters; the second part (Chapters 4–8) discusses the analysis of many short time series; and the third part (Chapters 9–12) concerns the analysis of long time series. Topics of missing data are treated in Chapter 13. The chart below displays the roadmap of all thirteen chapters, and readers can take a path to reach a topic of interest.

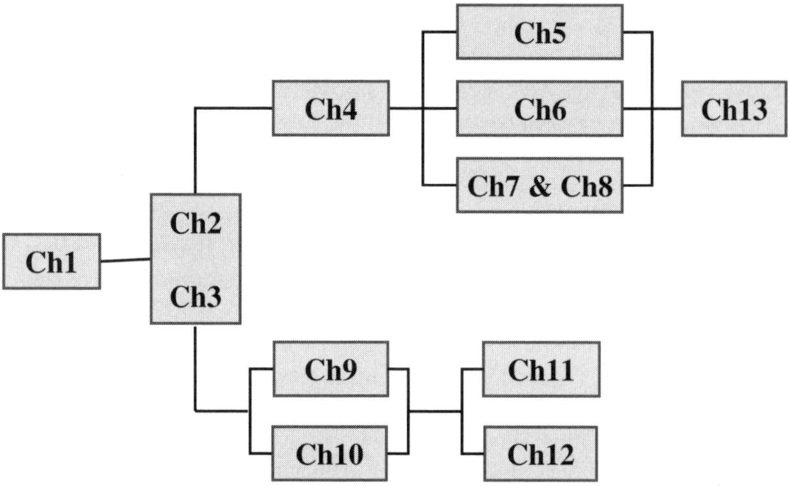

Chapter 2 gives an introduction to Jørgensen's theory of dispersion models, which presents a flexible class of parametric distributions, including those considered in the theory of generalized linear models (McCullagh and Nelder, 1989). In particular, two new regression models will be discussed. One is the model for compositional or continuous proportional data based on the simplex distribution, and the other is the model for directional (or circular or angular) data based on the von Mises distribution.

Chapter 3 focuses on the theory of inference functions, which is essential for many quasi-likelihood inference discussed in the book, including the approach of generalized estimating equations, the approach of quadratic inference functions, and the approach of Kalman estimating functions.

Chapter 4 outlines various strategies and approaches to the modeling of correlated data, especially longitudinal data from many short time series.

Marginal generalized linear models in Chapter 5 are an important class of regression models for correlated data, which attempts to model the population-average pattern of the data. These models only specify the first two moments of data distributions, rather than the joint probability distribution of the data, and consequently the correlation is treated as a nuisance in the analysis. Quasi-likelihood inference approaches derived from the inference functions will be discussed in detail, including generalized estimating equations and quadratic inference functions.

Chapter 6 discusses a class of joint generalized linear models based on full probability models for multi-dimensional outcomes. This joint modeling approach is applicable to analyze longitudinal, clustered, and spatial data with an equal number of repeated outcomes. The theory of simultaneous maximum likelihood inference is discussed to yield an efficient inference for the model parameters. Gaussian copulas are utilized to illustrate the procedure of joint modeling.

Chapters 7 and 8 are devoted to the theory of generalized linear mixed models in that random effects are used to address overdispersion, subject-specific heterogeneity, and within-cluster correlation. Chapter 7 mainly concerns likelihood-based inferences, including direct MLE, EM algorithm, and penalized quasi-likelihood and restricted maximum likelihood. Chapter 8 focuses on Bayesian inference based on Markov chain Monte Carlo (MCMC), in which analyzing multi-level correlated data is discussed. The Windows version of the BUGS (Bayesian Analysis Using Gibbs Sampling) software, in short `WinBUGS`, will be illustrated to implement the MCMC approach.

Chapter 9 is devoted to the theory of linear predictor, which provides the means of estimating random effects as well as the Kalman filter and smoothing. This serves as a preparation for the development of statistical inference in state space models considered in Chapters 11 and 12.

Chapter 10 gives an introduction to generalized state space models for long time series data. It reviews briefly the classical state space models for continuous-valued time series and some extensions.

Chapters 11 and 12 are devoted to modeling of discrete-valued time series. Chapter 11 concerns generalized state space models for time series of binomial observations, while Chapter 12 studies generalized state space models for time series of counts. Monte Carlo Kalman filter and smoother, Kalman estimating equations based on EM-algorithm, and Markov chain Monte Carlo algorithm will be discussed in the theory of parameter estimation.

Chapter 13 concentrates on the topic of missing data in the connection to the setting of many short time series. Topics include testing for missing data types, strategies of handling missing data processes of MAR type by multiple imputations and EM algorithm, and strategies of handling missing data processes of type NMAR.

2

Dispersion Models

2.1 Introduction

In the analysis of correlated data, it is relatively easy to recognize one-dimensional marginal distribution for each of the response vectors. In the example of Indonesian children's health study in Section 1.3.1, the univariate response at a given visit is the infection status, which takes two values with 1 representing the presence of infection and 0 otherwise. Obviously, the marginal distribution of such a binary response variable is Bernoulli or binomial with the size parameter equal to one. In some cases where marginal distributions are subtle to determine, one may apply some model diagnostic tools to check the assumption of marginal distributions. For example, univariate histograms, quantile-quantile plots, and some residual-based model diagnostics in univariate regression analysis, whichever is suitable, could be applied to draw some preliminary understanding of marginal distributions. In the GLMs, the diagnostic analysis of distributional assumptions is carried out through primarily validating the so-called mean and variance relationship. As far as a correlated data analysis concerns, the knowledge of marginal distributions is not yet developed enough to specify a full joint probability model for the data, and a proper statistical inference has to address the correlation among the components of the response vector. Failing to incorporate the correlation in the data analysis will, in general, result in a certain loss of efficiency in the estimation for the model parameters, which may cause misleading conclusions on statistical significance for some covariates.

There are two essential approaches to handling the correlation. One is to construct a full probability model that integrates the marginal distributions and the correlation coherently; within such a framework, the maximum likelihood estimation and inference can be then established. When the joint model is adequately specified, this approach is preferable, because the maximum likelihood method provides a fully efficient inference. Such an approach has been extensively investigated in the class of multivariate normal distributions. However, for many nonnormal data types, constructing a suitable joint probability distribution is not trivial, and relatively less effort on this matter has been made in the literature in comparison to other areas of research in statistics. In particular, the construction of multivariate discrete distributions, such as multivariate binomial distributions and multivariate Poisson distributions, is still under debate, particularly as to which of many versions of their multivariate extensions is desirable relative to the others. More details concerning this approach will be presented in Chapters 6 and 7. Two major classes of joint probability models are specified via, respectively, Gaussian copulas and random effects.

To avoid the difficulty of specifying a full probability model, the second approach takes a compromise; that is, it only specifies the first two moments of the data distribution. This approach constitutes the minimal requirements for a quasi-likelihood inference procedure. Although the resulting estimation is less efficient than the MLE, it enjoys the robustness against model misspecifications on higher moments. This quasi-likelihood inference would be the choice when robustness appears to be more appealing than efficiency in a given data analysis. A kind of such a quasi-likelihood approach, known as generalized estimating equations (GEE), will be discussed in Chapter 5.

To proceed, it is needed to first outline the marginal parametric distributions that will be used to develop either the full probability model approach or the quasi-likelihood approach. Marginal distributions are the essential pieces to formulate both inference approaches in correlated data analysis. To some extent, the breadth of marginal distributions determines the variety of data types that the proposed inference can handle. This means if one only considers marginal normal distributions, the resulting inference would be merely restricted to continuous data type.

This chapter is devoted to a review of the theory of dispersion models based primarily on Jørgensen's (1997) book, *The theory of dispersion models.* The dispersion models provide a rich class of one-dimensional parametric distributions for various data types, including those commonly considered in the GLM analysis. In effect, error distributions in the GLMs form a special subclass of the dispersion models, which are the *exponential dispersion models.* This means that the GLMs considered in this chapter, as well as in the entire book, encompass a wider scope of GLMs than those outlined in McCullagh and Nelder's (1989) book. Two special examples are the von Mises distribution for directional (circular or angular) data and the simplex distribution for compositional (or proportional) data, both of which are the dispersion models but not the exponential dispersion models.

According to McCullagh and Nelder (1989), the random component of a GLM is specified by an exponential dispersion (ED) family density of the following form:

$$p(y;\theta,\phi) = \exp\left[\frac{\{y\theta - \kappa(\theta)\}}{a(\phi)} + C(y,\phi)\right], y \in \mathcal{C}, \tag{2.1}$$

with parameters $\theta \in \Theta$ and $\phi > 0$, where $\kappa(\cdot)$ is the cumulant generating function and $\mathcal{C}$ is the support of the density. It is known that the first derivative of the cumulant function $\kappa(\cdot)$ gives the expectation of the distribution, namely $\mu = \mathrm{E}(Y) = \dot{\kappa}(\theta)$. Table 2.1 lists some ED distributions.

Table 2.1. Some commonly used exponential dispersion GLMs.

Distribution	Domain	Data type	Canonical link	Model
Normal	$(-\infty, \infty)$	Continuous	Identity	Linear model
Binomial	$\{0, 1, \ldots, n\}$	Binary or counts	Logit	Logistic model
Poisson	$\{0, 1, \ldots, \}$	Counts	Log	Loglinear model
Gamma	$(0, \infty)$	Positive continuous	Reciprocal	Reciprocal model

The systematic component of a GLM is then assumed to take the form:

$$g(\mu) = \mathbf{x}^T \boldsymbol{\beta} = \beta_0 + \beta_1 x_1 + \cdots + \beta_p x_p \tag{2.2}$$

where g is the link function, $\mathbf{x} = (1, x_1, \ldots, x_p)^T$ is a $(p+1)$-dimensional vector of covariates, and $\boldsymbol{\beta} = (\beta_0, \ldots, \beta_p)^T$ is a $(p+1)$-dimensional vector of regression coefficients. The *canonical link* function $g(\cdot)$ is such that $g(\mu) = \theta$, the canonical parameter.

The primary statistical tasks include estimation and inference for $\boldsymbol{\beta}$. Checking model assumptions is also an important task of regression analysis, which, however, is not the main focus of the book.

2.2 Dispersion Models

The normal distribution $N(\mu, \sigma^2)$ plays the central role in the classical linear regression regression. The density of $N(\mu, \sigma^2)$ is

$$p(y; \mu, \sigma^2) = \frac{1}{\sqrt{2\pi}\sigma} \exp\left\{-\frac{1}{2\sigma^2}(y-\mu)^2\right\}, \quad y \in \mathcal{R},$$

where $(y - \mu)^2$ can be regarded as an Euclidean distance that measures the discrepancy between the observed y and the expected μ. And this discrepancy measure is used to develop many regression analysis methods, such as the F-statistic for the assessment of goodness-of-fit for nested models.

Mimicking the normal density, Jørgensen (1987) defines a dispersion models (DM) by extending the Euclidean distance $(y-\mu)^2$ to a general discrepancy function $d(y;\mu)$. It is found that many commonly used parametric distributions, such as those in Table 2.1, are included as special cases of this extension. Moreover, each of such distributions will be determined uniquely by the discrepancy function d, and the resulting distribution is fully parameterized by two parameters μ and σ^2.

2.2.1 Definitions

A *(reproductive) dispersion model* $\text{DM}(\mu, \sigma^2)$ with *location parameter* μ and *dispersion parameter* σ^2 is a family of distributions whose probability density functions take the following form:

$$p(y;\mu,\sigma^2) = a(y;\sigma^2)\exp\left\{-\frac{1}{2\sigma^2}d(y;\mu)\right\}, \quad y \in \mathcal{C} \tag{2.3}$$

where $\mu \in \Omega$, $\sigma^2 > 0$, and $a \geq 0$ is a suitable normalizing term that is independent of the μ. Usually, $\Omega \subseteq \mathcal{C} \subseteq \mathcal{R}$. The fact that the normalizing term a does not involve μ will allow to estimate μ (or $\boldsymbol{\beta}$ in the GLM setting) separately from estimating σ^2, which gives rise to great ease in the parameter estimation. Such a nice property, known as the likelihood orthogonality, holds in the normal distribution, and it will remain in the dispersion models.

A bivariate function $d(\cdot;\cdot)$ is called the *unit deviance* defined on $(y,\mu) \in \mathcal{C} \times \Omega$ if it satisfies the following two properties:

i) It is zero when the observed y and the expected μ are equal, namely

$$d(y;y) = 0, \quad \forall y \in \Omega;$$

ii) It is positive when the observed y and the expected μ are different, namely

$$d(y;\mu) > 0, \quad \forall y \neq \mu.$$

Furthermore, a unit deviance is called *regular* if function $d(y;\mu)$ is twice continuously differentiable with respect to (y,μ) on $\Omega \times \Omega$ and satisfies

$$\frac{\partial^2 d}{\partial \mu^2}(y;y) = \left.\frac{\partial^2 d}{\partial \mu^2}(y;\mu)\right|_{\mu=y} > 0, \quad \forall y \in \Omega.$$

For a regular unit deviance, the variance function is defined as follows. The *unit variance function* $V : \Omega \to (0,\infty)$ is

$$V(\mu) = \frac{2}{\left.\frac{\partial^2 d}{\partial \mu^2}(y;\mu)\right|_{y=\mu}}, \quad \mu \in \Omega. \tag{2.4}$$

Some popular dispersion models are given in Table 2.2, in which the unit deviance d and variance function V can be found in a similar fashion to that presented in the following two examples.

Table 2.2. Unit deviance and variance functions of some dispersion models.

Distribution	Deviance d	$\mathcal{C}$	Ω	$V(\mu)$
Normal	$(y-\mu)^2$	$(-\infty, \infty)$	$(-\infty, \infty)$	1
Poisson	$2(y \log \frac{y}{\mu} - y + \mu)$	$\{0, 1, \ldots\}$	$(0, \infty)$	μ
Binomial	$2\left\{y \log \frac{y}{\mu} + (n-y) \log \frac{n-y}{n-\mu}\right\}$	$\{0, 1, \ldots, n\}$	$(0, 1)$	$\mu(1-\mu)$
Negative binomial	$2\left\{y \log \frac{y}{\mu} + (1-y) \log \frac{1-y}{1-\mu}\right\}$	$\{0, 1, \ldots\}$	$(0, \infty)$	$\mu(1+\mu)$
Gamma	$2\left(\frac{y}{\mu} - \log \frac{y}{\mu} - 1\right)$	$(0, \infty)$	$(0, \infty)$	μ^2
Inverse Gaussian	$\frac{(y-\mu)^2}{y\mu^2}$	$(0, \infty)$	$(0, \infty)$	μ^3
von Mises	$2\{1 - \cos(y-\mu)\}$	$(0, 2\pi)$	$(0, 2\pi)$	1
Simplex	$\frac{(y-\mu)^2}{y(1-y)\mu^2(1-\mu)^2}$	$(0, 1)$	$(0, 1)$	$\mu^3(1-\mu)^3$

Example 2.1 (Normal Distribution). In the normal distribution $N(\mu, \sigma^2)$, first the unit deviance function $d(y; \mu) = (y-\mu)^2$, $y \in \mathcal{C} = \mathcal{R}$, and $\mu \in \Omega = \mathcal{R}$. It is easy to see that this d function is non-negative and has the unique minimum 0 when $y = \mu$. This unit deviance is regular because it is twice continuously differentiable. Moreover, the first and second order derivatives of the d function *w.r.t.* μ are, respectively,

$$\frac{\partial d}{\partial \mu} = -2(y-\mu), \text{ and } \frac{\partial^2 d}{\partial \mu^2} = 2.$$

It follows that the unit variance function is $V(\mu) = \frac{2}{2} = 1$.

Example 2.2 (Poisson Distribution). To verify the results of the Poisson distribution given in Table 2.2, express the Poisson density with mean parameter μ as follows:

$$p(y; \mu) = \frac{\mu^y}{y!} e^{-\mu}, y \in \{0, 1, \ldots\};\ \mu \in \Omega = (0, \infty),$$

or equivalently

$$p(y; \mu) = \frac{1}{y!} \exp\{y \log \mu - \mu\}.$$

Note that the exponent $\{y \log \mu - \mu\}$ is not a deviance function because it does not equal to zero when $y = \mu$. To yield a deviance function, a new term $\{y \log y - y\}$ is added into the exponent, which results in

$$p(y; \mu) = \left\{ \frac{1}{y!} \exp(y \log y - y) \right\} \exp \left\{ -\frac{1}{2} 2(y \log y + y - y \log \mu + \mu) \right\}.$$

Comparing to the DM density in (2.3), one can identify the d function, the normalizing term, and the dispersion parameter, respectively,

$$\begin{aligned} d(y; \mu) &= 2(y \log \frac{y}{\mu} - y + \mu), \\ a(y) &= \frac{1}{y!} \exp\{y \log y - y\}, \\ \sigma^2 &= 1. \end{aligned}$$

To show this d function is a regular deviance function, it is sufficient to show it is convex with a unique minimum of zero. First, note that at a given mean value μ, the first and second order derivatives of the d *w.r.t.* y are

$$\frac{\partial d}{\partial y} = 2(\log y - \log \mu), \text{ and } \frac{\partial^2 d}{\partial y^2} = \frac{2}{y}.$$

Clearly, the first order derivative is negative when $y < \mu$ and positive when $y > \mu$, implying that the d is a convex function with a unique minimum 0 at $y = \mu$. Thus, the d function is a regular unit deviance for the Poisson distribution.

To find the unit variance function, note that the second order derivative $\frac{\partial^2 d}{\partial \mu^2} = 2\frac{y}{\mu^2}$, which immediately leads to $V(\mu) = \mu$ by the definition (2.4).

2.2.2 Properties

This section lists some useful properties of the dispersion models.

Proposition 2.3. *If a unit deviance d is regular, then*

$$\frac{\partial^2 d}{\partial y^2}(y; y) = \frac{\partial^2 d}{\partial \mu^2}(y; y) = -\frac{\partial^2 d}{\partial \mu \partial y}(y; y), \quad \forall y \in \Omega. \tag{2.5}$$

Proof. By the definition of a unit deviance,

$$d(y; y) = d(\mu; \mu) = 0 \text{ and } d(y; \mu) \geq 0, \ \forall y, \mu \in \Omega,$$

implying that $d(y; \cdot)$ has a unique minimum at y and similarly $d(\cdot; \mu)$ has a unique minimum at μ. Therefore,

$$\frac{\partial d}{\partial \mu}(y; y) = \frac{\partial d}{\partial y}(y; y) = 0. \tag{2.6}$$

The result of (2.5) holds by simply differentiating both equations in (2.6) *w.r.t.* y.

Proposition 2.4. *Taylor expansion of a regular unit deviance d near its minimum (μ_0, μ_0) is given by*

$$d(\mu_0 + x\delta; \mu_0 + m\delta) = \frac{\delta^2}{V(\mu_0)}(x - m)^2 + o(\delta^2),$$

where $V(\cdot)$ is the unit variance function.

Proof. It follows from equation (2.6) that

$$\begin{aligned} d(\mu_0 + x\delta; \mu_0 + m\delta) &= d(\mu_0, \mu_0) + \frac{\partial d}{\partial \mu}(\mu_0, \mu_0)(x\delta) + \frac{\partial d}{\partial y}(\mu_0, \mu_0)(m\delta) \\ &\quad \frac{1}{2}\frac{\partial d^2}{\partial \mu^2}(\mu_0, \mu_0)(\delta^2 x^2) + \frac{1}{2}2\frac{\partial d^2}{\partial \mu \partial y}(\mu_0, \mu_0)(\delta m) \\ &\quad + \frac{1}{2}\frac{\partial d^2}{\partial y^2}(\mu_0, \mu_0)(\delta^2 m^2) + o(\delta^2) \\ &= \frac{\delta^2}{V(\mu_0)}x^2 - \frac{\delta^2}{V(\mu_0)}2xm + \frac{\delta^2}{V(\mu_0)}m^2 + o(\delta^2) \\ &= \frac{\delta^2}{V(\mu_0)}(x - m)^2 + o(\delta^2). \end{aligned}$$

In some cases, the normalizing term $a(\cdot)$ has no closed form expression, which gives rise to the difficulty of estimating the dispersion parameter σ^2. The following proposition presents an approximation to the normalizing term $a(\cdot)$, resulting from the saddlepoint approximation of the density for small dispersion. Notation $a \simeq b$ exclusively stands for an approximation of a to b when the dispersion $\sigma^2 \to 0$, useful for small-dispersion asymptotics.

Proposition 2.5 (Saddlepoint approximation). *As the dispersion $\sigma^2 \to 0$, the density of a regular DM model can be approximated to be:*

$$p(y; \mu, \sigma^2) \simeq \{2\pi\sigma^2 V(y)\}^{-\frac{1}{2}} \exp\left\{-\frac{1}{2\sigma^2}d(y; \mu)\right\},$$

which equivalently says that as $\sigma^2 \to 0$, the normalizing term has a small dispersion approximation,

$$a(y; \sigma^2) \simeq \{2\pi\sigma^2 V(y)\}^{-1/2}, \tag{2.7}$$

with the unit variance function $V(\cdot)$.

The proof of this proposition is basically an application of the Laplace approximation given in, for example, Barndorff-Nielsen and Cox (1989, p.60). Also see Jørgensen (1997, p.28).

It follows from Propositions 2.4 and 2.5 that the small dispersion asymptotic normality holds, as stated in the following:

Proposition 2.6 (Asymptotic Normality). *: Let $Y \sim DM(\mu_0 + \sigma\mu, \sigma^2)$ be a dispersion model with uniformly convergent saddlepoint approximation, namely convergence in (2.7) is uniformly in y. Then*

$$\frac{Y - \mu_0}{\sigma} \xrightarrow{d} N(\mu, V(\mu_0)) \text{ as } \sigma^2 \to 0.$$

In other words, $DM(\mu_0 + \sigma\mu, \sigma^2) \overset{d}{\simeq} N(\mu_0 + \sigma\mu, \sigma^2 V(\mu_0))$ for small dispersion σ^2.

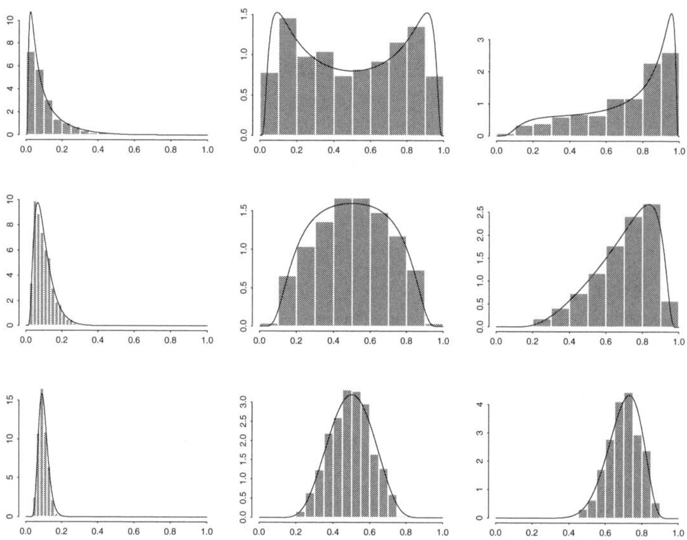

Fig. 2.1. Simplex density functions with mean $\mu = (0.1, 0.5, 0.7)$ from left to right and dispersion parameter $\sigma^2 = (4^2, 2^2, 1)$ from top to bottom. The solid lines represent the simplex densities with the histograms as the background. These histograms are based on 500 simulated data from respective densities.

To illustrate this small-dispersion asymptotic normality, Figure 2.1 displays the simplex distributions with different mean μ and dispersion σ^2 parameters. See the detail of a simplex distribution in Table 2.2. This figure clearly indicates that the smaller the dispersion is, the less deviation the simplex distribution is from the normality.

2.3 Exponential Dispersion Models

The class of dispersion models contains two important subclasses, namely *the exponential dispersion (ED) models* and *the proper dispersion (PD) models.*

The PD models are mostly of theoretical interest, so they are not discussed in this book. Readers may refer to the book of Jørgensen (1997) for relevant details.

This section focuses on the ED models, which have already been introduced in Section 2.1 as a family of GLMs' error distributions. The family of ED models includes continuous distributions such as normal, gamma, and inverse Gaussian, and discrete distributions such as Poisson, binomial, negative binomial, among others. To establish the connection of the ED model representation (2.1) to the DM, it is sufficient to show that expression (2.1) is a special form of (2.3). An advantage with the DM type of parametrization for the ED models is that both mean μ and dispersion parameters σ^2 are explicitly present in the density, whereas expression (2.1) hides the mean μ in the first order derivative $\mu = \dot{\kappa}(\theta)$. In addition, having a density form similar to the normal enables us to easily borrow the classical normal regression theory to the development of regression analysis for nonnormal data. One example is the analogue of the likelihood ratio test in the GLMs to the F-test for goodness-of-fit in the normal regression model.

To show an ED model, denoted by $\text{ED}(\mu, \sigma^2)$, as a special case of the DM, it suffices to find a unit deviance function d such that the density of the ED model can be expressed in the form of (2.3). First, denote $\lambda = 1/a(\phi)$. Then, the density in (2.1) can be rewritten as of the form:

$$p(y; \theta, \lambda) = c(y; \lambda) \exp[\lambda\{\theta y - \kappa(\theta)\}], \;\; y \in \mathcal{C} \tag{2.8}$$

where $c(\cdot)$ is a suitable normalizing term. Parameter $\lambda = 1/\sigma^2 \in \Lambda \subset (0, \infty)$ is called the *index parameter* and Λ is called the index set. To reparametrize this density (2.1) by the mean μ and dispersion σ^2, define the *mean value mapping:* $\tau : int\Theta \to \Omega$,

$$\tau(\theta) = \dot{\kappa}(\theta) \equiv \mu,$$

where $int(\Theta)$ is the interior of the parameter space Θ.

Proposition 2.7. *The mean mapping function* $\tau(\theta)$ *is strictly increasing.*

Proof. The property of the natural exponential family distribution leads to

$$\text{Var}(Y) = \lambda\ddot{\kappa}(\theta) > 0, \;\; \theta \in int\Theta.$$

In the mean time, because $\dot{\tau}(\theta) = \ddot{\kappa}(\theta)$, $\dot{\tau}(\cdot)$ is positive. This implies that $\tau(\theta)$ is a strictly increasing function in θ.

It follows that the inverse of the mean mapping function $\tau(\cdot)$ exists, denoted by $\theta = \tau^{-1}(\mu)$, $\mu \in \Omega$. Hence, the density in (2.8) can be reparametrized as follows,

$$p(y; \mu, \sigma^2) = c(y; \sigma^{-2}) \exp\left[\frac{1}{\sigma^2}\{y\tau^{-1}(\mu) - \kappa(\tau^{-1}(\mu))\}\right]. \tag{2.9}$$

Proposition 2.8. *The first order derivative of* $\tau^{-1}(\mu)$ *with respect to* μ *is* $1/V^*(\mu)$, *where* $V^*(\mu) = \dot{\tau}(\tau^{-1}(\mu))$.

Proof. Differentiating both sides of equation $\mu = \tau(\theta)$ gives

$$d\mu = \dot{\tau}(\theta)d\theta = \dot{\tau}(\tau^{-1}(\mu))d\theta = V^*(\mu)d\theta,$$

with $V^*(\mu) = \dot{\tau}(\tau^{-1}(\mu))$. This implies immediately that

$$\frac{d\tau^{-1}(\mu)}{d\mu} = \frac{d\theta}{d\mu} = \frac{1}{V^*(\mu)}.$$

Moreover, Proposition 2.9 below shows that the $V^*(\mu)$ is indeed the same as the unit variance function $V(\mu)$ given by the definition (2.4). The proof of this result will be given after the unit deviance function of the ED model is derived.

To derive the unit deviance function of the ED model, let

$$f(y;\mu) = y\tau^{-1}(\mu) - \kappa(\tau^{-1}(\mu)) = y\theta - \kappa(\theta).$$

Obviously, this f is not the unit deviance function since it does not equal to zero when $\mu = y$. One way to resolve this problem is to add a new term so that the resulting function is positive and equal to zero uniquely at $\mu = y$. Such a valley point corresponds effectively to the maximum of the density $p(y;\mu,\sigma^2)$.

Differentiating f with respect to μ and using Propositions 2.8 and 2.9, one can obtain

$$\dot{f}(y,\mu) = \frac{y-\mu}{V(\mu)}, \tag{2.10}$$

which is positive for $y > \mu$ and negative for $y < \mu$. This means that the f has a unique maximum, or equivalently, the $-f$ has a unique minimum at $\mu = y$. Therefore, it seems natural to define

$$\begin{aligned} d(y;\mu) &= 2\left[\sup_{\mu}\{f(y;\mu)\} - f(y;\mu)\right] \\ &= 2\left[\sup_{\theta\in\Theta}\{\theta y - \kappa(\theta)\} - y\tau^{-1}(\mu) + \kappa(\tau^{-1}(\mu))\right]. \end{aligned} \tag{2.11}$$

Clearly, this d function satisfies (i) $d(y;\mu) \geq 0$ for all $y \in \mathcal{C}$ and $\mu \in \Omega$, and (ii) $d(y;\mu)$ attains the minimum at $\mu = y$ because the supermum term is independent of μ. Thus, (2.11) gives a proper unit deviance function. Moreover, since it is continuously twice differentiable, it is also regular. As a result, the density of an ED model can be expressed as of the DM form:

$$p(y;\mu,\sigma^2) = a(y;\sigma^2)\exp\left\{-\frac{1}{2\sigma^2}d(y;\mu)\right\},$$

with the unit deviance function d given in (2.11) and the normalizing term given by

$$a(y;\sigma^2) = c(y;\sigma^{-2})\exp\left[\sigma^{-2}\sup_{\theta\in\Theta}\{y\theta - \kappa(\theta)\}\right].$$

Proposition 2.9. *For the unit deviance function (2.11), the corresponding unit variance function $V(\mu)$ given in (2.4) is $V^*(\mu) = \dot{\tau}(\tau^{-1}(\mu))$; that is, $V(\mu) = V^*(\mu)$.*

Proof. It follows from equations (2.10) and (2.11) that

$$\frac{\partial d}{\partial \mu} = -2\frac{\partial f}{\partial \mu} = -2\frac{y-\mu}{V^*(\mu)},$$

where $V^*(\mu) = \dot{\tau}(\tau^{-1}(\mu))$. Then, according to Proposition 2.3,

$$\frac{\partial^2 d}{\partial \mu^2} = -\frac{\partial^2 d}{\partial y \partial \mu} = \frac{2}{V^*(\mu)}.$$

Plugging this into the definition of the unit variance function (2.4) leads to

$$V(\mu) = \frac{2}{\frac{\partial^2 d}{\partial \mu^2}(y;\mu)\big|_{y=\mu}} = V^*(\mu).$$

Here are a few remarks for the ED models:

(1) Parameter μ is the mean of the distribution, namely $\mathrm{E}(Y) = \mu$.
(2) Variance of the distribution is

$$\mathrm{Var}(Y) = \sigma^2 V(\mu). \tag{2.12}$$

This mean-variance relationship is one of the key properties for the ED models, which will play an important role in the development of quasi-likelihood inference.

(3) An important variant of the reproductive ED model representation is the so-called *additive exponential dispersion model*, denoted by $\mathrm{ED}^*(\theta,\lambda)$, whose density takes the form

$$p^*(z;\theta,\lambda) = c^*(z;\lambda)\exp\{\theta z - \lambda\kappa(\theta)\}, \quad z \in \mathcal{C}. \tag{2.13}$$

Essentially the ED and ED* representations are equivalent under *the duality transformation* that converts one form to the other.
Suppose $Z \sim \mathrm{ED}^*(\theta,\lambda)$ and $Y \sim ED(\mu,\sigma^2)$. Then, the duality transformation performs

$$Z \sim \mathrm{ED}^*(\theta,\lambda) \Rightarrow Y = Z/\lambda \sim \mathrm{ED}(\mu,\sigma^2), \text{ with } \mu = \tau(\theta), \sigma^2 = 1/\lambda;$$
$$Y \sim \mathrm{ED}(\mu,\sigma^2) \Rightarrow Z = Y/\sigma^2 \sim \mathrm{ED}^*(\theta,\lambda), \text{ with } \theta = \tau^{-1}(\mu), \lambda = 1/\sigma^2.$$

Consequently, the mean and variance of $\mathrm{ED}^*(\theta,\lambda)$ are, respectively,

$$\mu^* = \mathrm{E}(Z) = \lambda\tau(\theta), \;\; \mathrm{Var}(Z) = \lambda V(\mu^*/\lambda).$$

Moreover, the normalizing term in the DM density (2.3) is

$$a^*(z;\sigma^2) = c^*(z;\sigma^{-2}) \exp\left[\sigma^{-2} \sup_{\theta\in\Theta}\{z\theta - \kappa(\theta)\}\right].$$

An important property for the ED models is the closure under convolution operation.

Proposition 2.10 (Convolution for the ED^* models). *Assume $Z_1, \ldots, Z_n$ are independent and $Z_i \sim ED^*(\theta, \lambda_i)$, $i = 1, \ldots, n$. Then the sum follows still an ED^* model:*

$$Z_+ = Z_1 + \cdots + Z_n \sim ED^*(\theta, \lambda_1 + \cdots + \lambda_n).$$

For example, consider two independent and identically distributed (*i.i.d.*) Poisson random variables $Z_i \sim \mathrm{ED}^*(\log\mu, 1), i = 1, 2$, where μ is the mean parameter and the canonical parameter $\theta = \log(\mu)$. Then, Proposition 2.10 implies that the sum $Z_+ = Z_1 + Z_2 \sim \mathrm{ED}^*(\log\mu, 2)$.

Proposition 2.11 (Convolution for the ED models). *Assume $Y_1, \ldots, Y_n$ are independent and*

$$Y_i \sim ED(\mu, \frac{\sigma^2}{w_i}), i = 1, \ldots, n,$$

where w_is are certain positive weights. Let $w_+ = w_1 + \cdots + w_n$. Then the weighted average follows still an ED model; that is,

$$\frac{1}{w_+}\sum_{i=1}^{n} w_i Y_i \sim ED(\mu, \frac{\sigma^2}{w_+}).$$

In particular, with $w_i = 1, i = 1, \ldots, n$ the sample average

$$\frac{1}{n}\sum_{i=1}^{n} Y_i \sim ED(\mu, \frac{\sigma^2}{n}).$$

For the example of two *i.i.d.* Poisson random variables with $Y_i \sim \mathrm{ED}(\mu, 1)$, $i = 1, 2$, their average $(Y_1 + Y_2)/2 \sim \mathrm{ED}(\mu, \frac{1}{2})$. Note that the resulting $\mathrm{ED}(\mu, \frac{1}{2})$ is no longer a Poisson distribution but it is still an ED distribution.

It is noticeable that although the class of the ED models is closed under the convolution operation, it is in general not closed under scale transformation. That is, cY may not follow an ED model even if $Y \sim \mathrm{ED}(\mu, \sigma^2)$, for a constant c. However, a subclass of the ED models, termed as the *Tweedie class*, is closed under this type of scale transformation. The Tweedie models will be discussed in Section 2.5.

Finally, the following property concerns sufficient and necessary conditions for the de-convolution for the ED models.

Definition 2.12 (Infinite Divisibility). *X is said to be infinitely divisible, if for any integer $n \in \{1, 2, \ldots\}$, there exist* i.i.d. *random variables $X_1, \ldots, X_n$ such that $X \stackrel{d}{=} X_1 + \cdots + X_n$. Notation $U \stackrel{d}{=} V$ means that two random variables U and V are identically distributed.*

Proposition 2.13 (Deconvolution for the ED*). *Suppose $Z \sim ED^*(\theta, \lambda)$. Then, Z is infinitely divisible if and only if the index parameter set $\Lambda = (0, \infty)$.*

This result holds simply because by Proposition 2.10 there exist $X_i \sim ED^*(\theta, \lambda/n), i = 1, \ldots, n$ such that

$$ED^*(\theta, \lambda) = ED^*(\theta, \lambda/n) + \cdots + ED^*(\theta, \lambda/n).$$

It is easy to see that gamma models are infinitely divisible, but binomial models are not infinitely divisible.

2.4 Residuals

Residual analysis is an important part of regression analysis. In the context of the dispersion models where the unit deviance functions d are highly nonlinear in comparison to the square normal deviance $(y - \mu)^2$ of the normal model, there are several other types of residuals besides the traditional Pearson residual $(y - \mu)$. Table 2.3 lists some proposed residuals in the GLMs. Among them, the Pearson and deviance residuals are most commonly used in practice, which are in fact implemented in statistical softwares such as `SAS`. For example, SAS PROC GENMOD uses the deviance residual in the analysis of outliers and influential data cases.

Table 2.3. Some types of residuals in the GLMs.

Type	Notation	Definition
Pearson residual	r_p	$\frac{y-\mu}{V^{1/2}(\mu)}$
Score residual	r_s	$-\frac{\partial d}{2\partial\mu} V^{1/2}(\mu)$
Dual score residual	r_d	$\frac{\partial d}{2\partial y} V^{1/2}(\mu)$
Deviance residual	r	$\pm d^{1/2}(y; \mu)$
Modified deviance residual	r^*	$\frac{r}{\sigma} + \frac{\sigma}{r} \log \frac{r_d}{r}$

Besides the residual analysis for model diagnosis, another important application of residuals is in the approximation of tail area probabilities with small dispersion. Calculating tail probabilities is often encountered, such as in the calculation of p-values. Most of cumulative distribution functions (CDFs) of the ED models have no closed form expressions, so a certain approximation to their CDF is useful.

Let $F(y; \mu, \sigma^2)$ be the CDF of an $\mathrm{ED}(\mu, \sigma^2)$. By Proposition 2.6, the small dispersion asymptotic normality gives

$$F(y; \mu, \sigma^2) \simeq \Phi(r_p/\sigma) \text{ for } \sigma^2 \text{ small},$$

where Φ is the CDF of the standard normal $N(0,1)$. This result is based on the Pearson residual r_p. Because it is a first-order linear approximation, this approximation may not be satisfactorily accurate when the unit deviance d is highly nonlinear.

Two formulas based on the so-called third-order approximation provide much more accurate approximations for the CDF of the DM model. One is Barndorff-Nielsen's formula given by,

$$F(y; \mu, \sigma^2) = \Phi(r^*)\{1 + O(\sigma^3)\},$$

where r^* is the modified deviance residual given in Table 2.3. The other is Lugannani-Rice's formula

$$F(y; \mu, \sigma^2) = \Phi^*(y; \mu, \sigma^2)\{1 + O(\sigma^3)\},$$

where

$$\Phi^*(y; \mu, \sigma^2) = \Phi(\frac{r}{\sigma}) + \sigma\phi(\frac{r}{\sigma})(\frac{1}{r} - \frac{1}{r_d}),$$

where r is the deviance residual and ϕ is the density of the standard normal $N(0,1)$.

2.5 Tweedie Class

Tweedie class is an important subclass of the ED models, which is closed under the scale transformation. Tweedie models are characterized by the unit variance functions in the form of the power function:

$$V_p(\mu) = \mu^p, \mu \in \Omega_p, \tag{2.14}$$

where $p \in R$ is a *shape* parameter.

It is shown that the ED model with the power unit variance function (2.14) always exists except $0 < p < 1$. A Tweedie model is denoted by $Y \sim Tw_p(\mu, \sigma^2)$ with mean μ and variance

$$\mathrm{Var}(Y) = \sigma^2\mu^p.$$

The following proposition gives the characterization of the Tweedie models.

Proposition 2.14 (Tweedie Characterization). *Let $ED(\mu, \sigma^2)$ be a reproductive ED model satisfying $V(1) = 1$ and $1 \in \Omega$. If the model is closed with respect to scale transformation, such that there exists a function $f : R_+ \times \Lambda^{-1} \to \Lambda^{-1}$ for which*

$$cED(\mu, \sigma^2) \sim ED[c\mu, f(c, \sigma^2)], \forall c > 0,$$

then

(a) $ED(\mu, \sigma^2)$ is a Tweedie model for some $p \in R\backslash(0, 1)$;
(b) $f(c, \sigma^2) = c^{2-p}\sigma^2$;
(c) the main domain $\Omega = R$ for $p = 0$ and $\Omega = (0, \infty)$ for $p \neq 0$;
(d) the model is infinitely divisible.

It follows immediately from Proposition 2.14 that

$$cTw_p(\mu, \sigma^2) = Tw_p(c\mu, c^{2-p}\sigma^2).$$

The importance of the Tweedie class is that it serves as a class of limiting distributions of the ED models, as described in the following proposition.

Definition 2.15. *The unit variance function V is said to be regular of order p at 0 (or at ∞), if $V(\mu) \sim c_0\mu^p$ as $\mu \to 0$(or $\mu \to \infty$) for certain $p \in \mathcal{R}$ and $c_0 > 0$.*

Proposition 2.16. *Suppose the unit variance function V is regular of order p at 0 or at ∞, with $p \notin (0, 1)$. For any $\mu > 0$ and $\sigma^2 > 0$,*

$$c^{-1}ED(c\mu, c^{2-p}\sigma^2) \xrightarrow{d} TW_p(\mu, c_0\sigma^2), \quad \text{as } c \to 0 \text{ or } \infty,$$

where the convergence is through values of c such that $c\mu \in \Omega$ and $c^{p-2}/\sigma^2 \in \Lambda$.

Refer to Jørgensen et al. (1994) for the proof of this result.

2.6 Maximum Likelihood Estimation

This section is devoted to maximum likelihood estimation in the GLMs based on the dispersion models. Therefore, the MLE theory given in, for example, McCullagh and Nelder (1989) are the special cases, because the ED family is a subclass of the DM family.

2.6.1 General Theory

Consider a cross-sectional dataset, $(y_i, \mathbf{x}_i), i = 1, \ldots, K$, where the y_i's are *i.i.d.* realizations of Y_i's according to $\mathrm{DM}(\mu_i, \sigma^2)$ and $g(\mu_i) = \mathbf{x}_i^T \boldsymbol{\beta}$. Let $\mathbf{y} = (y_1, \ldots, y_K)^T$ and $\boldsymbol{\mu} = (\mu_1, \ldots, \mu_K)^T$. The likelihood for the parameter vector $\boldsymbol{\theta} = (\boldsymbol{\beta}, \sigma^2)$ is given by

$$L(\boldsymbol{\theta}; \mathbf{y}) = \prod_{i=1}^{K} a(y_i; \sigma^2) \exp\left\{-\frac{1}{2\sigma^2} d(y_i; \mu_i)\right\}, \ \boldsymbol{\beta} \in \mathcal{R}^{p+1}, \sigma^2 > 0.$$

The log-likelihood is then

$$\begin{aligned} \ell(\boldsymbol{\theta}; \mathbf{y}) &= \sum_{i=1}^{K} \log a(y_i; \sigma^2) - \frac{1}{2\sigma^2} \sum_{i=1}^{K} d(y_i; \mu_i) \\ &= \sum_{i=1}^{K} \log a(y_i; \sigma^2) - \frac{1}{2\sigma^2} D(\mathbf{y}; \boldsymbol{\mu}), \end{aligned} \tag{2.15}$$

where $\mu_i = \mu_i(\boldsymbol{\beta})$ is a nonlinear function in $\boldsymbol{\beta}$ and $D(\mathbf{y}; \mu) = \sum_{i=1}^{K} d(y_i; \mu_i)$ is the sum of deviances depending on $\boldsymbol{\beta}$ only. This D is analogous to the sum of squared residuals in the linear regression model.

The score function for the regression coefficient $\boldsymbol{\beta}$ is

$$s(\mathbf{y}; \boldsymbol{\beta}) = \frac{\partial \ell(\boldsymbol{\theta})}{\partial \boldsymbol{\beta}} = -\frac{1}{2\sigma^2} \sum_{i=1}^{K} \frac{\partial d(y_i; \mu_i)}{\partial \mu_i} \frac{\partial \mu_i}{\partial \boldsymbol{\beta}}.$$

Denote the i-th linear predictor by $\eta_i = \mathbf{x}_i^T \boldsymbol{\beta}$, and denote the *deviance scores* by

$$\delta(y_i; \mu_i) = -\frac{1}{2} \frac{\partial d(y_i; \mu_i)}{\partial \mu_i}, \ i = 1, \ldots, K. \tag{2.16}$$

Note that

$$\frac{\partial \mu_i}{\partial \boldsymbol{\beta}} = \frac{\partial \mu_i}{\partial \eta_i} \frac{\partial \eta_i}{\partial \boldsymbol{\beta}} = \{\dot{g}(\mu_i)\}^{-1} \mathbf{x}_i,$$

where $\dot{g}(\mu)$ is the first order derivative of link function g *w.r.t* μ. Table 2.4 lists some commonly used link functions and their derivatives.

Then the score function for $\boldsymbol{\beta}$ takes the form

$$s(\mathbf{y}; \boldsymbol{\beta}) = \frac{1}{\sigma^2} \sum_{i=1}^{K} \mathbf{x}_i \frac{1}{\dot{g}(\mu_i)} \delta(y_i; \mu_i). \tag{2.17}$$

Moreover, the score equation leading to the maximum likelihood estimate of the $\boldsymbol{\beta}$ is

$$\sum_{i=1}^{K} \mathbf{x}_i \frac{1}{\dot{g}(\mu_i)} \delta(y_i; \mu_i) = 0. \tag{2.18}$$

Table 2.4. Some common link functions and derivatives. NB and IG stand for Negative binomial and Inverse Gaussian, respectively.

Model	Link g	Derivative $\dot{g}$	Domain Ω
Binomial or simplex	$\log\left(\frac{\mu}{1-\mu}\right)$	$\frac{1}{\mu(1-\mu)}$	$\mu \in (0,1)$
Poisson, NB, gamma, or IG	$\log(\mu)$	$\frac{1}{\mu}$	$\mu \in (0,\infty)$
Gamma	$\frac{1}{\mu}$	$-\frac{1}{\mu^2}$	$\mu \in (0,\infty)$
von Mises	$\tan(\mu/2)$	$\frac{1}{2}\sec^2(\mu/2)$	$\mu \in [-\pi,\pi)$

Note that this equation does not involve the dispersion parameter σ^2. Under some mild regularity conditions, the resulting ML estimator $\widehat{\boldsymbol{\beta}}_K$, which is the solution to the score equation (2.18), is consistent

$$\widehat{\boldsymbol{\beta}}_K \xrightarrow{p} \boldsymbol{\beta} \text{ as } K \to \infty,$$

and asymptotically normal with mean 0 and covariance matrix $\mathbf{i}^{-1}(\boldsymbol{\theta})$. Here $\mathbf{i}(\boldsymbol{\theta})$ is the Fisher information matrix given by

$$\begin{aligned}
\mathbf{i}(\boldsymbol{\theta}) &= -\mathrm{E}\{\dot{s}(\mathbf{Y};\boldsymbol{\beta})\} \\
&= \frac{1}{\sigma^2}\sum_{i=1}^{K} \mathbf{x}_i \frac{1}{\{\dot{g}(\mu_i)\}^2}\mathrm{E}\{-\dot{\delta}(Y_i;\mu_i)\}\mathbf{x}_i^T \\
&= \frac{1}{\sigma^2}\sum_{i=1}^{K} \mathbf{x}_i u_i^{-1}\mathbf{x}_i^T \\
&= \mathbf{X}^T U^{-1}\mathbf{X}/\sigma^2, \qquad (2.19)
\end{aligned}$$

where $\mathbf{X}$ is a $K \times (p+1)$ matrix with the i-th row being the $\mathbf{x}_i^T$, and U is a diagonal matrix with the i-th diagonal element u_i given by

$$u_i = \frac{\{\dot{g}(\mu_i)\}^2}{\mathrm{E}\{-\dot{\delta}(Y_i;\mu_i)\}}, \quad i = 1,\ldots,K. \qquad (2.20)$$

When the dispersion parameter σ^2 is present in the model, the ML estimation for the dispersion parameter σ^2 can be derived similarly, if the normalizing term $a(y;\sigma^2)$ is simple enough to allow such a derivation, such as the case of the normal distribution. However, in many cases, the term $a(\cdot)$ has no closed form expression and its derivative *w.r.t.* σ^2 may appear too complicated to be numerically solvable. In this case, two methods have been suggested to acquire the estimation for σ^2. The first method is to invoke the small dispersion asymptotic normality (Proposition 2.5), where subject to a constant,

$$\log a(y;\sigma^2) \simeq -\frac{1}{2}\log \sigma^2.$$

Applying this approximation in the log-likelihood (2.15) and differentiating the resulting approximate log-likelihood *w.r.t.* σ^2, one can obtain an equation as follows,

$$-\frac{K}{2\sigma^2} + \frac{1}{2\sigma^4}D(\mathbf{y};\boldsymbol{\mu}) = 0.$$

Solution to this equation gives an estimator of the dispersion parameter σ^2,

$$\widehat{\sigma}^2 = \frac{1}{K}D(\mathbf{y};\widehat{\boldsymbol{\mu}}) = \frac{1}{K}\sum_{i=1}^{K} d(y_i;\hat{\mu}_i). \tag{2.21}$$

This book refers this estimator to as *the Jørgensen estimator* of the dispersion parameter, which in fact is an average of the estimated unit deviances.

However, the Jørgensen estimator is not, in general, unbiased even if the adjustment on the degrees of freedom, $K - (p+1)$ is made to replace K. Moreover, this formula is recommended when the dispersion parameter σ^2 is small, say less than 5.

To obtain an unbiased estimator of the dispersion parameter σ^2, the second method utilizes a moment property given in the following proposition.

Proposition 2.17. *Let* $Y \sim DM(\mu, \sigma^2)$ *with a regular unit deviance* $d(y;\mu)$*. Then,*

$$E\{\delta(Y;\mu)\} = 0,$$
$$Var\{\delta(Y;\mu)\} = \sigma^2 E\{-\dot{\delta}(Y;\mu)\},$$

where $\dot{\delta}$ *is the first order derivative of the deviance score given in (2.16)* w.r.t. μ.

Proof. Differentiating both sides of equation $\int p(y;\mu,\sigma^2)dy = 1$ *w.r.t.* μ gives

$$-\frac{1}{2\sigma^2}\int \dot{d}(y;\mu)p(y;\mu,\sigma^2)dy = 0,$$

or $\mathrm{E}\{\dot{d}(Y;\mu)\} = 0$. Differentiating the above equation again *w.r.t.* μ, we obtain

$$-\frac{1}{2\sigma^2}\int \{\dot{d}(y;\mu)\}^2 p(y;\mu,\sigma^2)dy + \int \ddot{d}(y;\mu)p(y;\mu,\sigma^2)dy = 0,$$

or equivalently

$$\mathrm{E}\{\ddot{d}(Y;\mu)\} = \frac{1}{2\sigma^2}\mathrm{E}\{\dot{d}(Y;\mu)\}^2 = \frac{1}{2\sigma^2}\mathrm{Var}\{\dot{d}(Y;\mu)\}.$$

According to (2.16), this relation can be rewritten as follows,

$$\mathrm{Var}\{\delta(Y;\mu)\} = \sigma^2\mathrm{E}\{-\dot{\delta}(Y;\mu)\}.$$

Based on this result, one can consistently estimate the dispersion parameter σ^2 by the method of moments:

$$\widehat{\sigma}^2 = \frac{\sum_{i=1}^K (\delta_i - \bar{\delta})^2}{\sum_{i=1}^K (-\dot{\delta}_i)}, \tag{2.22}$$

where $\delta_i = \delta(y_i; \hat{\mu}_i)$, $\dot{\delta}_i = \dot{\delta}(y_i; \hat{\mu}_i)$ and $\bar{\delta} = \frac{1}{K}\sum_i \delta_i$.

2.6.2 MLE in the ED Models

Now return to the special case of the GLMs based on the ED models. For the unit deviance of the ED model given in (2.11), it is easy to see

$$\delta(y; \mu) = \frac{y - \mu}{V(\mu)}. \tag{2.23}$$

It follows that the score equation (2.18) becomes

$$\sum_{i=1}^K \mathbf{x}_i \frac{1}{\dot{g}(\mu_i)V(\mu_i)}(y_i - \mu_i) = 0.$$

Let $w_i = \dot{g}(\mu_i)V(\mu_i)$. Then the score equation can be re-expressed as of the form

$$\sum_{i=1}^K \mathbf{x}_i w_i^{-1}(y_i - \mu_i) = 0,$$

or in the matrix notation,

$$\mathbf{X}^T W^{-1}(\mathbf{y} - \boldsymbol{\mu}) = 0,$$

where $W = \text{diag}(w_1, \ldots, w_K)$. The following result is useful to calculate the Fisher information.

Proposition 2.18. *Suppose $Y \sim ED(\mu, \sigma^2)$. Then,*

$$E\{-\dot{\delta}(Y; \mu)\} = \frac{1}{V(\mu)},$$

where $\dot{\delta}(y; \mu)$ is the first order derivative of the deviance score $\delta(y; \mu)$ w.r.t. *μ.*

Proof. Differentiating δ in (2.16) *w.r.t.* μ gives

$$-\dot{\delta}(y; \mu) = \frac{1}{V(\mu)} + \frac{(y - \mu)\dot{V}(\mu)}{V^2(\mu)},$$

which leads to

$$\mathrm{E}\{-\dot{\delta}(Y; \mu)\} = \frac{1}{V(\mu)},$$

because $\mathrm{E}(Y) = \mu$ in the ED model.

In the Fisher information matrix $\mathbf{i}(\boldsymbol{\theta})$ of (2.19), $\mathbf{i}(\boldsymbol{\theta}) = \mathbf{X}^T U^{-1}\mathbf{X}/\sigma^2$, U is a diagonal matrix whose i-th diagonal element can be simplified as

$$u_i = \{\dot{g}(\mu_i)\}^2 V(\mu_i).$$

Furthermore, if the canonical link function $g = \tau^{-1}(\cdot)$ is chosen, then a further simplification leads to $w_i = 1$ and $u_i = 1/V(\mu_i)$ because in this case, $\dot{g}(\mu_i) = 1/V(\mu_i)$. So, the matrix W becomes the identity matrix and the matrix U is determined by the reciprocals of the variance functions.

It is interesting to note that the choice of the canonical link simplifies both score function and Fisher information. In summary, under the canonical link function, the score equation of an ED GLM is

$$\sum_{i=1}^{K} \mathbf{x}_i(y_i - \mu_i) = 0, \text{ or } \mathbf{X}^T(\mathbf{y} - \boldsymbol{\mu}) = 0,$$

and the Fisher information takes the form

$$\mathbf{i}(\boldsymbol{\theta}) = \mathbf{X}^T U^{-1}\mathbf{X}/\sigma^2$$

where $U = \text{diag}(u_1, \ldots, u_K)$, a diagonal matrix with variance function $V(\mu_i)$ as the i-th diagonal element.

Each ED model holds the so-called mean-variance relation, *i.e.* $\text{Var}(Y) = \sigma^2 V(\mu)$, which may be used to obtain a consistent estimator of the dispersion parameter σ^2 given as follows:

$$\widehat{\sigma}^2 = \frac{1}{K-p-1}\sum_{i=1}^{K} \hat{r}_{p,i}^2 = \frac{1}{K-p-1}\sum_{i=1}^{K}\left\{\frac{y_i - \hat{\mu}_i}{\sqrt{V(\hat{\mu}_i)}}\right\}^2,$$

where $\hat{r}_p$ is the Pearson residual listed in Table 2.3. This estimator is referred to as the Pearson estimator of the dispersion parameter σ^2. In fact, the relation given in Proposition 2.17 is equivalent to this mean-variance relation for the ED models, simply because of Proposition 2.18.

2.6.3 MLE in the Simplex GLM

The GLM for binary data or logistic regression model, the GLM for count data or log-linear regression model, and the GLM for positive continuous data or gamma regression model have been extensively illustrated in the literature. Interested readers can find examples of these ED GLMs easily in many references such as McCullagh and Nelder's (1989). This section supplies two non-ED GLMs based, respectively, on the simplex distribution and the von Mises distribution. Both are not available in the classical theory of GLMs.

In the ED GLMs, both score equation and Fisher information can be treated as a special case of weighted least squares estimation, due to the fact that its first order derivative of the unit deviance is $(y-\mu)/V(\mu)$, which is linear in y. However, this linearity no longer holds for a DM GLM outside the class of the ED GLMs. The simplex distribution is one of such examples. A simplex model $S^-(\mu;\sigma^2)$ has the density given by

$$p(y;\mu,\sigma^2)=[2\pi\sigma^2\{y(1-y)\}^3]^{-1/2}\exp\left\{-\frac{1}{2\sigma^2}d(y;\mu)\right\},\ y\in(0,1),\mu\in(0,1),$$

with the unit deviance function

$$d(y;\mu)=\frac{(y-\mu)^2}{y(1-y)\mu^2(1-\mu)^2},\ y\in(0,1),\mu\in(0,1),$$

where $\mu=\mathrm{E}(Y)$ is the mean. The unit variance function is $V(\mu)=\mu^3(1-\mu)^3$, obtained from (2.4).

For a non-ED GLM, the canonical link function no longer helps to simplify the weights u_i or the w_i, because the density does not explicitly involve the cumulant generating function $\kappa(\cdot)$ as in the ED GLM. For the simplex distribution, since $\mu\in(0,1)$, one may take the logit as the link function to formulate the systematic component:

$$\log\frac{\mu}{1-\mu}=\mathbf{x}^T\boldsymbol{\beta}.$$

According to Table 2.4, $\dot{g}(\mu)=\{\mu(1-\mu)\}^{-1}$. It follows from (2.18) that the score equation for the regression parameter $\boldsymbol{\beta}$ is

$$\sum_{i=1}^{K}\mathbf{x}_i\{\mu_i(1-\mu_i)\}\delta(y_i;\mu_i)=0, \tag{2.24}$$

where the deviance score is

$$\begin{aligned}\delta(y;\mu)&=-\frac{1}{2}\dot{d}(y;\mu)\\&=\frac{y-\mu}{\mu(1-\mu)}\left\{d(y;\mu)+\frac{1}{\mu^2(1-\mu)^2}\right\}.\end{aligned} \tag{2.25}$$

It is clear that this δ function is nonlinear in both y and μ. Solving nonlinear equation (2.24) can be done iteratively by the Newton-Raphson algorithm or quasi-Newton algorithm. The calculation of the Fisher information requires the knowledge of $\mathrm{E}\{-\dot{\delta}(Y_i;\mu_i)\}$. It is equivalent to deriving $\frac{1}{2}\mathrm{E}\ddot{d}(Y_i;\mu_i)$.

Differentiating $\dot{d}$ *w.r.t.* μ gives

$$\begin{aligned}\frac{1}{2}\ddot{d}(y;\mu)&=\frac{1}{\mu(1-\mu)}d(y;\mu)+\frac{1-2\mu}{\mu^2(1-\mu)^2}(y-\mu)d(y;\mu)\\&\quad+\frac{1}{\mu^3(1-\mu)^3}+\frac{1-2\mu}{\mu^4(1-\mu)^4}(y-\mu)\\&\quad-\frac{1}{\mu(1-\mu)}(y-\mu)\dot{d}(y;\mu)-\frac{2(2\mu-1)}{\mu^4(1-\mu)^4}(y-\mu).\end{aligned} \tag{2.26}$$

Hence,

$$
\begin{aligned}
\frac{1}{2}\mathrm{E}\{\ddot{d}(Y;\mu)\} &= \frac{1}{\mu(1-\mu)}\Big[\mathrm{E}\{d(Y;\mu)\} - \mathrm{E}\{(Y-\mu)\dot{d}(Y;\mu)\}\Big] \\
&\quad + \frac{1-2\mu}{\mu^2(1-\mu)^2}\mathrm{E}\{(Y-\mu)d(Y;\mu)\} + \frac{1}{\mu^3(1-\mu)^3} \\
&= \frac{3\sigma^2}{\mu(1-\mu)} + \frac{1}{\mu^3(1-\mu)^3},
\end{aligned} \tag{2.27}
$$

where the last equation holds by applying part (e) of Proposition 2.19 below. Therefore, the Fisher information is

$$
\mathbf{i}(\boldsymbol{\beta}) = \frac{1}{\sigma^2}\sum_{i=1}^{K} \mathbf{x}_i u_i^{-1} \mathbf{x}_i^T,
$$

where

$$
u_i = \frac{\mu_i(1-\mu_i)}{1+3\sigma^2\{\mu_i(1-\mu_i)\}^2}, \quad i = 1,\ldots,K.
$$

As seen in (2.26), the first order derivative of the deviance score $\dot{\delta}$ appears tedious, but its expectation in (2.27) is much simplified. Therefore, it is appealing to implement the Fisher-scoring algorithm in the search for the solution to the score equation (2.24). One complication in the application of Fisher-scoring algorithm is the involvement of the dispersion parameter σ^2. This can be resolved by replacing σ^2 with a $\sqrt{K}$-consistent estimate, $\hat{\sigma}^2$. A consistent estimate of such a type can be obtained by the method of moments. For example, the property (a) in Proposition 2.19 is useful to establish an estimate of σ^2 as follows:

$$
\hat{\sigma}^2 = \frac{1}{K-(p+1)}\sum_{i=1}^{K} d(y_i;\hat{\mu}_i). \tag{2.28}
$$

Proposition 2.19. *Suppose $Y \sim S^-(\mu;\sigma^2)$ with mean μ and dispersion σ^2. Then,*

(a) $E\{d(Y;\mu)\} = \sigma^2$;
(b) $E\{(Y-\mu)\dot{d}(Y;\mu)\} = -2\sigma^2$;
(c) $E\{(Y-\mu)d(Y;\mu)\} = 0$;
(d) $E\{\dot{d}(Y;\mu)\} = 0$;
(e) $\frac{1}{2}E\{\ddot{d}(Y;\mu)\} = \frac{3\sigma^2}{\mu(1-\mu)} + \frac{1}{\mu^3(1-\mu)^3}$;
(f) $Var\{d(Y;\mu)\} = 2\left(\sigma^2\right)^2$;
(g) $Var\{\delta(Y;\mu)\} = \frac{3\sigma^4}{\mu(1-\mu)} + \frac{\sigma^2}{\mu^3(1-\mu)^3}$.

The following lemma is needed in order to prove Proposition 2.19.

Lemma 2.20 (Jørgensen, 1997, P.191). *Consider a dispersion model $DM(\mu, \sigma^2)$ whose density takes the form:*

$$f(y; \mu, \lambda) = c_\alpha(\mu, \lambda) y^{\alpha-1} \exp\left\{-\frac{\lambda(y-\mu)^2}{2y\mu^{1-2\alpha}}\right\},$$

where $\lambda = 1/\sigma^2$ and the normalization constant is defined by

$$\frac{1}{c_\alpha(\mu, \lambda)} = 2K_\alpha(\lambda\mu^{2\alpha}) e^{\lambda\mu^{2\alpha}} \mu^\alpha.$$

Then the asymptotic expansion of $1/c_\alpha(\mu, \lambda)$ is given by, for large λ,

$$\left\{\frac{2\pi}{\lambda}\right\}^{\frac{1}{2}} \left\{1 + \frac{4\alpha^2-1}{8\lambda\mu^{2\alpha}} + \frac{(4\alpha^2-1)(4\alpha^2-9)}{2!(8\lambda\mu^{2\alpha})^2} + \frac{(4\alpha^2-1)(4\alpha^2-9)(4\alpha^2-25)}{3!(8\lambda\mu^{2\alpha})^3} + \cdots\right\}.$$

The proof of Proposition 2.19 is given as follows.

Proof. First prove part (b). Note that

$$0 = \mathrm{E}[(Y-\mu)] = \int_0^1 (y-\mu) p(y; \mu, \sigma^2) dy,$$

and differentiating both sides of the equation with respect to μ gives

$$0 = -1 - \frac{1}{2\sigma^2} \mathrm{E}[(Y-\mu)\dot{d}(Y; \mu)],$$

and hence $\mathrm{E}[(Y-\mu)\dot{d}(Y;\mu)] = -2\sigma^2$.

To prove part (a) and part (c), take the following transformations for both y and μ,

$$x = \frac{y}{1-y}, \quad \xi = \frac{\mu}{1-\mu}$$

and rewrite the two expectations in the following forms:

$$\begin{aligned}\mathrm{E}[d(Y;\mu)] &= \int_0^1 d(y;\mu) p(y;\mu,\sigma^2) dy \\ &= \sqrt{\frac{\lambda}{2\pi}} \frac{(1+\xi)^2}{\xi^2} \int_0^\infty \left\{x^{\frac{1}{2}} + (1-2\xi)x^{-\frac{1}{2}}\right. \\ &\quad \left. + \xi(\xi-2)x^{-\frac{3}{2}} + \xi^2 x^{-\frac{5}{2}}\right\} f(x;\xi,\lambda) dx,\end{aligned}$$

and

$$
\begin{aligned}
\mathrm{E}[(Y-\mu)d(Y;\mu)] &= \int_0^1 (y-\mu)d(y;\mu)p(y;\mu,\sigma^2)dy \\
&= \sqrt{\frac{\lambda}{2\pi}}\frac{1+\xi}{\xi^2}\int_0^\infty \Big\{x^{\frac{1}{2}} - 3\xi x^{-\frac{1}{2}} \\
&\quad +3\xi^2 x^{-\frac{3}{2}} - \xi^3 x^{-\frac{5}{2}}\Big\} f(x;\xi,\lambda)dx,
\end{aligned}
$$

where $\lambda = 1/\sigma^2$ and

$$
f(x;\xi,\lambda) = \exp\left\{-\frac{\lambda}{2}\frac{(1+\xi)^2}{\xi^2}\frac{(x-\xi)^2}{x}\right\}.
$$

Applying Lemma 2.20 leads to

$$
\int_0^\infty x^{\frac{1}{2}} f(x;\xi,\lambda)dx = \left(\frac{2\pi}{\lambda}\right)^{\frac{1}{2}} \frac{\xi^3 + \lambda\xi^2(1+\xi)^2}{\lambda(1+\xi)^3},
$$

$$
\int_0^\infty x^{-\frac{1}{2}} f(x;\xi,\lambda)dx = \left(\frac{2\pi}{\lambda}\right)^{\frac{1}{2}} \frac{\xi}{1+\xi},
$$

$$
\int_0^\infty x^{-\frac{3}{2}} f(x;\xi,\lambda)dx = \left(\frac{2\pi}{\lambda}\right)^{\frac{1}{2}} \frac{1}{1+\xi},
$$

and

$$
\int_0^\infty x^{-\frac{5}{2}} f(x;\xi,\lambda)dx = \left(\frac{2\pi}{\lambda}\right)^{\frac{1}{2}} \frac{\xi + \lambda(1+\xi)^2}{\lambda\xi(1+\xi)^3}.
$$

Plugging these results into the expressions above leads to

$$
\mathrm{E}\{d(Y;\mu)\} = 1/\lambda = \sigma^2 \quad \text{and} \quad \mathrm{E}\{(Y-\mu)d(Y;\mu)\} = 0.
$$

Part (d) is given by applying part (c) to (2.25) and then taking expectation. Also, part (e) is proved by applying parts (a), (b), and (c) to (2.27).

By part (a), to prove part (f), it is sufficient to show that

$$
\mathrm{E}\{d^2(Y;\mu)\} = 3\left(\sigma^2\right)^2.
$$

Simple algebra leads to

$$
\begin{aligned}
\mathrm{E}\{d^2(Y;\mu)\} &= \int_0^1 d^2(y;\mu)p(y;\mu,\sigma^2)dy \\
&= \sqrt{\frac{\lambda}{2\pi}}\frac{(1+\xi)^4}{\xi^4}\int_0^\infty \{x^{\frac{3}{2}} + (1-4\xi)x^{\frac{1}{2}} \\
&\quad +2\xi(3\xi-2)x^{-\frac{1}{2}} + 2\xi^2(3-2\xi)x^{-\frac{3}{2}} \\
&\quad +\xi^3(\xi-4)x^{-\frac{5}{2}} + \xi^4 x^{-\frac{7}{2}}\}f(x;\xi,\lambda)dx. \qquad (2.29)
\end{aligned}
$$

An application of Lemma 2.20 again results in

$$\int_0^\infty x^{\frac{3}{2}} f(x;\xi,\lambda)dx = \left(\frac{2\pi}{\lambda}\right)^{\frac{1}{2}} \frac{\lambda^2\xi^3(1+\xi)^4 + 3\lambda\xi^4(1+\xi)^2 + 3\xi^5}{\lambda^2(1+\xi)^5}$$

and

$$\int_0^\infty x^{-\frac{7}{2}} f(x;\xi,\lambda)dx = \left(\frac{2\pi}{\lambda}\right)^{\frac{1}{2}} \frac{\lambda^2(1+\xi)^4 + 3\lambda\xi(1+\xi)^2 + 3\xi^2}{\lambda^2\xi^2(1+\xi)^5}.$$

Based on these results, the integration (2.29) can be simplified as

$$\mathrm{E}\{d^2(Y;\mu)\} = 3\left(\sigma^2\right)^2.$$

Part (g) can be proved by applying part (e) in the relation between $\dot{d}$ and δ from Proposition 2.17.

In the application of the simplex GLM, one issue that deserves some attention is whether there is much difference between the normal linear model based on logit-transformed data, $\log\{y_i/(1-y_i)\}$, and the direct simplex GLM. The difference between the two models is the former models $\mathrm{E}[\log\{Y_i/(1-Y_i)\}]$ as a linear function of covariates, and the latter models $\mu_i = \mathrm{E}(Y_i)$ via $\log\{\mu_i/(1-\mu_i)\}$ as a linear function of covariates. Apparently the direct GLM approach gives rise to much ease in interpretation.

The following simulation study suggests that when the dispersion parameter σ^2 is large, the performance of the logit-transformed analysis may be questionable, if the data are really from a simplex distributed population.

The simulation study assumes the proportional data are generated independently from the following simplex distribution,

$$Y_i \sim S^-(\mu_i, \sigma^2),\ i = 1, \ldots, 150,$$

where the mean follows a GLM of the following form:

$$\mathrm{logit}(\mu_i) = \beta_0 + \beta_1 T_i + \beta_2 S_i.$$

Covariates T and S are presumably drug dosage levels indicated by $\{-1, 0, 1\}$ for each 50 subjects and illness severity score ranged in $\{0, 1, 2, 3, 4, 5, 6\}$ that is randomly assumed to each subject by a binomial distribution $\mathrm{B}(7, 0.5)$. The true values of regression coefficients are set as $\beta_0 = 0.5, \beta_1 = -0.5, \beta_2 = 0.5$, and the dispersion parameter $\sigma^2 = 0.5, 50, 200, 400$.

For each combination of parameters, the same simulated data was fit by the simplex GLM for the original responses and the normal linear model for logit-transformed responses. Two hundred replications were done for each case. Results are summarized in Table 2.5, including the averaged estimates, standard deviations of 200 replicated estimates, and standard errors of estimates calculated from the Fisher information.

Table 2.5. Summary of the simulation results for the comparison between the direct simplex GLM analysis and logit-transformed linear model analysis.

Parameter	Simplex GLM			Logit-Trans LM		
True	Mean	Std Dev	Std Err	Mean	Std Dev	Std Err
$\sigma^2 = 0.5$						
$\beta_0(0.5)$	.4996	.0280	.0254	.5089	.0288	.0263
$\beta_1(-0.5)$	−.5023	.0330	.0308	−.5110	.0345	.0322
$\beta_2(0.5)$	.5015	.0195	.0205	.5101	.0199	.0222
$\sigma^2 = 50$						
$\beta_0(0.5)$	.5062	.0983	.0960	.8057	.1769	.1752
$\beta_1(-0.5)$	−.5068	.1141	.1185	−.7998	.2065	.2148
$\beta_2(0.5)$	.5170	.0860	.0835	.8153	.1366	.1483
$\sigma^2 = 200$						
$\beta_0(0.5)$	.5060	.1145	.1021	1.0162	.2741	.2541
$\beta_1(-0.5)$	−.5262	.1346	.1263	−1.0479	.3218	.3114
$\beta_2(0.5)$	.5238	.0971	.0899	1.0430	.1919	.2150
$\sigma^2 = 400$						
$\beta_0(0.5)$	.5253	.0963	.1032	1.2306	.2767	.2980
$\beta_1(-0.5)$	−.5001	.1486	.1275	−1.1336	.3888	.3652
$\beta_2(0.5)$	.5165	.1000	.0909	1.1686	.2286	.2521

This simulation study indicates that *(i)* when the dispersion parameter σ^2 is small, the logit-transformed analysis appears fine, with little bias and little loss of efficiency, because of small-dispersion asymptotic normality; *(ii)* when the dispersion parameter is large, the estimation based on the logit-transformed analysis is unacceptable, in which bias increases and efficiency drops when the σ^2 increases.

One may try to make a similar comparison by simulating data from the normal distribution as well as from the beta distribution. Our simulation study suggested that in the case of normal data, the direct simplex GLM performed nearly as well as the normal model, with only a marginal loss of efficiency; in the case of beta data, the simplex GLM clearly outperformed the normal linear model. Interested readers can verify the findings through their own simulation studies.

Example 2.21 (Body Fat Index).

Penrose et al. (1985) reports a dataset consisting of 19 variables, including percentage of body fat, age, weight, height, and ten body circumference measurements (e.g., abdomen) for 252 men. This dataset is available at http://www.amstat.org/publications/jse/jse_data_archive.html. Body fat, a measure of health, is estimated through an underwater weighing technique. Percentage of body fat may be then calculated by either Brozek's equation or Siri's equation. Fitting body fat to the other measurements using GLM provides a convenient way of estimating body fat for men using only a scale and a measuring tape.

In this example, the simplex GLM is illustrated simply by fitting the the body fat index as a function of covariate age. Suppose the percentage of body fat $Y_i \sim S^-(\mu_i, \sigma^2)$, where

$$\log \frac{\mu_i}{1-\mu_i} = \beta_0 + \beta_1 \text{age}.$$

The Fisher-scoring algorithm was applied to obtain the estimates of the regression coefficients and the standard errors were calculated from the Fisher information. The results were summarized in Table 2.6, in which the dispersion parameter is estimated by the method of moments in (2.28). Clearly, from the results given in Table 2.6, age is an important predictor to the percentage of body fat in both Brozek's and Siri's equations. The dispersion σ^2 is found not small in this study, so it might be worrisome for the appropriateness of either a direct linear model analysis (with no transformation on the response) or logit-transformed linear model analysis. Some further investigations are needed to elucidate the choice of modeling approach in this data analysis.

Table 2.6. Results in the regression analysis of body fat percentage using the simplex GLM.

	Parameter		
Body-fat measure	Intercept (Std Err)	Age (Std Err)	σ^2
Brozek's	−2.7929(0.3304)	0.0193(0.0070)	55.9759
Siri's	−2.8258(0.3309)	0.0202(0.0070)	57.0353

2.6.4 MLE in the von Mises GLM

Angular data are a special case of circular data. Mardia (1972) has presented a general framework of estimation and inference in the models for circular data. Fisher (1993) gave an overview of the state-of-art of research in this field. Although the analysis of circular data is an old topic, there have been recent developments in applied areas, such as multi-dimensional circular data (Fisher, 1993; Rivest, 1997; Breckling, 1989), time series of circular observations (Accardi et al. 1987; Fisher and Lee 1994; Coles 1998), and longitudinal circular data (Artes et al. 2000; D'Elia et al. 2001).

The von Mises distribution is another example of the DM model but not of an ED model. The density of a von Mises distribution takes the form

$$p(y; \mu, \sigma^2) = \frac{1}{2\pi I_0(\lambda)} \exp\{\lambda \cos(y - \mu)\}, \; y \in [-\pi, \pi), \tag{2.30}$$

where $\mu \in [-\pi, \pi)$ is the mean, $\lambda = 1/\sigma^2 > 0$ is the index parameter, and $I_0(\lambda)$ is the modified Bessel function of the first kind of order 0, given by

$$I_0(\lambda) = \frac{1}{2\pi} \int_{-\pi}^{\pi} \exp\{\lambda \cos(y)\} dy.$$

It is easy to rewrite the von Mises density in the form of DM model with the unit deviance function given by

$$d(y; \mu) = 2\{1 - \cos(y - \mu)\}, \; y, \mu \in [-\pi, \pi),$$

whose first and second order derivatives *w.r.t.* μ are, respectively,

$$\dot{d} = -2\sin(y - \mu), \; \ddot{d} = 2\cos(y - \mu).$$

It follows that the unit variance function is $V(\mu) = 1$ for $\mu \in [-\pi, \pi)$ and the deviance score $\delta(y; \mu) = \sin(y - \mu)$.

Now consider a GLM for directional (circular or angular) data, where $Y_i \sim vM(\mu_i, \sigma^2)$, associated with p-element vector of covariates $\mathbf{x}_i$. According to Fisher and Lee (1992) or Fisher (1993, Section 6.4), a GLM for the mean direction $\mu_i = \mathrm{E}(Y_i|\mathbf{x}_i)$ may be formulated as follows:

$$\mu_i = \mu_0 + 2\arctan(x_{i1}\beta_1 + \cdots + x_{ip}\beta_p), \tag{2.31}$$

where μ_0 is an offset mean parameter representing the origin. If $Y_i^* = Y_i - \mu_0$ is taken as a surrogate response, then the corresponding mean direction is $\mu_i^* = \mu_i - \mu_0 = 2\arctan(\eta_i)$ with the origin of 0°. This implies

$$\tan(\mu_i^*/2) = \eta_i = \mathbf{x}_i^T \boldsymbol{\beta},$$

where the intercept term is not included, because of the 0° origin. Clearly, in this GLM, the link function $g(z) = \tan(z/2)$ and $\dot{g}(z) = \frac{1}{2}\sec^2(z/2)$, as shown in Table 2.4. To estimate the regression parameter $\boldsymbol{\beta}$, formula (2.18) is applied here to yield the following score equation:

$$
\begin{aligned}
s(\mathbf{y};\boldsymbol{\beta}) &= \lambda \sum_{i=1}^{K} \mathbf{x}_i \frac{1}{\dot{g}(\mu_i^*)} \delta(y_i^*; \mu_i^*) \\
&= 2\lambda \sum_{i=1}^{K} \mathbf{x}_i \left(\frac{1}{1+\eta_i^2} \right) \sin(y_i^* - \mu_i^*) \\
&= 2\lambda \sum_{i=1}^{K} \mathbf{x}_i \left(\frac{1}{1+\eta_i^2} \right) \sin(y_i - \mu_0 - 2\mathrm{arctan}(\mathbf{x}_i^T \boldsymbol{\beta})),
\end{aligned}
$$

where the identity of $\sec^2(\arctan(a)) = 1 + a^2$ is used. The MLE of $\boldsymbol{\beta}$ is the solution to the score equation

$$
s(\mathbf{y};\boldsymbol{\beta}) = 0. \tag{2.32}
$$

To find the Fisher Information for $\widehat{\boldsymbol{\beta}}$, first note that the surrogate response $Y_i^* \sim vM(\mu_i^*, \sigma^2)$, and then

$$
\mathrm{E}\{-\dot{\delta}(Y_i^*; \mu_i^*)\} = \mathrm{E}\{\cos(Y_i^* - \mu_i^*)\} = \frac{I_1(\lambda)}{I_0(\lambda)},
$$

where $I_1(\lambda)$ is the first order modified Bessel function of the first kind given by

$$
I_1(\lambda) = \frac{1}{2\pi} \int_{-\pi}^{\pi} \cos(y) \exp\{\lambda \cos(y)\} dy.
$$

Denote the mean resultant length by $A_1(\sigma^2) = I_1(\sigma^{-2})/I_0(\sigma^{-2})$. Then the weights u_i in (2.20) are found as

$$
u_i = \frac{(1+\eta_i^2)^2}{4A_1(\sigma^2)}, \quad i = 1, \ldots, K.
$$

Moreover, the Fisher Information for the $\widehat{\boldsymbol{\beta}}$ is $\mathbf{i}(\boldsymbol{\beta}) = \mathbf{X}^T U^{-1} \mathbf{X}/\sigma^2$, with $U = \mathrm{diag}(u_1, \ldots, u_K)$.

To estimate the parameter μ_0 and the dispersion parameter σ^2, the MLE may be also employed. The log likelihood is proportional to

$$
\ell(\boldsymbol{\theta}) \propto -K \log I_0(\lambda) + \lambda \sum_{i=1}^{K} \cos(y_i - \mu_0 - 2\mathrm{arctan}(\eta_i)),
$$

and the scores for μ_0 and λ are, respectively,

$$
\begin{aligned}
s(\mathbf{y};\mu_0) &= \lambda \sum_{i=1}^{K} \sin(y_i - \mu_0 - 2\mathrm{arctan}(\eta_i)), \\
s(\mathbf{y};\lambda) &= -K \frac{\dot{I}_0(\lambda)}{I_0(\lambda)} + \sum_{i=1}^{K} \cos(y_i - \mu_0 - 2\mathrm{arctan}(\eta_i)).
\end{aligned}
$$

Also,

$$
\begin{aligned}
-\mathrm{E}\{\dot{s}_{\mu_0}(\mathbf{y};\mu_0)\} &= K\lambda A_1(\lambda),\\
-\mathrm{E}\{\dot{s}_{\boldsymbol{\beta}}(\mathbf{y};\mu_0)\} &= 4\lambda A_1(\lambda)\sum_{i=1}^{K}\mathbf{x}_i\frac{1}{(1+\eta_i^2)^2},\\
-\mathrm{E}\{\dot{s}_{\lambda}(\mathbf{y};\lambda)\} &= K\left\{\frac{\dot{I}_1(\lambda)}{I_0(\lambda)} - A_1^2(\lambda)\right\},\\
-\mathrm{E}\{\dot{s}_{\boldsymbol{\beta}}(\mathbf{y};\lambda)\} &= 0.
\end{aligned}
$$

It is easy to show that $\dot{I}_0(\lambda) = I_1(\lambda)$ and $\dot{I}_1(\lambda) = \frac{1}{2}\{I_1(\lambda) + I_0(\lambda)\}$. Let $\hat{\mu}_0$ and $\hat{\lambda}$ be the MLE. Then $(\hat{\mu}_0, \widehat{\boldsymbol{\beta}}, \hat{\lambda})$ will be the solution to the following joint score equations:

$$
\begin{aligned}
\begin{pmatrix} s(\mathbf{y};\mu_0)\\ s(\mathbf{y};\boldsymbol{\beta})\\ s(\mathbf{y};\lambda)\end{pmatrix} &= \begin{pmatrix} \lambda\sum_{i=1}^K \mathrm{diag}[1,\mathbf{x}_i][1, 2(1+\eta_i^2)^{-1}]^T \sin(y_i-\mu_0-2\mathrm{arctan}(\eta_i))\\ -KA_1(\lambda)+\sum_{i=1}^K\cos(y_i-\mu_0-2\mathrm{arctan}(\eta_i))\end{pmatrix}\\
&= \mathbf{0}.
\end{aligned}
\tag{2.33}
$$

The corresponding Fisher information matrix is

$$
\mathbf{i}(\mu_0,\boldsymbol{\beta},\lambda) = \begin{pmatrix} \lambda\sum_{i=1}^K \mathrm{diag}[1,\mathbf{x}_i][1,u_i^{-1}]^T[1,u_i^{-1}]\mathrm{diag}^T[1,\mathbf{x}_i] & 0\\ 0 & K\left\{\frac{1}{2}(A_1(\lambda)+1)-A_1^2(\lambda)\right\}\end{pmatrix}.
$$

One may use the iterative Fisher-scoring algorithm to solve jointly the score equation (2.33) for the MLE, which involves inverting the above Fisher information matrix at current values of the parameters. Alternatively, one may solve equations (2.32), and the following (2.34) and (2.35) in cycle,

$$
\hat{\mu}_0 = \arctan\left(\bar{S}/\bar{C}\right) \tag{2.34}
$$

$$
\hat{\lambda} = A_1^{-1}\left\{\frac{1}{K}\sum_{i=1}^K\cos(y_i-\mu_0-2\mathrm{arctan}(\eta_i))\right\}, \tag{2.35}
$$

where $A_1^{-1}\{\cdot\}$ is the inverse function of A_1, and

$$
\begin{aligned}
\bar{S} &= \frac{1}{K}\sum_{i=1}^K \sin(y_i - 2\mathrm{arctan}(\eta_i)),\\
\bar{C} &= \frac{1}{K}\sum_{i=1}^K \cos(y_i - 2\mathrm{arctan}(\eta_i)).
\end{aligned}
$$

It is known in the literature that when the sample size K is small, the MLE of σ^2 or λ appears to have some noticeable bias. Alternatively, one may use the moment property, $\text{Var}(Y) = 1 - A_1(\lambda)$, to obtain a consistent moment estimator,

$$\hat{\lambda}_{\text{mom}} = A_1^{-1} \left\{ 1 - \frac{1}{K-p-1} \sum_{i=1}^{K} (y_i - \hat{\mu}_i)^2 \right\}. \tag{2.36}$$

An `R` package `CircStats` provides functions to plot circular data (e.g., function `circ.plot`) and compute many quantities given above, such as $I_0(\lambda)$, $I_1(\lambda)$, and even $I_p(\lambda)$ for any integer p. In this package, another useful function is `circ.kappa`, which provides a bias correction for the MLE estimation for the index parameter $\lambda = 1/\sigma^2$. Interested readers can follow Problem 2.5 in Problem Set 2 (available at the book webpage) to gain some numerical experience with the analysis of circular data.

3

Inference Functions

3.1 Introduction

The theory of inference functions or estimating functions may be viewed as a generalization of the maximum likelihood theory. In a usual estimation problem, an estimate (for example, the MLE) is obtained as a solution to an equation of the form

$$\Psi(\text{data}; \boldsymbol{\theta}) = 0,$$

where $\boldsymbol{\theta}$ is the set of parameters of interest. This equation may be derived from a fully specified parametric model, say a logistic regression model with Ψ being the score function and $\boldsymbol{\theta}$ being the regression coefficient $\boldsymbol{\beta}$. As far as estimation concerns, the key device needed is a sort of equation from which an estimate of the parameter can be found. In other words, if one is able to directly come up with a "sensible" function Ψ without using any specific underlying probability distribution, an estimation procedure can proceed based only on this given function Ψ. This approach has been seen in several settings, including the least squares estimation for a linear regression model in which only moments are assumed for error terms, and M-estimation in the robust statistics literature. Another important application of this approach is the so-called quasi-likelihood estimation for regression coefficients in the context of GLMs, when the distribution assumption is violated due, for instance, to overdispersion. The next section will present a detailed discussion about the quasi-likelihood inference in GLMs.

The term *equation of estimation* was first used in Fisher (1935). Kimball (1946) presented a non-trivial example of inference function, where estimating equations are proposed to construct confidence regions for the parameters in Gumbel distributions. Later, McLeish and Small (1988) generalized Kimball's idea of *stable* estimating equations to establish the theory of sufficiency and ancillarity for inference functions.

The theory of optimal inference functions was first studied by Godambe (1960). In the same year Durbin (1960) introduced the notion of unbiased linear inference function in the time series analysis. Since then, inference functions have been drawing much attention to researchers in different areas, such as robust statistics, GLMs, and econometrics. This chapter will focus on a few elements of the inference function theory, in order to prepare the introduction to Liang and Zeger's generalized estimating equations (GEE) approach and quadratic inference functions (QIF) in the correlated data analysis. Readers can find more details of the inference functions from Godambe (1991), Heyde (1997), Hardin and Hilbe (2003), and Hall (2005), among others.

3.2 Quasi-Likelihood Inference in GLMs

It is not rare that practitioners encounter situations where the probability mechanism (e.g., density functions) by which the data are generated cannot be fully specified due to reasons such as the fact that the underlying biological theory is not yet fully understood or no substantial experience of analyzing similar data from previous studies is available. As a consequence, investigators were only able to impose assumptions on some aspects of the probability mechanism such as moments, but not on the full parametric distributions. Another scenario is that investigators knew from some preliminary analyses that certain aspects of the parametric model they intended to use for the data analysis were violated. One example is overdispersion in the GLM analysis. Overdispersion basically violates the mean-variance relation induced from a proper probability model, which prohibits investigators from using a specific parametric distribution for the data. Overdispersion may emerge from different data collection procedures, one of which is that the response variable is recorded as an aggregation of dependent variables.

To elucidate, let us consider a Poisson log-linear regression model for count data. In a standard GLM analysis, count responses are typically assumed to follow a Poisson distribution with mean μ, which is further assumed to take a log-linear form as follows,

$$\log(\mu) = \mathbf{x}^T \boldsymbol{\beta}.$$

The assumption of a Poisson distribution for the data implies the mean-variance relation of the following form

$$\text{Var}(Y) = \mu,$$

since the dispersion parameter $\sigma^2 = 1$ and the unit variance function is $V(\mu) = \mu$. This relation says that the variation of the assumed Poisson counts should take the same magnitude of its mean. In many practical cases, the variance of data appears to be substantially larger (or smaller) than its mean, referred to as *overdispersion (or underdispersion)*; hence, the mean-variance relation is no longer valid. One way to deal with overdispersed count data is to introduce a dispersion parameter σ^2 that inflates the Poisson variance as given by

$$\text{Var}(Y) = \sigma^2\mu, \ \sigma^2 > 1.$$

Obviously, this response variable Y satisfying such a new mean-variance relation is no longer Poisson distributed.

In the cases discussed above, the maximum likelihood estimation approach may not be applicable due to the unavailability of full density functions. Wedderburn (1974) proposed an idea of quasi-likelihood estimation for regression coefficients in the setting of GLMs. Also see McCullagh (1983). Suppose that part of the objective in data analysis can be addressed by the following regression model, specified only by the first two moments:

$$g(\mu) = \mathbf{x}^T\boldsymbol{\beta} \ \text{ and } \ \text{Var}(Y) = \sigma^2 V(\mu).$$

Similar model specifications have been considered in the least squares theory of classical linear regression models, where the first two moments of error terms are assumed. The least squares estimation for the regression coefficients can be carried out without a fully specified distribution for error terms.

In the context of GLMs, the ordinary least squares approach generally does not provide a consistent estimator for the coefficient $\boldsymbol{\beta}$. Instead, it is suggested to directly solve the equation that is originally derived from the MLE setting,

$$\Psi(\mathbf{y};\boldsymbol{\beta}) = \sum_{i=1}^{K} \frac{\partial \mu_i(\boldsymbol{\beta})^T}{\partial \boldsymbol{\beta}} \Sigma_i^{-1}(y_i - \mu_i(\boldsymbol{\beta})) = 0,$$

where $\Sigma_i = \sigma^2 V(\mu_i)$ is the variance of Y_i. Note that the utility of this estimating equation for $\widehat{\boldsymbol{\beta}}$ only requires the assumptions about the first two moments; namely, models for $\mu_i = \text{E}(Y_i)$ and $\Sigma_i = \text{Var}(Y_i)$. This inference function $\Psi(\mathbf{y};\boldsymbol{\beta})$ is referred to as a quasi-score function. It is also worth noting that this estimating equation $\Psi(\mathbf{y};\boldsymbol{\beta}) = 0$ does not estimate the dispersion parameter σ^2, because this parameter is factorized out of the expression and cancelled. Therefore, in the variance Σ_i what really matters is the variance function $V(\cdot)$, rather than the variance itself.

Why does this idea work for the estimation in the GLMs? It works because this quasi-score function preserves the two key properties that the real score function has:

(a) The quasi-score function is unbiased, namely $\text{E}\{\Psi(\boldsymbol{\beta})\} = 0$. This unbiasedness ensures the consistency of the resulting estimator $\hat{\boldsymbol{\beta}}$.
(b) The following identity holds,

$$\text{E}\left\{-\frac{\partial \Psi(\mathbf{Y};\boldsymbol{\beta})}{\partial \boldsymbol{\beta}}\right\} = \text{Var}\{\Psi(\mathbf{Y};\boldsymbol{\beta})\}.$$

This equality ensures that the resulting estimator $\hat{\boldsymbol{\beta}}$ will have the asymptotic covariance matrix equal to that of the MLE or the inverse of Fisher information matrix. In other words, the estimator produced from this estimating equation, although no explicit parametric models are assumed, will achieve the same estimation efficiency as that of the MLE that is known to be fully efficient. This optimality property presents a remarkable attraction to illustrate the usefulness of the theory of inference functions.

More discussions will be given in the subsequent sections of this chapter under a general framework of inference functions. Without assuming a true underlying parametric model, likelihood function is unavailable. However, it is possible to yield a function similar to the likelihood, called *quasi-likelihood*, that corresponds to the given quasi-score function $\Psi(\cdot)$ by simply taking integration *w.r.t.* μ; that is,

$$l_q(\mathbf{y};\mu) = \sum_{i=1}^{K} \int_{y_i}^{\mu_i} \frac{y_i - t}{V(t)} dt,$$

where $V(\cdot)$ is the unit variance function. Apparently, the quasi-score function $\Psi(\mathbf{y};\cdot)$ can be retrieved by

$$\begin{aligned}\Psi(\mathbf{y};\boldsymbol{\beta}) &= \frac{\partial l_q(\mathbf{y};\mu(\boldsymbol{\beta}))}{\partial \boldsymbol{\beta}} \\ &= \sum_{i=1}^{K} \dot{\mu}_i(\boldsymbol{\beta})^T V^{-1}(\mu_i)(y_i - \mu_i(\boldsymbol{\beta})).\end{aligned}$$

In the application of inference functions, two central technical questions are:

(i) which function $\Psi(\cdot)$ is sensible for parameter estimation among many candidate functions, and
(ii) under which criteria the optimality of a chosen inference function can be reasonably assessed.

Answers to these questions will be provided in the rest of this chapter.

3.3 Preliminaries

Consider a family of parametric statistical models

$$\mathcal{P} = \{p_{\boldsymbol{\theta}} : \boldsymbol{\theta} \in \Theta\}, \Theta \subseteq \mathcal{R}^p.$$

Let $\mathcal{X}$ be the sample space, defined as the collection of all possible samples.

Definition 3.1. *A function* $\Psi : \mathcal{X} \times \Theta \to \mathcal{R}^p$ *is called an* inference function *or* estimating function *if* $\Psi(\cdot;\boldsymbol{\theta})$ *is measurable for any* $\boldsymbol{\theta} \in \Theta$ *and* $\Psi(\mathbf{x};\cdot)$ *is continuous in a compact subspace of* Θ *containing the true parameter* $\boldsymbol{\theta}_0$ *for any sample* $\mathbf{x} \in \mathcal{X}$.

Let Ψ be an $\mathcal{R}^p$-valued vector function with components $(\psi_1, \ldots, \psi_p)^T$, where p is equal to the dimension of the parameter space Θ, or equal to the number of parameters to be estimated. Obviously, in order to estimate the p-dimensional parameter vector $\boldsymbol{\theta}$, it is necessary to have an inference function that contains at least p components of linearly independent inference functions ψ_js. When the number of inference function components is larger than the dimension of the parameter vector, according to Hansen (1982), the parameter vector $\boldsymbol{\theta}$ is said to be *over-identified.* Let us first consider the regular case of non-over-identification, and then deal with the case of over-identification in Section 3.5.

With a given inference function, Ψ, and a sample $\mathbf{x} \in \mathcal{X}$, one can establish an estimating equation given by

$$\Psi(\mathbf{x}; \boldsymbol{\theta}) = \mathbf{0}, \tag{3.1}$$

and as a result, an estimate $\widehat{\boldsymbol{\theta}} = \widehat{\boldsymbol{\theta}}(\mathbf{x})$ of parameter $\boldsymbol{\theta}$ is obtained as an solution to this equation (3.1).

Definition 3.2. *Two inference functions Ψ and φ are said to be equivalent, denoted by $\Psi \sim \varphi$, if they give the same estimate of $\boldsymbol{\theta}$ for any given sample $\mathbf{x} \in \mathcal{X}$.*

For example, one may use a given inference function Ψ to construct a new one of the following form:

$$\varphi_0(\mathbf{x}; \boldsymbol{\theta}) = C(\boldsymbol{\theta})\Psi(\mathbf{x}; \boldsymbol{\theta}), \ \mathbf{x} \in \mathcal{X}, \boldsymbol{\theta} \in \Theta$$

where $C(\boldsymbol{\theta})$ is a $p \times p$ matrix of full rank and independent of sample $\mathbf{x}$. Clearly, $\varphi_0 \sim \Psi$.

Definition 3.3. *An inference function Ψ is said to be* unbiased *if it has mean zero,*

$$E_{\boldsymbol{\theta}}\{\Psi(\mathbf{X}; \boldsymbol{\theta})\} = \mathbf{0}, \ \forall \ \boldsymbol{\theta} \in \Theta.$$

In the rest of this chapter, suppose that sample $\mathbf{x} = (x_1, \ldots, x_K)^T$ constitutes *i.i.d.* observations drawn from a parametric $p(x; \boldsymbol{\theta})$. Note that observation x_i may be a vector of, for example, correlated measurements from i subject. For this data structure, we consider an *additive inference function* given by

$$\Psi_K(\mathbf{x}; \boldsymbol{\theta}) = \sum_{i=1}^{K} \Psi(x_i; \boldsymbol{\theta}), \tag{3.2}$$

where $\Psi(\cdot)$ is called the *kernel* inference function. Then, an estimate of $\boldsymbol{\theta}$, $\widehat{\boldsymbol{\theta}}_K = \widehat{\boldsymbol{\theta}}_K(\mathbf{x})$ is defined as an solution to the equation $\Psi_K(\mathbf{x}; \boldsymbol{\theta}) = 0$; that is,

$$\sum_{i=1}^{K} \Psi(x_i; \widehat{\boldsymbol{\theta}}_K) = \Psi_K(\mathbf{x}; \widehat{\boldsymbol{\theta}}_K) = \mathbf{0}.$$

It is easy to see that in the case of the additive inference function, the unbiasedness of the Ψ_K is guaranteed by the unbiasedness of the kernel Ψ, $\mathrm{E}_{\boldsymbol{\theta}}\{\Psi(X_i;\boldsymbol{\theta})\} = 0, \ \forall\, \boldsymbol{\theta} \in \Theta$.

Let $\boldsymbol{\theta}_0 \in \Theta$ be the true parameter value, and let

$$\lambda(\boldsymbol{\theta}) = \mathrm{E}_{\boldsymbol{\theta}_0}\Psi(X;\boldsymbol{\theta}) = \int \Psi(x;\boldsymbol{\theta})p(x;\boldsymbol{\theta}_0)dx.$$

Thus, the unbiasedness of Ψ at the true value $\boldsymbol{\theta}_0$ holds if and only if $\lambda(\boldsymbol{\theta}_0) = 0$.

Theorem 3.4 (Consistency). *Suppose an additive inference function Ψ_K is unbiased at the $\boldsymbol{\theta}_0$. If $\lambda(\boldsymbol{\theta})$ has a unique zero at $\boldsymbol{\theta}_0$, then there exists a sequence of roots to equation $\Psi_K(\mathbf{x};\boldsymbol{\theta}) = \mathbf{0}$, $\{\widehat{\boldsymbol{\theta}}_K\}$, such that*

$$\widehat{\boldsymbol{\theta}}_K \xrightarrow{p} \boldsymbol{\theta}_0, \ \textit{under } P_{\boldsymbol{\theta}_0}.$$

Readers who are interested in the rigorous proof may refer to van der Vaart and Wellner (1996, Section 3) and Godambe (1991).

To appreciate the importance of the unbiasedness condition to establish consistency, a heuristic proof of this result is outlined below for the 1-dimensional case, i.e., $p = 1$.

Proof. Since function $\lambda(\theta)$ is continuous in a compact set containing θ_0, there must exist a (small) $\delta_0 > 0$, such that

$$\lambda(\theta) > 0, \theta \in (\theta_0 - \delta_0, \theta_0), \ \text{and } \lambda(\theta) < 0, \theta \in (\theta_0, \theta_0 + \delta_0),$$

or

$$\lambda(\theta) < 0, \theta \in (\theta_0 - \delta_0, \theta_0), \ \text{and } \lambda(\theta) > 0, \theta \in (\theta_0, \theta_0 + \delta_0).$$

Note that the conditions of $\lambda(\theta_0) = 0$ and the uniqueness at θ_0 essentially rule out the possibility that the $\lambda(\theta)$ is either always positive (concave) or always negative (convex) for all $\theta \in (\theta_0 - \delta_0, \theta_0 + \delta_0)$ except at $\theta = \theta_0$.

Let us consider the first scenario, and the second case can be argued in a similar fashion. The law of large numbers says

$$\frac{1}{K}\Psi_K(\mathbf{x};\theta) \xrightarrow{a.s.} \lambda(\theta) \text{ under } P_{\theta_0},$$

which implies that for large K and for any $0 < \delta < \delta_0$,

$$\Psi_K(\mathbf{x};\theta_0 - \delta) > 0, \ \text{and } \Psi_K(\mathbf{x};\theta_0 + \delta) < 0.$$

Here $\xrightarrow{a.s.}$ stands for almost sure convergence. By the continuity of probability measure, a root $\hat{\theta}_K(\delta)$ in the interval $(\theta_0 - \delta, \theta_0 + \delta)$ will have

$$P_{\theta_0}\left\{|\hat{\theta}_K(\delta) - \theta_0| < \delta\right\} \to 1, \text{as } K \to \infty,$$

which leads to

$$P_{\theta_0}\left\{|\hat{\theta}_K - \theta_0| < \delta\right\} \to 1, \text{as } K \to \infty,$$

where $\hat{\theta}_K$ is the root closest to the true value θ_0.

It is worth commenting that the existence of a solution to estimating equation $\Psi_K(\mathbf{x};\theta) = 0$ can be relaxed to the existence of the minimum for generalized method of moments (GMM)

$$Q(\mathbf{x};\boldsymbol{\theta}) = \Psi_K^T(\mathbf{x};\boldsymbol{\theta}) C_\Psi^{-1} \Psi_K(\mathbf{x};\boldsymbol{\theta}),$$

where $C_\Psi = \mathrm{Var}_{\boldsymbol{\theta}_0}\{\Psi_K^T(\mathbf{X};\boldsymbol{\theta})\}$; that is,

$$\widetilde{\boldsymbol{\theta}}_K = \arg\min_{\boldsymbol{\theta}\in\Theta} Q(\mathbf{x};\boldsymbol{\theta}),$$

which always exists if the parameter space Θ is compact. It is easy to prove that the minimizer $\widetilde{\boldsymbol{\theta}}_K$ and the root $\widehat{\boldsymbol{\theta}}_K$ are stochastically equivalent, namely $\widetilde{\boldsymbol{\theta}}_K = \widehat{\boldsymbol{\theta}}_K + o_p(1)$ as $K \to \infty$.

3.4 Optimal Inference Functions

Now consider a class of inference functions, denoted by $\mathcal{G}$, within which the resultant estimators are consistent and asymptotically normally distributed. Among the inference functions in class $\mathcal{G}$, define an *optimal* inference function as the one that leads to an estimator with the smallest asymptotic covariance and hence has the highest asymptotic efficiency. In the case that class $\mathcal{G}$ includes the score function, the optimal inference function is apparently the score function that gives the MLE. The objective is then whether there are any other inference functions in the $\mathcal{G}$ that produce estimators of equal efficiency to that of the MLE.

One relevant question is what are the conditions under which optimal inference functions exist. This question is answered in Godambe's series of seminal papers (Godambe 1960 and 1976; Godambe and Thompson, 1974 and 1978).

To begin, let us first consider a simple case where the parameter $\boldsymbol{\theta}$ is one-dimensional ($p = 1$).

Definition 3.5. *An inference function Ψ is said to be regular, if it satisfies the following conditions:*

(a) $E_\theta \Psi(X;\theta) = 0, \ \forall \theta \in \Theta$;
(b) $\frac{\partial \Psi(x;\theta)}{\partial \theta}$ *exists,* $\forall x \in \mathcal{X}$;
(c) The order of integration and differentiation may be interchangeable

$$\frac{\partial}{\partial \theta} \int_{\mathcal{X}} f(x)\Psi(x;\theta)p(x;\theta)dx = \int_{\mathcal{X}} f(x)\frac{\partial}{\partial \theta}\left\{\Psi(x;\theta)p(x;\theta)\right\}dx$$

for any bounded function $f(x)$ that is independent of θ;
(d) $0 < E_\theta\left\{\Psi^2(X;\theta)\right\} < \infty$;
(e) $0 < \left\{E_\theta \left|\frac{\partial \Psi(X;\theta)}{\partial \theta}\right|\right\}^2 < \infty$.

For now on, let $\mathcal{G}$ be the class of all regular inference functions.

Definition 3.6. *A statistical model is said to be regular if its score function* $u(x;\theta) = \frac{\partial \log p(x;\theta)}{\partial \theta}$ *is a regular inference function; that is,* $u(x;\theta) \in \mathcal{G}, \theta \in \Theta$ *an open interval.*

Under a regular model, the *Fisher information* for a single observation is defined by

$$\mathbf{i}(\theta) = -\mathrm{E}_\theta \left\{ \frac{\partial^2 \log p(X;\theta)}{\partial \theta^2} \right\} = -\mathrm{E}_\theta \left\{ \frac{\partial u(X;\theta)}{\partial \theta} \right\}.$$

It is known that Fisher information $\mathbf{i}(\theta)$ holds the following identity relation:

$$\mathbf{i}(\theta) = -\mathrm{E}_\theta \left\{ \frac{\partial u(X;\theta)}{\partial \theta} \right\} = \mathrm{E}\{u(X;\theta)\}^2.$$

Thus, function $\mathbf{i} : \Theta \to (0, \infty)$ excludes 0.

Let the first and second order derivatives of inference function Ψ *w.r.t.* θ be

$$\dot{\Psi}(\theta) = \frac{\partial \Psi(x;\theta)}{\partial \theta}, \quad \text{and} \quad \ddot{\Psi}(\theta) = \frac{\partial \Psi^2(x;\theta)}{\partial \theta^2}.$$

For a regular function $\Psi(x;\theta)$, the *variability*, V_Ψ, of Ψ is defined as

$$V_\Psi = \mathrm{E}_\theta \left\{ \Psi^2(X;\theta) \right\} = \mathrm{Var}\left\{ \Psi(X;\theta) \right\},$$

and the *sensitivity*, S_Ψ, of Ψ is defined as

$$S_\Psi(\theta) = \mathrm{E}_\theta \left\{ \frac{\partial \Psi(X;\theta)}{\partial \theta} \right\} = \mathrm{E}_\theta \left\{ \dot{\Psi}(X;\theta) \right\}.$$

An inference function Ψ is said to be θ-*sensitive* if the sensitivity $S_\Psi(\theta) > 0$. In the presence of a nuisance parameter τ, *say*, an inference function Ψ is said to be τ-*insensitive* if $\mathrm{E}\dot{\Psi}_\tau(X;\theta,\tau) = 0$. See, for example, Jørgensen and Knudsen (2004).

Definition 3.7. *For a regular inference function* $\Psi \in \mathcal{G}$*, the* Godambe information *is defined by*

$$\mathbf{j}_\Psi(\theta) = \frac{S_\Psi^2(\theta)}{V_\Psi(\theta)}, \quad \theta \in \Theta. \tag{3.3}$$

The larger the Godambe information is, the more efficient an estimator is. In other words, an inference function with small variability but large sensitivity is desirable. Note that comparison of inference functions based only on the variability or only on the sensitivity will be misleading. This is because equivalent inference functions can have rather different variabilities or sensitivities. The Godambe information takes the form of a ratio and therefore overcomes such a problem. As a result, Godambe information is unique among equivalent inference functions.

Theorem 3.8 (Asymptotic Normality). *Consider a regular statistical model. Let Ψ be a regular inference function, and let $\{\widehat{\theta}_K\}$ be a sequence of roots to the additive estimating equation:*

$$\sum_{i=1}^{K} \Psi(x_i; \widehat{\theta}_K) = 0, \quad K \geq 1.$$

Suppose that

(i) $\{\widehat{\theta}_K\}$ is consistent, i.e. $\widehat{\theta}_K \stackrel{P_{\theta_0}}{\to} \theta_0$; and
(ii) the second order derivative w.r.t. *θ is bounded, namely*

$$|\ddot{\Psi}(x;\theta)| < M(x), \text{ for } \theta \in (\theta_0 - c, \theta_0 + c)$$

for a certain constant c and a P_θ-measurable function $M(x)$ such that $E_\theta\{M(X)\} < \infty$.

Then,

$$\sqrt{K}(\hat{\theta}_K - \theta_0) \stackrel{d}{\to} N(0, \mathbf{j}^{-1}(\theta_0)), \text{ under } P_{\theta_0}. \tag{3.4}$$

A sketch of the proof of this theorem is given as follows.

Proof. Given that $\hat{\theta}_K$ is consistent to the true value θ_0, a linear Taylor expansion of $\Psi(x_i; \hat{\theta}_k)$ at the θ_0 leads to

$$\begin{aligned}\sqrt{K}(\hat{\theta}_K - \theta_0) &\approx \frac{\frac{1}{\sqrt{K}} \sum_{i=1}^K \Psi(x_i;\theta_0)}{-\frac{1}{K}\sum_{i=1}^K \dot{\Psi}(x_i;\theta_0)} \\ &\stackrel{d}{\to} \frac{N(0, V_\Psi(\theta_0))}{-S_\Psi(\theta_0)}, \text{ as } K \to \infty.\end{aligned}$$

The higher-order terms than the linear in the Taylor expansion can be controlled at the rate of $o_p(K^{-1/2})$. This is because for large K,

$$\begin{aligned}\frac{1}{\sqrt{K}} \sum_{i=1}^{K} |\ddot{\Psi}(x_i;\xi_K)(\widehat{\theta}_K - \theta_0)^2| &\leq \frac{1}{\sqrt{K}} \sum_{i=1}^{K} M(x_i)(\widehat{\theta}_K - \theta_0)^2 \\ &= O_p(1) o_p(1) \\ &= o_p(1),\end{aligned}$$

where ξ_K is a value between $\hat{\theta}_K$ and θ_0. By Slutsky's theorem (Arnold, 1990), the asymptotic normality holds and the asymptotic variance of $\hat{\theta}_K$ is the inverse of the Godambe information, $S_\Psi^2(\theta_0)/V_\Psi(\theta_0)$.

It follows immediately from the result (3.4) that the asymptotic variance of the estimator $\hat{\theta}_K$ is the same as the variance of the *normalized* inference function, $\bar{\Psi}(\theta) = \Psi(\theta)/\{-S_\Psi(\theta)\}$; that is, $V_{\bar{\Psi}}(\theta_0)$.

Theorem 3.9 (Godambe Inequality). *Assume an inference function $\Psi \in \mathcal{G}$. Then*

$$\mathbf{j}_\Psi(\theta) \le \mathbf{i}(\theta), \ \forall \ \theta \in \Theta,$$

where the equality holds if and only if $\Psi \sim u$, namely Ψ is equivalent to the score function.

Proof. Since Ψ is unbiased,

$$\int_{\mathcal{X}} \Psi(x;\theta)p(x;\theta)dx = 0, \ \forall \theta \in \Theta.$$

Differentiating the two sides of the above equation *w.r.t.* θ, and interchanging the order of integration and differentiation, one can obtain

$$\begin{aligned} 0 &= \int_{\mathcal{X}} \dot{\Psi}(x;\theta)p(x;\theta)dx + \int_{\mathcal{X}} \Psi(x;\theta)\dot{p}(x;\theta)dx \\ &= \mathrm{E}_\theta \dot{\Psi}(X;\theta) + \mathrm{E}_\theta \left\{ \Psi(X;\theta)u(X;\theta) \right\}, \end{aligned} \tag{3.5}$$

because $\partial \log p(x;\theta)/\partial\theta = \dot{p}(x;\theta)/p(x;\theta)$. It follows from the Cauchy-Schwartz inequality that

$$\left\{ \mathrm{E}_\theta \dot{\Psi}(X;\theta) \right\}^2 = \left[\mathrm{E}_\theta \left\{ \Psi(X;\theta)u(X;\theta) \right\} \right]^2 \le \mathrm{E}_\theta \Psi^2(X;\theta) \mathrm{E}_\theta u^2(X;\theta), \ \forall \theta.$$

Therefore,

$$\frac{\left\{ \mathrm{E}_\theta \dot{\Psi}(X;\theta) \right\}^2}{\mathrm{E}_\theta \Psi^2(X;\theta)} \le \mathrm{E}_\theta u^2(X;\theta), \ \forall \theta$$

or equivalently,

$$\mathbf{j}_\Psi(\theta) \le \mathbf{i}(\theta), \ \forall \theta.$$

In the meanwhile, the equality holds if and only if there exist two non-random coefficients such that

$$\Psi(x;\theta) = a(\theta) + b(\theta)u(x;\theta),$$

but $a(\theta) = 0$ due to the unbiasedness. Thus, $\Psi \sim u$.

Due to the fact that inference functions are similar to the score function, many results established in the likelihood inference may be transplanted to inference functions.

Now consider a special subclass, $\mathcal{G}_c$, of regular inference functions defined as

$$\Psi_c(\theta) = \sum_{i=1}^{K} c_i(\theta)\Psi_i(x_i;\theta), \ \theta \in \Theta \subset \mathcal{R}, \tag{3.6}$$

where $\Psi_i \in \mathcal{G}$ and $c_i(\theta)$ is a non-random function of θ such that the sequence of roots, $\{\hat{\theta}_K, K \ge 1\}$, to the estimating equation $\Psi_c(\theta) = 0$ is consistent.

The collection $\mathcal{G}_c \subset \mathcal{G}$ is referred to as the *Crowder class* of regular inference functions.

The following optimality theorem attributed to Crowder (1987) gives the optimal linear inference function in class $\mathcal{G}_c$.

Theorem 3.10 (Crowder Optimality). *Consider a regular inference function $\Psi_c(\theta) \in \mathcal{G}_c$. Then, the optimal inference function in the class $\mathcal{G}_c$, which has the largest Godambe information, is the one with the $c_i(\cdot)$ function taking a ratio of the sensitivity over the variability, namely*

$$c_i(\theta) = \frac{E_\theta\{\dot{\Psi}_i(\theta)\}}{Var_\theta\{\Psi_i(\theta)\}} = \frac{S_{\Psi_i}(\theta)}{V_{\Psi_i}(\theta)}, \quad \theta \in \Theta. \tag{3.7}$$

Proof. For an inference function in the Crowder class $\mathcal{G}_c$, it is easy to show that the Godambe information is

$$\mathbf{j}_c(\theta) = \frac{\left\{\sum_{i=1}^K c_i(\theta)\mathrm{E}_\theta(\dot{\Psi}_i)\right\}^2}{\sum_{i=1}^K c_i^2(\theta)\mathrm{Var}_\theta(\Psi_i)}.$$

In particular, when the $c_i(\cdot)$ takes the form of (3.7), the resulting Godambe information becomes

$$\mathbf{j}_c^*(\theta) = \sum_{i=1}^K \frac{\mathrm{E}_\theta^2(\dot{\Psi}_i)}{\mathrm{Var}_\theta(\Psi_i)}.$$

To show the optimality, it suffices to prove that $j_c^*(\theta) - j_c(\theta) \geq 0$, for any non-random functions c_i. This is equivalent to proving that

$$\sum_{i=1}^K \frac{\mathrm{E}_\theta^2(\dot{\Psi}_i)}{\mathrm{Var}_\theta(\Psi_i)} \sum_{i=1}^K c_i^2(\theta)\mathrm{Var}_\theta(\Psi_i) \geq \left\{\sum_{i=1}^K c_i(\theta)\mathrm{E}_\theta(\dot{\Psi}_i)\right\}^2,$$

which always holds according to the Cauchy-Schwartz inequality.

3.5 Multi-Dimensional Inference Functions

Consider a p-element additive inference function $\Psi_K = \sum_{i=1}^K \Psi(\mathbf{x}_i;\boldsymbol{\theta})$, where $\boldsymbol{\theta} = (\theta_1,\ldots,\theta_p)^T \in \mathcal{R}^p$ and $\Psi(\mathbf{x}_i;\boldsymbol{\theta}) = (\psi_1(\mathbf{x}_i;\boldsymbol{\theta}),\ldots,\psi_p(\mathbf{x}_i;\boldsymbol{\theta}))^T$, $\mathbf{x}_i \in \mathcal{X}$. Similar to univariate inference functions, a regular multi-dimensional inference function is defined as follows. A p-element inference function $\Psi(\mathbf{x};\boldsymbol{\theta})$ is said to be *regular* if it satisfies:

(a) $\mathrm{E}_{\boldsymbol{\theta}}\Psi(\mathbf{X};\boldsymbol{\theta}) = \mathbf{0}$, $\forall\, \boldsymbol{\theta} \in \Theta$.
(b) $\frac{\partial\Psi(\mathbf{x};\boldsymbol{\theta})}{\partial\theta_j}$ exists, $\forall\, \mathbf{x} \in \mathcal{X}$, $j = 1,\ldots,p$.

(c) The order of integration and differentiation may be interchangeable

$$\frac{\partial}{\partial \theta_j} \int_{\mathcal{X}} f(\mathbf{x}) \Psi(\mathbf{x}; \boldsymbol{\theta}) p(\mathbf{x}; \boldsymbol{\theta}) d\mathbf{x} = \int_{\mathcal{X}} f(\mathbf{x}) \frac{\partial}{\partial \theta_j} \left\{ \Psi(\mathbf{x}; \boldsymbol{\theta}) p(\mathbf{x}; \boldsymbol{\theta}) \right\} d\mathbf{x},$$

for $j = 1, \ldots, p$ and for any bounded function $f(\mathbf{x})$ that is independent of $\boldsymbol{\theta}$.

(d) $\mathrm{E}_{\boldsymbol{\theta}} \left\{ \psi_j(\mathbf{X}; \boldsymbol{\theta}) \psi_k(\mathbf{X}; \boldsymbol{\theta}) \right\}$ exists, and a $p \times p$ matrix

$$\mathbf{V}_{\Psi}(\boldsymbol{\theta}) = \mathrm{E}_{\boldsymbol{\theta}} \left\{ \Psi(\mathbf{X}; \boldsymbol{\theta}) \Psi^T(\mathbf{X}; \boldsymbol{\theta}) \right\}$$

is positive-definite. $\mathbf{V}_{\Psi}(\boldsymbol{\theta})$ is called the *variability matrix*.

(e) A $p \times p$ matrix

$$\mathbf{S}_{\Psi}(\boldsymbol{\theta}) = \mathrm{E}_{\boldsymbol{\theta}} \left\{ \nabla_{\boldsymbol{\theta}} \Psi(\mathbf{X}; \boldsymbol{\theta}) \right\}$$

is non-singular. $\mathbf{S}_{\Psi}(\boldsymbol{\theta})$ is referred as the *sensitivity matrix.*

Here the $\nabla_{\boldsymbol{\theta}}$ denotes the gradient operator on function f with respect to $\boldsymbol{\theta}$, defined by

$$\nabla_{\boldsymbol{\theta}} f(\boldsymbol{\theta}) = \left(\frac{\partial f(\boldsymbol{\theta})}{\partial \theta_1}, \ldots, \frac{\partial f(\boldsymbol{\theta})}{\partial \theta_p} \right)^T.$$

This implies that the (j,k)-th element of matrix $\mathbf{S}_{\Psi}(\boldsymbol{\theta})$ is $\frac{\partial \psi_j(\boldsymbol{\theta})}{\partial \theta_k}$, $j,k = 1, \ldots, p$. Let $\mathcal{G}$ be the class of all p-dimensional regular inference functions. Similarly, denote the score function by

$$\mathbf{u}(\mathbf{x}; \boldsymbol{\theta}) = \nabla_{\boldsymbol{\theta}} \log p(\mathbf{x}; \boldsymbol{\theta}).$$

If $\mathbf{u} \in \mathcal{G}$, then the Fisher information matrix for a single observation is

$$\mathbf{i}(\boldsymbol{\theta}) = -\mathrm{E}_{\boldsymbol{\theta}} \left\{ \nabla_{\boldsymbol{\theta}} \mathbf{u}(\mathbf{X}; \boldsymbol{\theta}) \right\} = \mathrm{E}_{\boldsymbol{\theta}} \left\{ \mathbf{u}(\mathbf{X}; \boldsymbol{\theta}) \mathbf{u}^T(\mathbf{X}; \boldsymbol{\theta}) \right\}.$$

Also, for a given regular inference function $\Psi \in \mathcal{G}$, the Godambe information matrix takes the form

$$\mathbf{j}_{\Psi}(\boldsymbol{\theta}) = \mathbf{S}_{\Psi}^T(\boldsymbol{\theta}) \mathbf{V}_{\Psi}^{-1}(\boldsymbol{\theta}) \mathbf{S}_{\Psi}(\boldsymbol{\theta}). \tag{3.8}$$

Let $\{\widehat{\boldsymbol{\theta}}_K, K \geq 1\}$ be a sequence of roots to the estimating equations

$$\Psi_K(\boldsymbol{\theta}) = \sum_{i=1}^{K} \Psi(\mathbf{x}_i; \boldsymbol{\theta}) = \mathbf{0}, \ K \geq 1$$

where $\Psi \in \mathcal{G}$.

Similar to the univariate case, under the condition that the Ψ is unbiased, one can establish the consistency $\widehat{\boldsymbol{\theta}}_K \xrightarrow{p} \boldsymbol{\theta}_0$ under $P_{\boldsymbol{\theta}_0}$ as $K \to \infty$, and the asymptotic normality as described in the following theorem.

Theorem 3.11 (Multivariate Asymptotic Normality). *If $\widehat{\boldsymbol{\theta}}_K$ is consistent, and in a small neighborhood, $\mathcal{N}(\boldsymbol{\theta}_0)$, centered at the true value $\boldsymbol{\theta}_0$,*

$$||\ddot{\Psi}(\mathbf{x};\boldsymbol{\theta})|| < M(\mathbf{x}), \ \boldsymbol{\theta} \in \mathcal{N}(\boldsymbol{\theta}_0),$$

with a $P_{\boldsymbol{\theta}}$-measurable function $M(\mathbf{x})$ such that $E_{\boldsymbol{\theta}_0}\{M(\mathbf{X})\} < \infty$, then

$$\sqrt{K}(\widehat{\boldsymbol{\theta}}_K - \boldsymbol{\theta}_0) \xrightarrow{d} MVN_p(\mathbf{0}, \mathbf{j}_\Psi^{-1}(\boldsymbol{\theta}_0)), \ \textit{under } P_{\boldsymbol{\theta}_0},$$

where $\mathbf{j}_\Psi(\boldsymbol{\theta})$ is the Godambe information given by (3.8).

Theorem 3.12 (Multivariate Godambe Inequality). *Consider a regular inference function $\Psi \in \mathcal{G}$. Then*

$$\mathbf{j}_\Psi(\boldsymbol{\theta}) \preceq \mathbf{i}(\boldsymbol{\theta}), \quad \forall\, \boldsymbol{\theta} \in \Theta,$$

and the equality holds if and only if $\Psi \sim \mathbf{u}$, the score function.

Here the inequality symbol "$\preceq$" means the Löwner's partial ordering in the space of non-negative definite matrices. That is, for two matrices A and B,

$$A \preceq B \text{ if and only if } B - A \text{ is non-negative definite.}$$

Proof. A similar derivation to equation (3.5) *w.r.t.* vector $\boldsymbol{\theta}$ leads to

$$\mathbf{0} = \mathbf{S}_\Psi(\boldsymbol{\theta}) + \text{cov}[\Psi(\mathbf{X};\boldsymbol{\theta}), \mathbf{u}(\mathbf{X};\boldsymbol{\theta})]. \tag{3.9}$$

Note that the normalized inference function and the normalized score function are given by, respectively,

$$\bar{\Psi}(\mathbf{X};\boldsymbol{\theta}) = \{-\mathbf{S}_\Psi^{-1}(\boldsymbol{\theta})\}\Psi(\mathbf{X};\boldsymbol{\theta}),$$
$$\bar{\mathbf{u}}(\mathbf{X};\boldsymbol{\theta}) = \{\mathbf{i}^{-1}(\boldsymbol{\theta})\}\Psi(\mathbf{X};\boldsymbol{\theta}).$$

Thus, the identity relation (3.9) can be rewritten as follows:

$$\mathbf{i}^{-1}(\boldsymbol{\theta}) = \text{cov}[\bar{\Psi}(\mathbf{X};\boldsymbol{\theta}), \bar{\mathbf{u}}(\mathbf{X};\boldsymbol{\theta})].$$

This implies the following results:

(a) $\bar{\mathbf{u}}(\mathbf{X};\boldsymbol{\theta})$ and $\bar{\Psi}(\mathbf{X};\boldsymbol{\theta}) - \bar{\mathbf{u}}(\mathbf{X};\boldsymbol{\theta})$ are uncorrelated;
(b) their information matrices satisfy $\mathbf{j}_\Psi(\boldsymbol{\theta}) \preceq \mathbf{i}(\boldsymbol{\theta})$; and
(c) the score function $\bar{\mathbf{u}}(\mathbf{X};\boldsymbol{\theta})$ contains more information about the parameter $\boldsymbol{\theta}$ than the inference function $\bar{\Psi}(\mathbf{X};\boldsymbol{\theta})$ in any linear subspace of the inference functions, in the sense that

$$\text{Var}^{-1}\{\mathbf{a}^T\bar{\mathbf{u}}(\mathbf{X};\boldsymbol{\theta})\} \geq \text{Var}^{-1}\{\mathbf{a}^T\bar{\Psi}(\mathbf{X};\boldsymbol{\theta})\}, \text{ for any } \mathbf{a} \in \mathcal{R}^p.$$

Part (a) holds because

$$\begin{aligned}\mathrm{cov}(\bar{\mathbf{u}}, \bar{\Psi} - \bar{\mathbf{u}}) &= \mathrm{cov}(\bar{\mathbf{u}}, \bar{\Psi}) - \mathrm{cov}(\bar{\mathbf{u}}, \bar{\mathbf{u}}) \\ &= \mathbf{i}^{-1}(\boldsymbol{\theta}) - \mathbf{i}^{-1}(\boldsymbol{\theta}) \\ &= \mathbf{0}.\end{aligned}$$

Part (b) holds because $\mathrm{Var}\{\bar{\Psi}(\mathbf{X};\boldsymbol{\theta}) - \bar{\mathbf{u}}(\mathbf{X};\boldsymbol{\theta})\} \succeq 0$ (i.e., it is a non-negative definite matrix) and

$$\begin{aligned}\mathrm{cov}(\bar{\Psi} - \bar{\mathbf{u}}, \bar{\Psi} - \bar{\mathbf{u}}) &= \mathrm{cov}(\bar{\Psi}, \bar{\Psi} - \bar{\mathbf{u}}) - \mathrm{cov}(\bar{\mathbf{u}}, \bar{\Psi} - \bar{\mathbf{u}}) \\ &= \mathbf{j}_{\Psi}^{-1}(\boldsymbol{\theta}) - \mathbf{i}^{-1}(\boldsymbol{\theta}).\end{aligned}$$

Thus, $\mathbf{j}_{\Psi}^{-1}(\boldsymbol{\theta}) - \mathbf{i}^{-1}(\boldsymbol{\theta}) \succeq 0$ leads to $\mathbf{j}_{\Psi}(\boldsymbol{\theta}) \preceq \mathbf{i}(\boldsymbol{\theta})$.

Part (c) is true because letting $\bar{\Psi} = \bar{\mathbf{u}} + (\bar{\Psi} - \bar{\mathbf{u}})$, we obtain the variance

$$\begin{aligned}\mathrm{Var}(\mathbf{a}^T\bar{\Psi}) &= \mathrm{Var}(\mathbf{a}^T\bar{\mathbf{u}}) + \mathrm{Var}\{\mathbf{a}^T(\bar{\Psi} - \bar{\mathbf{u}})\} \\ &= \mathrm{Var}(\mathbf{a}^T\bar{\mathbf{u}}) + \mathbf{a}^T\mathrm{Var}(\bar{\Psi} - \bar{\mathbf{u}})\mathbf{a} \\ &\geq \mathrm{Var}(\mathbf{a}^T\bar{\mathbf{u}}),\end{aligned}$$

which holds for all $\mathbf{a} \in \mathcal{R}^p$.

Theorem 3.13 (Multivariate Crowder Optimality). *Consider a regular inference function $\Psi_K(\boldsymbol{\theta}) \in \mathcal{G}_c$ defined by*

$$\Psi_c(\mathbf{x};\boldsymbol{\theta}) = \sum_{i=1}^{K} C_i(\boldsymbol{\theta})\psi_i(x_i;\boldsymbol{\theta}),\ \ \boldsymbol{\theta} \in \Theta \subseteq \mathcal{R}^p$$

where $C_i(\boldsymbol{\theta})$ is a non-random matrix of $\boldsymbol{\theta}$ such that the sequence of roots to equation $\Psi_c(\mathbf{x};\boldsymbol{\theta}) = 0$, $K \geq 1$, is consistent. Then, the optimal inference function in the Crowder class $\mathcal{G}_c$ is the one with the matrix $C_i(\cdot)$ given by

$$C_i(\boldsymbol{\theta}) = E_{\boldsymbol{\theta}}\{\dot{\Psi}_i(X_i;\boldsymbol{\theta})\}^T\, Var_{\boldsymbol{\theta}}^{-1}\{\Psi_i(X_i;\boldsymbol{\theta})\}.$$

Proof. Since the proof is similar to the one given in Theorem 3.12, only an outline is given. Let $\Psi_c \in \mathcal{G}_c$ be a general inference function, and let Ψ_o be the inference function with the given C_is. First, show that $\mathrm{cov}(\bar{\Psi}_c, \bar{\Psi}_o) = -\mathbf{S}_{\Psi_o}^{-1}$, where $\bar{\Psi}_c$ and $\bar{\Psi}_o$ are the corresponding normalized inference functions, and $\mathbf{S}_{\Psi_o}$ is the sensitivity matrix of the Ψ_o. Then, Part (b) in the proof of Theorem 3.12 can be similarly established here, which leads to the Crowder optimality.

3.6 Generalized Method of Moments

Now let us consider the case of over-identification, in which a q-element $\Psi_K(\mathbf{x};\boldsymbol{\theta})$ contains more components of inference functions than the number of parameters in $\boldsymbol{\theta} \in \mathcal{R}^p$; that is, $q > p$.

Hansen (1982) introduced the generalized method of moments (GMM), which may be regarded as a class of inference functions constructed through the moments of the underlying statistical models. In particular, the number of moment conditions (preferably orthogonal) used in the construction of an inference function is usually greater than the number of parameters to be estimated. If each moment condition forms one component of the inference function, then the resulting inference function will have more equations than the unknowns, and hence the parameter vector $\boldsymbol{\theta}$ is said to be over-identified. Obviously, it can not simply set

$$\Psi(\mathbf{x};\boldsymbol{\theta}) = 0,$$

as this equation has no non-trivial solution. To elucidate, let us consider a simple example as follows: Let $X_1, \ldots, X_K$ be a random sample from Poisson distribution with mean μ. Two moment conditions are given by

$$\begin{aligned} \mathrm{E}(X_i) &= \mu, \\ \mathrm{E}(X_i^2) &= \mu(1+\mu). \end{aligned}$$

Using these moment conditions, one may construct two unbiased inference functions for parameter μ,

$$\begin{aligned} \psi_1(\mu) &= \sum_{i=1}^{K}(X_i - \mu), \\ \psi_2(\mu) &= \sum_{i=1}^{K}\{X_i^2 - \mu(1+\mu)\}. \end{aligned}$$

Joining two functions sets up an estimating equation,

$$\Psi_K(\mu) = \frac{1}{K}\begin{bmatrix} \psi_1(\mu) \\ \psi_2(\mu) \end{bmatrix} = 0.$$

Clearly, this equation has no root, because the solution to $\psi_1(\mu)$ is uniquely the sample mean $\bar{X}$ and the solution to $\psi_2(\mu)$ is uniquely the sample variance S^2, and the probability that the sample mean and variance are the same is zero. However, there is a consistent solution to an equation in a form of, say linear combination of the two functions, $\psi_3(\mu) = a(x)\psi_1(\mu) + b(x)\psi_2(\mu) = 0$, because $\psi_3(\mu)$ is unbiased.

Taking a slightly different approach, Hansen (1982) suggested to find the estimator of the $\boldsymbol{\theta}$ by minimizing a quadratic objective function as follows,

$$\widehat{\boldsymbol{\theta}}_K^{(w)} = \arg\min_{\boldsymbol{\theta}} \Psi_K^T(\mathbf{x};\boldsymbol{\theta})W^{-1}(\boldsymbol{\theta})\Psi_K(\mathbf{x};\boldsymbol{\theta}), \tag{3.10}$$

where $W(\boldsymbol{\theta})$ is a certain suitable $q \times q$ weighting matrix. Under some regularity conditions such as unbiasedness, this estimator $\widehat{\boldsymbol{\theta}}_K^{(w)}$ is consistent (Lee, 1996,

P. 26). Obviously, estimator $\widehat{\boldsymbol{\theta}}_K^{(w)}$ depends on the choice of weighting matrix W.

Moreover, Hansen (1982) proved that the optimal weighting matrix W_{opt}, in the sense that the resulting estimator, $\widehat{\boldsymbol{\theta}}_{opt,K}$ of (3.10) has the smallest asymptotic covariance among all estimators $\widehat{\boldsymbol{\theta}}_K^{(w)}$ for all W, is effectively equal to the variance of the inference function Ψ; that is, $W_{\text{opt}} = \text{Var}\{\Psi_K(\mathbf{X};\boldsymbol{\theta})\}$ leads to the optimal estimator of $\boldsymbol{\theta}$ (Chamberlain, 1987). In other words,

$$\widehat{\boldsymbol{\theta}}_{\text{opt},K} = \arg\min_{\boldsymbol{\theta}} \Psi_K^T(\mathbf{x};\boldsymbol{\theta}) W_{\text{opt}}^{-1} \Psi_K(\mathbf{x};\boldsymbol{\theta}).$$

This $\widehat{\boldsymbol{\theta}}_{\text{opt},K}$ is the most efficient among all estimators of the form (3.10).

In addition, according to Hansen (1982), under some regularity conditions, this estimator is not only consistent but also asymptotically normally distributed with mean zero and asymptotic covariance given by the inverse of the Godambe information; that is,

$$\sqrt{K}(\widehat{\boldsymbol{\theta}}_{\text{opt},K} - \boldsymbol{\theta}) \xrightarrow{d} \text{MVN}_p(\mathbf{0}, \mathbf{j}^{-1}(\boldsymbol{\theta})),$$

where $\mathbf{j}(\boldsymbol{\theta})$ is the Godambe information of Ψ_K. Therefore, the optimal objective function,

$$Q(\mathbf{x};\boldsymbol{\theta}) = \Psi_K^T(\mathbf{x};\boldsymbol{\theta}) W_{\text{opt}}^{-1}(\boldsymbol{\theta}) \Psi_K(\mathbf{x};\boldsymbol{\theta}). \tag{3.11}$$

To apply this method, one has to plug in a consistent estimate of the covariance matrix W_{opt} in the function (3.11).

Theorem 3.14 (Hansen, 1982). *Suppose* $\widehat{W}_{opt,K}$ *is a* $\sqrt{K}$*-consistent estimate of* W_{opt}*, and* $\{\widehat{\boldsymbol{\theta}}_K\}$ *is a GMM estimator such that*

$$\widehat{\boldsymbol{\theta}}_K = arg\min_{\boldsymbol{\theta}} \Psi_K^T(\mathbf{x};\boldsymbol{\theta}) \widehat{W}_{opt,K}^{-1} \Psi_K(\mathbf{x};\boldsymbol{\theta}).$$

Then, under some mild regularity conditions and $q > p$*, the estimated quadratic function*

$$\widehat{Q}(\mathbf{x};\widehat{\boldsymbol{\theta}}) = \Psi_K^T(\mathbf{x};\widehat{\boldsymbol{\theta}}) \widehat{W}_{opt,K}^{-1} \Psi_K(\mathbf{x};\widehat{\boldsymbol{\theta}}) \xrightarrow{d} \chi^2(q-p), \ \ as\ K \to \infty.$$

This theorem provides a Rao's score-type test in the context of inference functions for the goodness-of-fit. Note that this test is only valid if $q > p$; otherwise, the limiting χ^2 distribution does not hold. Refer to Lindsay and Qu (2003) and Qu et al. (2000) for more discussions on quadratic inference function based test and related issues.

According to Newey and McFadden (1994), the GMM estimators are preferred to MLE estimators in the aspects of robustness, analytic tractability, and numerical stability, despite the fact that MLE is asymptotically more efficient when the model is correctly specified. GMM estimators are robust because they are based on a limited set of moment conditions without assuming parametric distributions. For GMM consistency, only these moment conditions need to be correctly specified, unlike MLEs that require correct specification of every conceivable moment condition, i.e., GMM estimators are robust with respect to distributional misspecification.

However, the price for the gain of robustness is the loss in efficiency with respect to MLE estimators. Furthermore, in some cases, the MLE estimators may not be available due to inability to deduce the likelihood function. The GMM estimators may still be feasible even in situations where the MLE is not possible.

4

Modeling Correlated Data

4.1 Introduction

The first part of this chapter is devoted to a general discussion about strategies of modeling correlated data, and the second part introduces several methods to simulate correlated data that are essential in simulation studies. As shown in Chapter 1, a set of correlated data comprises a collection of repeated triplets, $(y_{ij}, \mathbf{x}_{ij}, t_{ij})$, $j = 1, \ldots, n_i$ and $i = 1, \ldots, K$, where $\mathbf{x}_{ij} = (1, x_{ij1}, \ldots, x_{ijp})^T$ is a $(p+1)$-element vector of covariates. Here variable t_{ij} may index time for longitudinal data, or index spatial location for spatial data, or index other features of a sampling protocol from which correlated observations are collected. To express the data in matrix notation, $(\mathbf{y}_i, \mathbf{X}_i, \mathbf{t}_i)$, $i = 1, \ldots, K$, let

$$\mathbf{y}_i = (y_{i1}, \ldots, y_{in_i})^T, \ \mathbf{X}_i = (\mathbf{x}_{i1}, \ldots, \mathbf{x}_{in_i}), \ \mathbf{t}_i = (t_{i1}, \ldots, t_{in_i})^T.$$

A parametric modeling framework assumes that $\mathbf{y}_i$ is a realization of $\mathbf{Y}_i$ drawn from a certain population of the form,

$$\mathbf{Y}_i | (\mathbf{X}_i, \mathbf{t}_i) \stackrel{ind.}{\sim} p(\mathbf{y} | \mathbf{X} = \mathbf{X}_i, \mathbf{t} = \mathbf{t}_i; \boldsymbol{\theta}), \ i = 1, \ldots, K,$$

where $\boldsymbol{\theta}$ is the parameter of interest. The primary objective is to estimate and infer the model parameter $\boldsymbol{\theta}$. In the regression analysis of correlated data, the parameter $\boldsymbol{\theta}$ typically consists of two subsets, $\boldsymbol{\beta}$ and Γ, where $\boldsymbol{\beta}$ is the parameter vector involved in a regression model for the mean of the population, and Γ represents the other model parameters needed for the specification of a full parametric distribution $p(\cdot|\cdot)$, including those in the correlation structure.

Explicitly specifying such a parametric distribution for nonnormal data is not trivial. In spite of the fact that the multivariate normal distribution has been widely used in the analysis of continuous vector outcomes, it cannot model some data types, such as correlated discrete and categorical data. As far as regression analysis concerns, in order to handle correlated data, inevitably there is a need for the extension of the univariate GLM theory to be multi-dimensional, in which one can utilize rich 1-dimensional marginal distributions of the GLM to accommodate various data types.

In the construction of a multivariate distribution for correlated data, marginal densities are assumed as

$$Y_{ij}|\mathbf{x}_{ij}, t_{ij} \sim \mathrm{DM}(\mu_{ij}, \sigma_{ij}^2),$$

where a regression analysis further specifies the location parameter μ_{ij} to follow a GLM,

$$g(\mu_{ij}) = \eta(\mathbf{x}_{ij}, t_{ij}; \boldsymbol{\beta}), \; j = 1, \ldots, n_i. \tag{4.1}$$

In general, the dispersion parameter is also indexed by (i, j), as it may be also dependent on covariates and modeled by, for example,

$$\log(\sigma_{ij}^2) = \zeta(\mathbf{x}_{ij}, t_{ij}; \varsigma).$$

Depending on how the marginal GLM is specified, the parameter $\boldsymbol{\beta}$ may appear in different forms. Several commonly used marginal models are proposed in the literature. For the convenience of exposition, discussions will be restricted to longitudinal data, where t_{ij} represents a time index. Also, time t_{ij} may be included as a covariate in the covariate $\mathbf{x}_{ij}$, if time plays a particular role in the model specification.

(a) (Marginal GLM Model) When model (4.1) takes the form

$$\eta(\mathbf{x}_{ij}, t_{ij}; \boldsymbol{\beta}) = \mathbf{x}_{ij}^T \boldsymbol{\beta},$$

parameter $\boldsymbol{\beta}$ is the vector of regression coefficients. Here parameter $\boldsymbol{\beta}$ is interpreted as the population-average effect, since it is constant over time as well as across subjects.

(b) (Marginal Generalized Additive Model) When model (4.1) takes an additive form as follows,

$$\eta(\mathbf{x}_{ij}, t_{ij}; \boldsymbol{\beta}) = \theta_0 + \theta_1(x_{ij1}) + \cdots + \theta_p(x_{ijp}),$$

$\boldsymbol{\beta}$ denotes the set of nonparametric regression functions $\theta_0, \theta_1(\cdot), \ldots, \theta_p(\cdot)$. In this formulation, when one covariate is time t_{ij}, the resulting model characterizes a nonlinear time-varying profile of the data, which is particularly desirable in longitudinal data analysis.

(c) (Semi-Parametric Marginal Model) When model (4.1) includes both parametric and nonparametric predictors, for example,

$$\eta(\mathbf{x}_{ij}, t_{ij}; \boldsymbol{\beta}) = \theta_0(t_{ij}) + \mathbf{x}_{ij}^T \Upsilon,$$

the parameter $\boldsymbol{\beta}$ contains both the nonparametric function $\theta_0(\cdot)$ and the regression coefficients Υ. In this model, the population-average effect of a covariate is adjusted by a nonlinear time-varying baseline effect.

(d) (Time-Varying Coefficient Marginal Model) When model (4.1) follows a GLM with time-varying coefficients,

$$\eta(\mathbf{x}_{ij}, t_{ij}; \boldsymbol{\beta}) = \mathbf{x}_{ij}^T \boldsymbol{\beta}(t_{ij}),$$

parameter $\boldsymbol{\beta} = \boldsymbol{\beta}(t)$ represents a vector of regression coefficient functions in time. This model characterizes time-varying effects of covariates, which is useful in longitudinal data analysis. This is because in longitudinal studies, time-varying effects of covariates, rather than population-average constant effects, are often of interest.

(e) (Single-Index Marginal Model) When model (4.1) is specified as follows,

$$\eta(\mathbf{x}_{ij}, t_{ij}; \boldsymbol{\beta}) = \theta_0(t_{ij}) + \theta_1\left(\mathbf{x}_{ij}^T \Upsilon\right),$$

the parameter $\boldsymbol{\beta}$ includes the nonparametric functions $\theta_0(\cdot)$ and $\theta_1(\cdot)$ and the vector of coefficients Υ. This model is particularly useful for dimension reduction in the presence of a large number of covariates.

(f) A certain combination of models (a)-(e).

All these models essentially specify the first moment of the joint distribution $p(\cdot|\cdot)$; that is, the mean vector $\mathrm{E}(\mathbf{Y}_i|\mathbf{x}_i) = (\mu_{i1}, \ldots, \mu_{in_i})^T$ is modeled by assuming each μ_{ij} follow one of the above marginal models. Which model form to choose depends mainly on objectives of data analysis and features of the data.

A much harder task here is to specify higher moments of the joint distribution $p(\cdot)$ or even the joint distribution itself. This is the key to join marginal models under a certain suitable association structure in correlated data analysis. Note that the multivariate normal is the distribution that can be fully determined when mean $\boldsymbol{\mu}$ (the first moment) and covariance matrix Σ (the second moment) are given. In nonnormal distributions, it is generally difficult to determine a joint distribution based only on few low-order moments. Three approaches have been suggested in the literature to develop statistical inference procedures under different strategies of modeling association. They are the quasi-likelihood (QL) modeling approach, the conditional modeling approach, and the joint modeling approach.

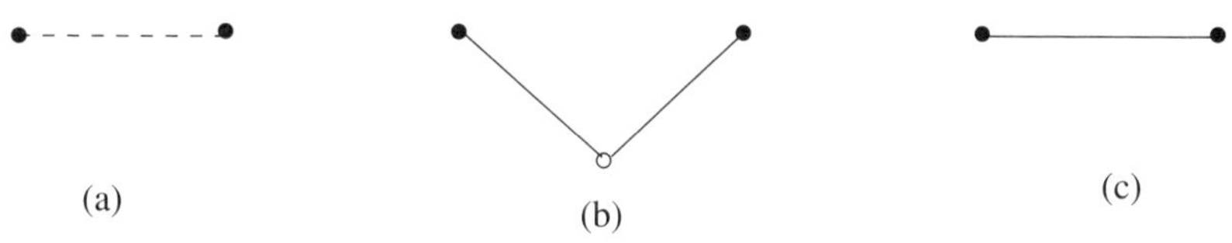

Fig. 4.1. Diagrams for three types of association models.

Essentially, the three approaches assume the same marginal model for the first moments, but differ according to the modeling of correlation. Figure 4.1 shows these three scenarios, each representing one model of association. In these panels, a solid dot represents a observation and a circle represents a latent variable. A solid line represents a legitimate association and a dashed line represents a working (hypothetical) association. By a legitimate association, it means an association induced in a proper multivariate distribution, and it is referred, otherwise, to as working correlation. Panel (a) indicates that the association of two observations is modeled by a working correlation, which is assumed in the QL inference. Panel (b) indicates that the association of two observations arises from their sharing of a common latent variable, which is the assumption made in the conditional modeling approach. Panel (c) indicates that the association of two observations is directly modeled by a joint probability model, which is the approach of joint modeling.

4.2 Quasi-Likelihood Approach

The quasi-likelihood (QL) approach does not rely on the specification of a full density function $p(\cdot)$, but just requires the availability of the first two moments, namely the mean and covariance matrix of the data. The covariance matrix is necessary in order to incorporate correlation of correlated outcomes in inference for the improvement of efficiency with regard to parameters in the first moment model. The mean and covariance constitute the minimal requirements for a QL modeling approach.

A QL approach requires an explicit formulation of covariance $\Sigma_i = \text{cov}(\mathbf{Y}_i|\mathbf{X}_i, \mathbf{t}_i)$. It is known that a covariance matrix may be written in the form:

$$\Sigma_i = \text{diag}\left[\sqrt{\text{Var}(Y_{i1})}, \ldots, \sqrt{\text{Var}(Y_{in_i})}\right] \Gamma \,\text{diag}\left[\sqrt{\text{Var}(Y_{i1})}, \ldots, \sqrt{\text{Var}(Y_{in_i})}\right], \tag{4.2}$$

where $\Gamma = (\gamma_{ts})$ is the correlation matrix of $\mathbf{Y}_i$. Note that γ_{ts} is the Pearson correlation coefficient between Y_{it} and Y_{is}. A relevant question is whether the covariance matrix, in which γ_{ts} primarily measures pairwise linear dependence, is suitable to characterize the strength and nature of association for correlated nonnormal data. The Pearson correlation is commonly known as a linear dependence measure for normally distributed variates. To measure dependence between nonnormal variates, there are some better tools than Pearson correlation. For example, *odds ratio* (OR) is a measure of association for categorical variates. For a pair of correlated binary variables (Y_{it}, Y_{is}), the OR is given by

$$OR = \frac{P(Y_{it} = 1, Y_{is} = 1)P(Y_{it} = 0, Y_{is} = 0)}{P(Y_{it} = 1, Y_{is} = 0)P(Y_{it} = 0, Y_{is} = 1)}.$$

The odds ratio provides a better interpretation for the association because it essentially contrasts probabilities of concordance to probabilities of discordance. On the other hand, the Pearson correlation of (Y_{it}, Y_{is}) is

$$\text{corr}(Y_{it}, Y_{is}) = \frac{P(Y_{it} = 1, Y_{is} = 1) - \mu_{it}\mu_{is}}{\{\mu_{it}(1 - \mu_{it})\mu_{is}(1 - \mu_{is})\}^{1/2}},$$

where μ_{it} and μ_{is} are the probabilities of success, respectively. To ensure the condition $|\text{corr}(Y_{it}, Y_{is})| < 1$, it immediately follows that

$$\max(0, \mu_{it} + \mu_{is} - 1) < P(Y_{it} = 1, Y_{is} = 1) < \min(\mu_{it}, \mu_{is}),$$

where the two limits are effectively the Fréchet lower and upper bounds (Joe, 1997), respectively. An obvious disadvantage of the Pearson correlation is that it depends on the means μ_{it} and μ_{is}, which hence depends on the regression coefficient parameter $\boldsymbol{\beta}$, when these marginal means are modeled in regression analysis. This often leads to restriction on the parameter space of $\boldsymbol{\beta}$ and inconvenience in modeling correlated data. In contrast, the odds ratio between the two variables is

$$\text{OR}(Y_{it}, Y_{is}) = \frac{\text{E}(Y_{it}Y_{is})\text{E}\{(1 - Y_{it})(1 - Y_{is})\}}{\text{E}\{Y_{it}(1 - Y_{is})\}\text{E}\{(1 - Y_{it})Y_{is}\}}, \tag{4.3}$$

which can vary in $(0, \infty)$ and hence is no longer constrained by their means.

Some nonlinear dependence measures, such as Spearman's ρ and Kendall's τ, may also be considered, as long as related interpretations are meaningful.

The use of covariance matrix (or Pearson correlation) is mostly for convenience, because it is mathematically well defined for all distributions with finite second moments. This generality allows researchers to develop QL inference in a unified fashion. In addition, incorporating the covariance matrix in QL inference is closely related to the weighted least squares method, which has been extensively studied in the literature.

On the line of Pearson correlation to characterize pairwise linear dependence, an alternative measure is defined as follows. For a given pair of variables Y_{it} and Y_{is}, with respective continuous cumulative distribution functions (CDF) G_{it} and G_{is}. Consider a normal-score transformation of these variables as follows:

$$Z_{ij} = \Phi^{-1}\{G_{ij}(Y_{ij})\}, \ j = t, s$$

where Φ^{-1} is the inverse of the standard normal CDF. Clearly, both transformed variables Z_{it} and Z_{is} are normally distributed. Then, one may compute the Pearson linear dependence for the Z_{it} and Z_{is}, and the resulting dependence measure is denoted by ν,

$$\nu(Y_{it}, Y_{is}) = \text{corr}(Z_{it}, Z_{is}) = \text{corr}[\Phi^{-1}\{G_{it}(Y_{it})\}, \Phi^{-1}\{G_{is}(Y_{is})\}]. \tag{4.4}$$

We have a few remarks:

(a) $\Phi^{-1}\{G(\cdot)\}$ is the so-called normal-score transformation. Note that when the Y_{it} and Y_{is} are normally distributed, the measure ν will give the Pearson correlation.

(b) This transformation is monotonic, which implies that although $\nu(Y_{it}, Y_{is})$ and $\text{corr}(Y_{it}, Y_{is})$ may have different magnitudes, the directions of both correlations are the same. That is, if $\text{corr}(Y_{it}, Y_{is})$ is positive (negative), then $\nu(Y_{it}, Y_{is})$ will be positive (negative), and *vice versa*. In addition, if $\text{corr}(Y_{it}, Y_{is}) > \text{corr}(Y'_{it}, Y'_{is})$, then $\text{corr}(Y_{it}, Y_{is}) > \text{corr}(Y'_{it}, Y'_{is})$.

(c) When the CDF G is discrete, the transformed variables Z_{it} and Z_{is} will not be normally distributed. However, the definition of the ν-measure remains mathematically valid provided that the second moment exists. In other words, the ν-measure still provides a dependence measure for the normal scores.

QL inference focuses on the marginal modeling of the first moments and treats the second moments such as covariance matrix Σ_i as nuisance parameters (with no modeling). An important advantage of QL inference is that misspecification of Σ_i will not affect the consistency of the estimator of $\boldsymbol{\beta}$, as long as the proposed QL inference function is unbiased. To implement a QL approach, an estimate of the covariance matrix Σ_i, which will be plugged in the QL inference function, has to be positive definite and consistent. Some examples that QL fails to produce reasonable estimates of $\boldsymbol{\beta}$ have been discussed in the literature; e.g., Crowder (1986), Sutradhar and Das (1999), Wang and Carey (2003), and Chaganty and Joe (2004).

Fitzmaurice et al. (1993) points out that to improve the efficiency of QL inference, the covariance matrix Σ_i needs to be specified as close to the true structure as possible, especially when covariates vary over different occasions during data collection. The unstructured correlation is not always the best. According to Liang and Zeger (1986), some common types of correlation structures used in the building of the covariance matrix Σ_i are as follows.

(1) (*Independence*) The independence correlation structure assumes that all pairwise correlation coefficients are zero:

$$\operatorname{corr}(Y_{it}, Y_{is}) = 0, \ t \neq s, \ \text{or} \ \nu(Y_{it}, Y_{is}) = 0, \ t \neq s.$$

(2) (*Unstructured*) An unstructured correlation structure assumes that all pairwise correlation coefficients are different parameters:

$$\operatorname{corr}(Y_{it}, Y_{is}) = \gamma_{ts} = \gamma_{st}, \ t \neq s, \ \text{or} \ \nu(Y_{it}, Y_{is}) = \gamma_{ts} = \gamma_{st}, \ t \neq s.$$

(3) (*Interchangeability*) Called also *compound symmetry*, the interchangeability correlation structure assumes that all pairwise correlation coefficients are equal and hence the components in the response vector $\mathbf{Y}$ are exchangeable (not ordered),

$$\operatorname{corr}(Y_{it}, Y_{is}) = \gamma, \ t \neq s, \ \text{or} \ \nu(Y_{it}, Y_{is}) = \gamma, \ t \neq s.$$

(4) (*AR-1*) The autoregressive correlation structure of order 1 assumes that the correlation coefficients decay exponentially over time, and the responses are ordered in time and more correlated if they are closer to each other in time than if they are more distant,

$$\operatorname{corr}(Y_{it}, Y_{is}) = \gamma^{|t-s|}, \ t \neq s, \ \text{or} \ \nu(Y_{it}, Y_{is}) = \gamma^{|t-s|}, \ t \neq s.$$

(5) (*m-dependence*) This correlation structure is generated by a moving-average process of order m, in which it assumes that the responses are uncorrelated if they are apart more than m units in time, or $|t-s|>m$,

$$\text{corr}(Y_{it}, Y_{is}) = \gamma_{ts}, \text{for } |t-s| \le m, \text{ or } \nu(Y_{it}, Y_{is}) = \gamma_{ts}, \text{for } |t-s| \le m.$$

To determine which correlation structure among those listed above would be suitable for a given correlated data, a preliminary residual analysis may be invoked, which may be proceeded in the following steps:

Step I: Fit correlate data by a marginal GLM (4.1) under the independence correlation structure, and outpout fitted values $\hat{\mu}_{it}$.

Step II: Calculate the following Pearson-type residuals, which presumably carry over the information of correlation that was originally ignored in Step I:

$$r_{it} = \frac{y_{it} - \widehat{\mu}_{it}}{\sqrt{V(\widehat{\mu}_{it})}}, t = 1, \ldots, n_i, i = 1, \ldots, K,$$

where $V(\cdot)$ is the unit variance function chosen according to the marginal model.

Step III: Compute the pairwise Pearson correlations $\widehat{\gamma}_{ts}$ using residuals available in Step II for each pair of fixed indices (t, s), which produces a sample correlation matrix $R = (\widehat{\gamma}_{ts})$.

Step IV: Examine matrix R to see if there is any pattern in the sample correlation coefficients over time that matches with one of those listed above.

The above preliminary examination of correlation structure is limited mostly to equally spaced repeated observations. When data are collected at unequally space occasions, the means of auto-correlation function is no longer useful. The *variogram* proposed by Diggle (1990) provides an alternative function to depict the association among irregularly spaced repeated measurements. For a time series $Y(t)$, *say*, the variogram is defined by

$$\hat{\zeta}(u) = \frac{1}{2}\text{E}\{Y(t) - Y(t-u)\}^2, \; u \ge 0. \tag{4.5}$$

To acquire the sample variogram of the residuals r_{it}, first calculate pairs of (v_{ijk}, u_{ijk}), where v_{ijk} is the half squared difference

$$v_{ijk} = \frac{1}{2}(r_{ij} - r_{ik})^2$$

and u_{ijk} is the corresponding time difference between the two residuals used in the calculation of the v_{ijk}; that is,

$$u_{ijk} = t_{ij} - t_{ik}.$$

When times t_{ij} are highly regular, the sample variogram is just given by the average of all of the v_{ijk}, corresponding to a fixed u. Otherwise, the sample variogram is estimated from the data (v_{ijk}, u_{ijk}) by a nonparametric smoothing technique, such as kernel smoothing or spline smoothing.

The variogram function is not ideal to explore the association of categorical repeated measurements directly. The *lorelogram* proposed by Heagerty and Zeger (1998) is useful to describe the association among categorical responses based on log-odds ratios. The lorelogram is defined as the function

$$\text{LOR}(t_j, t_k) = \log \text{OR}(Y_j, Y_k),$$

where the OR between Y_j and Y_k is given in equation (4.3). The sample lorelogram is estimated in a similar way to that in the sample variogram; that is, use the sample proportions across subjects to replace the theoretical probabilities (or expectations) in the definition of the OR. Figure 5.3 shows a sample lorelogram in the analysis of multiple sclerosis trial data in Chapter 5.

Another issue related to the determination of a correlation structure is the trade-off between the number of nuisance parameters and the closeness to the true underlying structure. The question is which one–simpler correlation structure with, *say*, one nuisance parameter, or closer to the true structure with many nuisance parameters–would give better efficiency? Some simulation studies (e.g., Fitzmaurice et al. 1993) have unveiled that the latter case, choosing a correlation structure close to the true one, seems preferred, especially when covariates are time-varying in the longitudinal data analysis. Modeling covariance or correlation structure of correlated data has drawn much attention in the recent literature.

As usual, a residual analysis based decision of the correlation structure is preliminary and subjective. A more rigorous approach to making decision would be based on a certain model selection criterion, such as Akaike information criterion (AIC), if available. In Chapter 5, a model selection procedure is derived using QIF in the framework of QL inference.

4.3 Conditional Modeling Approaches

Two kinds of conditional modeling approaches will be discussed in this section. The first kind is a latent variable based conditional modeling approach, and the other kind is a transitional model based conditional approach.

4.3.1 Latent Variable Based Approach

One way to overcome the difficulty of directly specifying the joint distribution is to consider conditional distributions, which are essentially one dimensional (Laird and Ware, 1982). Suppose there exists a latent variable $\mathbf{b}$, indicated as the circle in Panel (b) of Figure 4.1, such that conditional on the $\mathbf{b}$, the components in $\mathbf{Y} = (Y_1, \ldots, Y_n)^T$ are independent, namely

$$\mathbf{Y} = (Y_1, \ldots, Y_n)^T | \mathbf{b} \sim p(y_1|\mathbf{b}) \cdots p(y_n|\mathbf{b}). \tag{4.6}$$

Then, the joint distribution $p(\cdot)$ will be obtained by integrating out the latent variable $\mathbf{b}$ as follows:

$$\begin{aligned} p(\mathbf{y}|\mathbf{X}, \mathbf{t}) &= \int_{\mathcal{B}} p(\mathbf{y}, \mathbf{b}|\mathbf{X}, \mathbf{t}) d\mathbf{b} \\ &= \int_{\mathcal{B}} p(\mathbf{y}|\mathbf{b}, \mathbf{X}, \mathbf{t}) p(\mathbf{b}|\mathbf{X}, \mathbf{t}) d\mathbf{b} \\ &= \int_{\mathcal{B}} \prod_{i=1}^{n} p(y_i|\mathbf{b}, \mathbf{X}, \mathbf{t}) p(\mathbf{X}, \mathbf{t}) d\mathbf{b}, \end{aligned} \tag{4.7}$$

where the joint density $p(\mathbf{y}|\mathbf{b}, \mathbf{X}, \mathbf{t})$ is fully specified by the one-dimensional conditional distributions $p(y_i|\mathbf{b}, \mathbf{X}, \mathbf{t})$ under the assumption of conditional independence.

In this modeling approach, the latent variable $\mathbf{b}$ plays a critical role. The following example presents one scenario in which the latent variables are involved in the construction of a bivariate Poisson distribution.

Example 4.1 (Bivariate Poisson Distribution). To illustrate the conditional approach, let Y_1 and Y_2 be two correlated Poisson variates that are defined as follows,

$$Y_1 = Z_1 + Z_{12}, \; Y_2 = Z_2 + Z_{12},$$

where Z_1, Z_2, Z_{12} are independent latent variables following Poisson distributions with mean μ_1, μ_2 and μ_{12}, respectively. Then, given Z_{12}, Y_1 and Y_2 are conditionally independent. Note that by convolution, marginally $Y_1 \sim Po(\mu_1 + \mu_{12})$ and $Y_2 \sim Po(\mu_2 + \mu_{12})$. The covariance of (Y_1, Y_2) is

$$\begin{aligned} \text{cov}(Y_1, Y_2) &= \text{cov}(E(Y_1|Z_{12}), E(Y_2|Z_{12})) \\ &= \text{cov}(Z_{12}, Z_{12}) \\ &= \mu_{12} > 0, \end{aligned}$$

where $\text{Ecov}(Y_1, Y_2 \mid Z_{12}) = 0$. As a result, this latent variable approach generates a positively correlated bivariate Poisson vector. The joint probability mass function is

$$\begin{aligned} P(Y_1 = k_1, Y_2 = k_2) &= \sum_{k_3=0}^{\infty} P(Y_1 = k_1, Y_2 = k_2 | Z_{12} = k_3) P(Z_{12} = k_3) \\ &= \left(\frac{\mu_1^{k_1}}{k_1!} e^{-\mu_1} \right) \left(\frac{\mu_2^{k_2}}{k_2!} e^{-\mu_2} \right) \times \\ &\quad \sum_{k_3=0}^{\min\{k_1,k_2\}} \frac{k_1! k_2!}{(k_1 - k_3)!(k_2 - k_3)! k_3!} \left(\frac{\mu_{12}}{\mu_1 \mu_2} \right)^{k_3} e^{-\mu_{12}}. \end{aligned}$$

In general, in order to apply this conditional modeling approach, one needs to specify two model components: 1-dimensional conditional distributions $p(y_{ij}|\mathbf{b}_i), j = 1, \ldots, n_i$ and the distribution of the latent variables, $p(\mathbf{b}_i)$. Here the conditional density $p(y_{ij}|\mathbf{b}_i)$ may be specified by a dispersion model $DM(\mu_{ij}^b, \sigma_{ij}^b)$ (see Stiratelli et al. 1984). The latent variables $\mathbf{b}_i$s are referred to as *random effects*, which will be studied in detail in Chapters 7 and 8. With the availability of the joint distribution in (4.7), the full maximum likelihood estimation and inference can be developed. Albert (1999) points out that this modeling approach is particularly useful for analyzing longitudinal data in which there is a sizable number of missing observations either due to missed visits, loss to follow-up, or death: missing at random (MAR) and/or missing not at random (MNAR).

Some challenges associated with the conditional modeling approach are:

(a) If the dimension of the latent variable $\mathbf{b}_i$ is high, say larger than 5, numerical evaluation of the integral (4.7) in the calculation of the joint density $p(\mathbf{y}|\mathbf{X}, \mathbf{t})$ can be intricate. Some computationally intensive methods such as Monte Carlo EM algorithm and Markov Chain Monte Carlo (MCMC) algorithm may be invoked to overcome this difficulty.

(b) Specification of the distribution, $p(\mathbf{b})$, of the latent variable $\mathbf{b}$ will affect the form of the joint distribution $p(\mathbf{y}|\mathbf{X}, \mathbf{t})$ and hence affect the resulting MLE. Some studies (e.g., Butler and Louis, 1992; Verbeke and Lesaffre, 1997) have shown that the random-effects distribution has little effect on fixed-effects estimation, due largely to the fact that the random effects contribute only to the second moments, not to the first moments. Nevertheless, the asymptotic Fisher information matrix would become problematic if the distribution of random effects is misspecified. More importantly, the predictive distribution, $p(\mathbf{b}|\mathbf{y}, \mathbf{x}, \mathbf{t})$, of random effects is strongly dependent on the assumed distribution of random effects. See more discussions in Zhang (2006) for details.

(c) This conditional approach relies on the conditional independence assumption (4.6), which needs to be validated. A simple question is how many latent variables, one or ten, *say*, and in which form, would be appropriate to approve the conditional independence. Little study has been done in the literature for answers to this question.

4.3.2 Transitional Model Based Approach

Another conditional modeling approach makes a detour of the latent variable $\mathbf{b}$ and directly specifies the conditional distribution of one response on others, resulting the so-called transition models. This class of models is analogous to the time-series autoregressive model, which allows practitioners to examine the effect of covariates on the transitional patterns across the expectations of the responses over time. See Cox (1970), Muenz and Rubinstein (1985), and

Zeger and Qaqish (1988). Transition models are primarily useful to analyze serially correlated longitudinal data. That is, given a collection of equally spaced longitudinal observations, $(y_{it}, \mathbf{x}_{it})$, one may assume that the conditional (or transitional) density $f(y_{it}|y_{it-1}, \ldots, y_{it-q})$ follows an exponential dispersion model where the mean may take an autoregressive model. More specifically, let us consider the case of binary longitudinal responses. A logistic transition model of order q assumes that conditional on $y_{it-1}, y_{it-2}, \ldots, y_{it-q}$, the current response Y_{it} follows a Bernoulli distribution with the probability of success given as follows:

$$\text{logit} P[Y_{it} = 1|y_{it-1}, y_{it-2}, \ldots, y_{it-q}] = \mathbf{x}_{it}^T \boldsymbol{\beta} + \sum_{j=1}^{q} \theta_j y_{it-j} \tag{4.8}$$

where $\mathbf{x}_{itl}$ are subject-specific and time-dependent covariates, and q is the order of Markov dependence. The regression coefficients from models given by (4.8) can be interpreted as the effects of covariates on the probability of a binary event adjusting for the past history of the process.

Diggle et al. (2002) and Fahrmeir and Kaufmann (1987) extended these models to ordinal and repeated categorical data, respectively. In both cases, they considered models in discrete time with observations at regularly spaced intervals. Kalbfleisch and Lawless (1985) and Kosorok and Chao (1996) further considered the design and analysis of continuous-time transitional processes from such data observed in discrete time. These approaches allow for transitional inferences from highly irregularly-spaced observations by making modeling assumptions that flexibly relate instantaneous probabilities of state transitions to discrete-time transition probabilities.

The above specification for transition models as given by equation (4.8) can be generalized to include a broader class of models in which the observed response at time t, Y_{it}, is modeled conditionally as an explicit function of its history (past responses) $\mathcal{H}_{it} = (Y_{i1}, \ldots, Y_{it-1})$ and covariates $\mathbf{x}_{it}$. Cook and Ng (1997) and Albert and Waclawiw (1998) developed two approaches that allow for the transition probabilities in binary processes to follow a random-effects distribution.

In fitting transitional models for a weak stationary Gaussian process, the marginal distribution of Y_{it} can be fully determined from the conditional model without additional unknown parameters. However, when the marginal distribution of Y_{it} is not fully specified by the conditional model, one can estimate $\boldsymbol{\beta}$ and θ_js by maximizing the conditional log-likelihood $\ell = \sum_i \log L_i$, whose i-th piece for one subject is given by

$$\begin{aligned} L_i(\boldsymbol{\beta}, \boldsymbol{\theta}) &= f(Y_{iq+1}, \ldots, Y_{in_i}|Y_{i1}, \ldots, Y_{iq}; \boldsymbol{\beta}, \boldsymbol{\theta}) \\ &= \prod_{t=q+1}^{n_i} f(Y_{it}|Y_{it-1}, \ldots, Y_{it-q}; \boldsymbol{\beta}, \boldsymbol{\theta}). \end{aligned}$$

The conditional score equation is then given by equating the first-order derivatives of the conditional log-likelihood function to zero, i.e., $\dot{\ell} = \mathbf{0}$. Albert and Waclawiw (1998) demonstrated that this approach is robust against misspecified transition, a property also shared with the GEE-based marginal models. However, they warn that if q is large relative to n_i, then the use of transitional models with the above conditional likelihood could be inefficient and if the conditional mean is correctly specified but the conditional variance is not, then they recommend use of the empirical sample variance estimates in order to obtain a consistent inference about the model parameters.

This book focuses only on the latent variable based conditional approach. This choice is based on the fact that there are no good software packages currently available for transition models.

4.4 Joint Modeling Approach

The joint modeling approach attempts to directly construct the joint distribution $p(\cdot)$. In the 1-dimensional case, the DM family or the ED family presents a rich class of parametric distributions for different data types under a unified framework. To develop a multivariate analysis of correlated data similar to the univariate GLM theory, multi-dimensional DM or ED families are inevitable. In the literature, some *ad hoc* solutions have proposed to construct various multivariate distributions, such as Bahadur's representation for multivariate binomial distributions, the stochastic representations for multivariate Poisson distributions (e.g., Example 4.1), and multivariate gamma distributions. Unfortunately, none of these existing methods can be readily extended for a unified construction of multivariate DM models.

Directly constructing multivariate DM models is of both theoretical and practical interest. A unified framework for the multivariate DM models will lead to an extension of the univariate GLM theory to the multivariate GLM. With applications of the extended theory, correlated data can be modeled and analyzed under a proper probability model, in which MLE can be developed in a unified fashion.

Song (2000a) studied a unified DM framework generated by Gaussian copulas, where the proposed multivariate DM families satisfy the following properties, some of which are similar to those of the multivariate normal.

(a) The proposed family of multivariate DMs is only parametrized by the location (or mean) $\boldsymbol{\mu}$ and a dependence structure Γ, similar to the multivariate normal. The dependence parameters in the distribution can be interpreted in light of the ν-measure given in (4.4).
(b) The proposed multivariate DM is reproducible or marginally closed. That is, its marginal distributions have the same distribution type as that of the joint distribution.
(c) Parameters in the dependence matrix Γ can characterize both positive and negative associations. The bivariate Poisson distribution given by the

stochastic representation in Example 4.1 can only allow positive association. This imposes limitation for the modeling of dependence structure.

(d) The proposed family of multivariate DMs includes the multivariate normal as a special case, similar to the 1-dimension DM family.

Although this book focuses only on the development of the joint modeling approach based on Gaussian copulas, other types of parametric copulas such as Archimedean copulas may be applied in a similar way to carry out the analysis of correlated data. With the availability of the multivariate DM families, one can analyze correlated data in a fashion similar to what has been developed in setting of the multivariate normal distribution. Details are supplied later in Chapter 6.

5

Marginal Generalized Linear Models

For the ease of exposition, the presentation of this chapter is based on longitudinal data, and transplanting the core material to analyzing other types of correlated data can be done with little effort.

Marginal generalized linear models (MGLMs) are useful to conduct regression analysis of longitudinal data in the form of many short time series (i.e., small n_i large K). They arise from the formulation of quasi-likelihood modeling approach. This chapter focus on two quasi-likelihood inferences on regression coefficients, namely Liang and Zeger's (1986) *generalized estimating equations* (GEE) and Qu et al.'s (2000) quadratic inference function (QIF). As discussed in Chapter 4, MGLMs are used to study the population-average pattern or trend over time for longitudinal data, where the serial correlation is treated as a nuisance.

An underlying assumption for the use of MGLMs is that the subjects/clusters from which the data are collected are relatively homogeneous, in the sense that the variation in the response is mostly due to different levels of covariates. This may be true when subjects/clusters are sampled under a well designed and controlled study protocol. If subjects/clusters were sampled from a population that contains variation beyond what available covariates can explain, then the use of MGLMs should be cautious. Instead, the conditional modeling approach may be appealing.

It is still under debate whether a correctly (or nearly correctly) specified correlation structure is necessary for the application of quasi-likelihood inference in MGLMs. This debate has a lot to do with the efficiency of quasi-likelihood inference. Nevertheless, quasi-likelihood inference is advantageous for its robustness against the model misspecification on the correlation structure, and enjoys its simplicity, as it requires only correctly specifying the first two moments of the underlying probability distribution of the data.

The theory of inference functions in Chapter 3 provides the theoretical basis to establish Liang and Zeger's GEE, Prentice's (1988) GEE2, and Qu et al.'s QIF. All of them may be regarded as special cases in a general framework of inference functions.

5.1 Model Formulation

An MGLM comprises three model components, as follows:

(a) (Marginal Random Component) Assume that response Y_{ij} marginally follow a one-dimensional dispersion model:

$$Y_{ij}|(\mathbf{x}_{ij}, t_{ij}) \sim \mathrm{DM}(\mu_{ij}, \sigma_{ij}^2) \tag{5.1}$$

where μ_{ij} and σ_{ij}^2 are respectively the location and dispersion parameters, possibly depending on covariates $\mathbf{x}_{ij}$ or/and time t_{ij}.

(b) (Marginal Systematic Component) Let $\mu_{ij} = \mathrm{E}(Y_{ij}|\mathbf{x}_{ij}, t_{ij})$ be the marginal expectation. Assume that both μ_{ij} and σ_{ij}^2 follow, respectively, GLMs given by,

$$g(\mu_{ij}) = \mathbf{x}_{ij}^T \boldsymbol{\beta}, \tag{5.2}$$

$$\log(\sigma_{ij}^2) = \mathbf{z}_{ij}^T \boldsymbol{\alpha} \tag{5.3}$$

where $\mathbf{z}_{ij}$ may be a subset of $\mathbf{x}_{ij}$, and the log link function on the second model is to ensure the positivity for the dispersion parameter. This book only discusses the marginal GLM (5.2)-(5.3), and for the other types of marginal models, refer to Davidian and Giltinan (1995) and Wu and Zhang (2006), and relevant references therein. For convenience, in the following the covariate vector $\mathbf{x}_{ij}$ contains time covariate t_{ij} or a function of t_{ij} whenever such a covariate is present in the model.

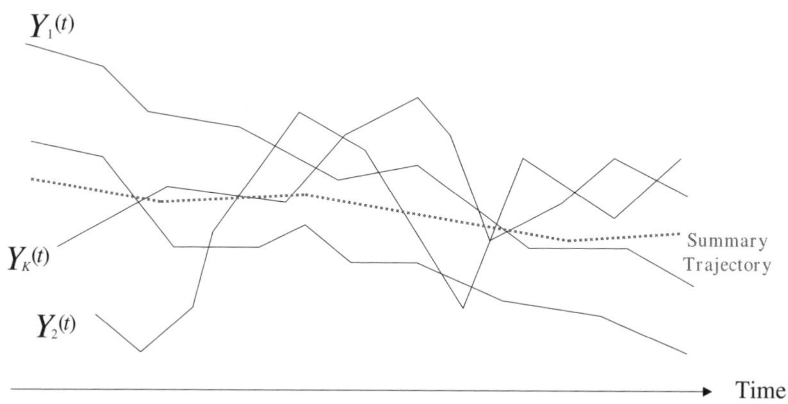

Fig. 5.1. A diagram of population-average summarization at each time point, while the other covariates are withheld. The dotted line represents the summary curve and each solid line represents an individual trajectory of observed time series.

The marginal model (5.2) may be essentially interpreted as an average summary over the responses in the subgroup of subjects that share common values of covariates $\mathbf{x}$, say the covariate of time, while the other covariates are withheld. Figure 5.1 presents a diagram, in which the dotted line presents a summary curve of many individual time-series trajectories, each observed for one subject.

(c) (Correlation Structure Component) The Pearson correlation between Y_{ij} and Y_{ik} is a function of both t_{ij} and μ_{ij}, parametrized by a parameter vector $\boldsymbol{\gamma}$,

$$\operatorname{corr}(Y_{ij}, Y_{ik}) = \rho(t_{ij}, t_{ik}, \mu_{ij}, \mu_{ik}; \boldsymbol{\gamma}) \tag{5.4}$$

where $\rho(\cdot)$ is a known function. Five common types of function $\rho(\cdot)$ have been given in Section 4.2. They are independence, unstructured, interchangeability (compound symmetry), AR-1, and m-dependence.

Because the MGLM model features the average summary of the data, it is also called *the population-average model* in the literature. The set of model parameters is $\boldsymbol{\theta} = (\boldsymbol{\beta}, \boldsymbol{\alpha}, \boldsymbol{\gamma})$, which is unknown and needs to be estimated. In this chapter, the estimation will be given via quasi-likelihood approaches.

In the MGLM formulation, because of the lack of a joint probability model, it separately specifies marginal location, marginal dispersion, and correlation structure. For normal longitudinal data, these three model components are enough to build a multivariate normal distribution, but for nonnormal longitudinal data, in general they are insufficient to fully determine a multivariate dispersion model. For example, some higher-order associations than the pairwise correlation may be needed.

With regard to choosing a proper function $\rho(\cdot)$ in the correlation structure component (5.4), a preliminary residual analysis may be conducted to find some useful clues. Refer to Section 4.1 for relevant discussions. In equation (5.4), the Pearson correlation is used as the dependence measure, which is desirable for the normal data. However, for nonnormal data, other types of association measures such as ν-measure may better describe the association in the data. The use of Pearson correlation brings technical ease to the development of a quasi-likelihood inference similar to the weighted least squares estimation approach, which has been well studied in the statistical literature.

5.2 GEE: Generalized Estimating Equations

Theory of inference functions in Chapter 3 is now applied to develop generalized estimating equations (GEE), a quasi-likelihood estimation for parameter $\boldsymbol{\theta}$. First consider the case where the dispersion parameter is constant, $\sigma_{ij}^2 = \sigma^2$. Let $\mathcal{G}$ be the collection of all regular inference functions, which contains the score function of the underlying joint model for the data. According to Chapter 3, the score function is the optimal inference function in the class $\mathcal{G}$, in which the optimality effectively corresponds to a full likelihood-based inference. However, this full likelihood inference is not permissible, given that only the first two moments of the MGLM are specified. Thus, two key steps are needed to establish quasi-likelihood inference:

(a) identify a suitable subclass $\mathcal{G}^* \subset \mathcal{G}$, which possibly does not contain the score function, and then
(b) find the optimal inference function within this subclass $\mathcal{G}^*$.

To identify $\mathcal{G}^*$, it is helpful to first observe the estimating function derived from a naive analysis under the independence correlation structure. Using this naive estimating function as a reference, one can propose a subclass $\mathcal{G}^*$ of inference functions that contains such a naive estimating function. This will result in the optimal inference function that exceeds the performance of the naive estimation method.

5.2.1 General Theory

Under the assumption that all Y_{ij} are independent and $Y_{ij} \sim \mathrm{DM}(\mu_{ij}, \sigma^2)$, the data can be thought essentially of as a cross-sectional dataset with sample size $\sum_{i=1}^K n_i$. A similar derivation to equation (2.18) leads to a quasi-score function *w.r.t.* parameter $\boldsymbol{\beta}$ given by

$$\mathbf{s}(\mathbf{Y}; \boldsymbol{\beta}) = \frac{1}{\sigma^2} \sum_{i=1}^K \left(\frac{\partial \boldsymbol{\mu}_i}{\partial \boldsymbol{\beta}} \right)^T \boldsymbol{\delta}_i(\mathbf{y}_i; \boldsymbol{\mu}_i),$$

where $\boldsymbol{\mu}_i = (\mu_{i1}, \ldots, \mu_{in_i})^T$, $\boldsymbol{\delta}_i = (\delta_{i1}, \ldots, \delta_{in_i})^T$, with $\delta_{ij} = -\frac{1}{2}\dot{d}(y_{ij}; \mu_{ij})$, and

$$\mathbf{D}_i^T = \left(\frac{\partial \boldsymbol{\mu}_i}{\partial \boldsymbol{\beta}} \right)^T = \mathbf{X}_i \left[\mathrm{diag}\{\dot{g}(\mu_{i1}), \ldots, \dot{g}(\mu_{in_i})\} \right]^{-1}. \tag{5.5}$$

It follows from Proposition 2.13 that $\mathrm{E}(\delta_{ij}) = 0$, for all i, j, so $\mathrm{E}\{\mathbf{s}(\mathbf{Y}; \boldsymbol{\beta})\} = 0$. This implies that inference function $\mathbf{s}(\mathbf{Y}; \boldsymbol{\beta})$ is unbiased. According to Theorem 3.1, under some mild regularity conditions the solution, $\widetilde{\boldsymbol{\beta}}$, to the following equation

$$\Psi_0(\mathbf{Y}; \boldsymbol{\beta}) = \sum_{i=1}^K \left(\frac{\partial \boldsymbol{\mu}_i}{\partial \boldsymbol{\beta}} \right)^T \boldsymbol{\delta}_i(\mathbf{y}_i; \boldsymbol{\mu}_i) = 0, \tag{5.6}$$

will be a consistent estimator, even though the independence assumption does not hold in the longitudinal data.

Next, to specify class $\mathcal{G}^*$, which contains the $\Psi_0(\cdot)$ as an element, one may consider a class of linear unbiased inference functions based on the deviance score vector $\boldsymbol{\delta}_i = (\mathbf{y}_i; \boldsymbol{\mu}_i)$ of the form:

$$\Psi(\mathbf{Y}; \boldsymbol{\beta}) = \sum_{i=1}^K \mathbf{C}_i(\boldsymbol{\theta}) \boldsymbol{\delta}_i(\mathbf{y}_i; \boldsymbol{\mu}_i),$$

where $\mathbf{C}_i$ is an arbitrary $p \times n_i$ nonstochastic weighting matrix that may depend on parameters other than the $\boldsymbol{\beta}$. Denote this class by $\mathcal{G}_\delta = \{\Psi(\mathbf{Y}; \boldsymbol{\beta})\}$. Clearly, the Ψ_0 belongs to the class $\mathcal{G}_\delta$ with $\mathbf{C}_i = \left(\frac{\partial \boldsymbol{\mu}_i}{\partial \boldsymbol{\beta}} \right)^T$. The Crowder optimality theory (Theorem 3.10) implies that the optimal inference function in class $\mathcal{G}_\delta$ is the one with the following weighting matrix

$$\begin{aligned}\mathbf{C}_i(\boldsymbol{\beta}) &= \mathrm{E}\left\{\frac{\partial \boldsymbol{\delta}_i(\mathbf{Y}_i;\boldsymbol{\mu}_i)}{\partial \boldsymbol{\beta}}\right\}^T [\mathrm{Var}\{\boldsymbol{\delta}_i(\mathbf{Y}_i;\boldsymbol{\mu}_i)\}]^{-1} \\ &= -\mathbf{D}_i^T \mathrm{diag}[\mathrm{E}\{-\dot{\delta}(Y_{i1};\mu_{i1})\},\ldots,\mathrm{E}\{-\dot{\delta}(Y_{in_i};\mu_{in_i})\}]\,[\mathrm{Var}\{\boldsymbol{\delta}_i(\mathbf{Y}_i;\boldsymbol{\mu}_i)\}]^{-1},\end{aligned}$$

where $\mathbf{D}_i$ is defined in (5.5).

Let $\mathbf{r}_i = (r_{i1},\ldots,r_{in_i})^T$ be a vector of the modified score residuals (see Table 2.3) with the jth element being $r_{ij} = \sqrt{V(\mu_{ij})} r_{s,ij} = V(\mu_{ij})\delta_{ij}$. Then, $\mathbf{r}_i = \mathrm{diag}\left\{V(\mu_{i1}),\cdots,V(\mu_{in_i})\right\}\boldsymbol{\delta}_i(\mathbf{y}_i;\boldsymbol{\mu}_i)$. Thus, the optimal inference function can be expressed as

$$\Psi_{op}(\mathbf{Y};\boldsymbol{\beta}) = \sum_{i=1}^K \mathbf{D}_i^T \mathbf{A}_i \left[\mathrm{Var}(\mathbf{r}_i)\right]^{-1} \mathbf{r}_i, \tag{5.7}$$

where

$$\begin{aligned}\mathbf{A}_i &= \mathrm{diag}[\mathrm{E}\{-\dot{\delta}(Y_{i1};\mu_{i1})\},\ldots,\mathrm{E}\{-\dot{\delta}(Y_{in_i};\mu_{in_i})\}]\mathrm{diag}\left\{V(\mu_{i1}),\cdots,V(\mu_{in_i})\right\} \\ &= \sigma^{-2}\mathrm{diag}\left\{\mathrm{Var}(\delta_{i1})V(\mu_{i1}),\cdots,\mathrm{Var}(\delta_{in_i})V(\mu_{in_i})\right\}.\end{aligned}$$

The last equality holds because of Proposition 2.17. It is easy to see that the optimal inference function Ψ_{op} is unbiased since $\mathrm{E}(\mathbf{r}_i) = 0$.

The difficulty with the utility of the optimal inference function Ψ_{op} is that the variance-covariance matrix $\mathrm{Var}(\mathbf{r}_i)$ is unknown. To overcome this, Liang and Zeger (1986) suggested replacing the $\mathrm{Var}(\mathbf{r}_i)$ by a *working* covariance matrix defined as follows,

$$\Sigma_i = \mathbf{G}_i^{1/2}\mathbf{R}(\boldsymbol{\gamma})\mathbf{G}_i^{1/2},$$

where $\mathbf{G}_i = \mathrm{diag}\left\{\mathrm{Var}(r_{i1}),\cdots,\mathrm{Var}(r_{in_i})\right\}$, and $\mathbf{R}(\boldsymbol{\gamma})$ is an $n_i \times n_i$ correlation matrix that is fully characterized by a q-dimensional vector of parameters $\boldsymbol{\gamma}$. This $\mathbf{R}(\boldsymbol{\gamma})$ is referred to as a *working correlation matrix*. Clearly, when $\mathbf{R}(\boldsymbol{\gamma})$ is the true correlation matrix of $\mathbf{r}_i$, the resulting inference function is the optimal Ψ_{op}, and when $\mathbf{R}(\boldsymbol{\gamma})$ is the independence correlation matrix (i.e., the indentity matrix), the inference function reduces to the naive Ψ_0. In addition, Proposition 2.17 gives the variance of a modified score residual r_{ij} as follows:

$$\mathrm{Var}(r_{ij}) = V^2(\mu_{ij})\mathrm{Var}\{\delta_{ij}(Y_{ij};\mu_{ij})\} = \sigma^2 V^2(\mu_{ij})\mathrm{E}\left\{-\dot{\delta}(Y_{ij};\mu_{ij})\right\}. \tag{5.8}$$

As a result, an estimating equation is given by

$$\Psi(\mathbf{Y};\boldsymbol{\beta}) = \sum_{i=1}^K \mathbf{D}_i^T \mathbf{A}_i \Sigma_i^{-1}\mathbf{r}_i(\mathbf{y}_i;\boldsymbol{\mu}_i) = \mathbf{0}, \tag{5.9}$$

which is termed as *the generalized estimating equation (GEE)* by Liang and Zeger (1986). Consequently, the estimate, $\widehat{\boldsymbol{\beta}}$, of parameter $\boldsymbol{\beta}$ is defined as the solution to the GEE (5.9).

This inference function $\Psi(\mathbf{y};\boldsymbol{\beta})$ may be viewed as a multivariate extension of the quasi-score function first proposed by Wedderburn (1974). The complication in such an extension is rooted in the fact that this function depends not only on the parameter $\boldsymbol{\beta}$ of interest, but also on some nuisance parameters $\boldsymbol{\gamma}$ and σ^2. It is suggested in the literature that the dependence on $(\boldsymbol{\gamma},\sigma^2)$ can be resolved by replacing these nuisance parameters in the Σ_i with, respectively, their $K^{1/2}$-consistent estimators, $\widehat{\boldsymbol{\gamma}}$ and $\widehat{\sigma}^2$. It is shown in Liang and Zeger (1986) that, under some mild conditions, the resulting estimator $\widehat{\boldsymbol{\beta}}(\widehat{\boldsymbol{\gamma}},\widehat{\sigma}^2)$ is asymptotically equally efficient to the estimator $\widehat{\boldsymbol{\beta}}(\boldsymbol{\gamma},\sigma^2)$ with the true values of the parameters $\boldsymbol{\gamma}$ and σ^2. Note that in some distributions, such as binomial and Poisson, dealing with the nuisance parameters becomes slightly simpler because the dispersion parameter σ^2 is known.

In the following presentation of asymptotics, both $\boldsymbol{\gamma}$ and σ^2 are assumed to be known. So, denote $\widehat{\boldsymbol{\beta}}=\widehat{\boldsymbol{\beta}}(\boldsymbol{\gamma},\sigma^2)$. Note that the GEE inference function Ψ is unbiased. Under some mild regularity conditions, the estimator $\widehat{\boldsymbol{\beta}}$ is consistent, and $K^{1/2}(\widehat{\boldsymbol{\beta}}-\boldsymbol{\beta})$ is asymptotically multivariate Gaussian with zero mean and covariance matrix of the form $\lim_K K\mathbf{j}^{-1}(\boldsymbol{\beta})$, where $\mathbf{j}(\boldsymbol{\beta})$ is the Godambe information matrix (also called the sandwich covariance estimator) given by

$$\mathbf{j}(\boldsymbol{\beta})=\mathbf{S}(\boldsymbol{\beta})^T\mathbf{V}(\boldsymbol{\beta})^{-1}\mathbf{S}(\boldsymbol{\beta}).$$

The sensitivity matrix $\mathbf{S}(\boldsymbol{\beta})$ is given by

$$\begin{aligned}\mathbf{S}(\boldsymbol{\beta})=\mathrm{E}\{\nabla_{\boldsymbol{\beta}}\Psi(\mathbf{Y};\boldsymbol{\beta})\}&=\sum_{i=1}^{K}\mathbf{D}_i^T\mathbf{A}_i\Sigma_i^{-1}\mathrm{E}\left\{\nabla_{\boldsymbol{\beta}}\mathbf{r}_i(\mathbf{Y}_i;\boldsymbol{\beta})\right\}\\&=-\sum_{i=1}^{K}\mathbf{D}_i^T\mathbf{A}_i\Sigma_i^{-1}\mathbf{A}_i\mathbf{D}_i.\end{aligned} \tag{5.10}$$

And the variability matrix is given by

$$\mathbf{V}(\boldsymbol{\beta})=\mathrm{E}\left\{\Psi(\mathbf{Y};\boldsymbol{\beta})\Psi^T(\mathbf{Y};\boldsymbol{\beta})\right\}=\sum_{i=1}^{K}\mathbf{D}_i^T\mathbf{A}_i\Sigma_i^{-1}\mathrm{Var}(\mathbf{r}_i)\Sigma_i^{-1}\mathbf{A}_i\mathbf{D}_i, \tag{5.11}$$

where $\mathrm{Var}(\mathbf{r}_i)$ is related to $\mathrm{Var}(\boldsymbol{\delta}_i)$ via the following form

$$\mathrm{Var}(\mathbf{r}_i)=\mathrm{diag}\left\{V(\mu_{i1}),\cdots,V(\mu_{in_i})\right\}\mathrm{Var}(\boldsymbol{\delta}_i)\mathrm{diag}\left\{V(\mu_{i1}),\cdots,V(\mu_{in_i})\right\}.$$

Here the variance matrix of $\boldsymbol{\delta}_i$, $\mathrm{Var}(\boldsymbol{\delta}_i)$, may be estimated by $\boldsymbol{\delta}_i\boldsymbol{\delta}_i^T$.

It is worth pointing out that the insensitivity of the GEE (5.9) to the nuisance parameters $\boldsymbol{\gamma}$ and σ^2 ensures that the Godambe information $\mathbf{j}(\boldsymbol{\beta})$ above is only marginally affected by the efficiency of the nuisance parameter estimators, $\hat{\boldsymbol{\gamma}}$ and $\hat{\sigma}^2$. This insensitivity is seen by the fact that both $\mathrm{E}\{\nabla_{\boldsymbol{\gamma}}\Psi(\mathbf{Y};\boldsymbol{\beta})\}=\mathbf{0}$ and $\mathrm{E}\{\nabla_{\sigma^2}\Psi(\mathbf{Y};\boldsymbol{\beta})\}=0$. This is because these nuisance parameters involve only in terms $\mathbf{A}_i$ and Σ_i, and $\mathrm{E}\{\mathbf{r}_i(\mathbf{Y}_i;\boldsymbol{\mu}_i)\}=\mathbf{0}$.

5.2.2 Some Special Cases

This section presents three examples of the GEE with, respectively, the exponential dispersion (ED) distribution, the simplex distribution, and the von Mises distribution.

Example 5.1 (ED GEE).

Liang and Zeger's GEE (1986) becomes a special case of the estimating equations (5.9) when the marginal distribution is an $\mathrm{ED}(\mu_{ij}, \sigma^2)$. In this case, the modified score residual $r_{ij} = y_{ij} - \mu_{ij}$, the regular raw residual, which implies that $\mathrm{Var}(\mathbf{r}_i) = \mathrm{Var}(\mathbf{Y}_i)$. Thus, the working covariance matrix is

$$\begin{aligned}\Sigma_i &= \mathrm{diag}^{1/2}\left\{\mathrm{Var}(Y_{i1}), \cdots, \mathrm{Var}(Y_{in_i})\right\}\mathbf{R}(\gamma)\mathrm{diag}^{1/2}\left\{\mathrm{Var}(Y_{i1}), \cdots, \mathrm{Var}(Y_{in_i})\right\}\\ &= \mathbf{G}_i^{1/2}\mathbf{R}(\gamma)\mathbf{G}_i^{1/2},\end{aligned}$$

with

$$\mathbf{G}_i = \sigma^2 \mathrm{diag}\left\{V(\mu_{i1}), \cdots, V(\mu_{in_i})\right\}.$$

It is easy to show that for the ED model matrix $\mathbf{A}_i$ reduces to the identity matrix. As a result, the optimal estimating equation becomes

$$\Psi_{op}(\mathbf{Y};\boldsymbol{\beta}) = \sum_{i=1}^{K} \mathbf{D}_i^T \mathrm{Var}(\mathbf{Y}_i)^{-1}(\mathbf{y}_i - \boldsymbol{\mu}_i) = \mathbf{0},$$

which is of the same form as the optimal weighted least squares estimating function. The corresponding GEE then takes the form

$$\Psi(\mathbf{Y};\boldsymbol{\beta}) = \sum_{i=1}^{K} \mathbf{D}_i^T \Sigma_i^{-1}(\mathbf{y}_i - \boldsymbol{\mu}_i) = \mathbf{0}, \tag{5.12}$$

where only the nuisance parameter $\boldsymbol{\gamma}$ is involved, as the dispersion parameter σ^2 is factorized out of the equation.

The sensitivity and variability matrices can be simplified, respectively, as

$$\mathbf{S}(\boldsymbol{\beta}) = -\sum_{i=1}^{K} \mathbf{D}_i^T \Sigma_i^{-1}\mathbf{D}_i$$

and

$$\mathbf{V}(\boldsymbol{\beta}) = \sum_{i=1}^{K} \mathbf{D}_i^T \Sigma_i^{-1}\mathrm{Var}(\mathbf{y}_i)\Sigma_i^{-1}\mathbf{D}_i.$$

Hence, the Godambe information matrix is

$$\mathbf{j}(\boldsymbol{\beta}) = \left\{\sum_{i=1}^{K} \mathbf{D}_i^T \Sigma_i^{-1}\mathbf{D}_i\right\}\left\{\sum_{i=1}^{K} \mathbf{D}_i^T \Sigma_i^{-1}\mathrm{Var}(\mathbf{Y}_i)\Sigma_i^{-1}\mathbf{D}_i\right\}^{-1}\left\{\sum_{i=1}^{K} \mathbf{D}_i^T \Sigma_i^{-1}\mathbf{D}_i\right\}, \tag{5.13}$$

where the variance matrix $\mathrm{Var}(\mathbf{Y}_i)$ is estimated by $(\mathbf{y}_i - \boldsymbol{\mu}_i)(\mathbf{y}_i - \boldsymbol{\mu}_i)^T$.

Example 5.2 (Simplex GEE).

Song and Tan (2000) studied a GEE with the simplex distribution margin, which also produces a special case of (5.9). Suppose marginally $Y_{ij} \sim S^{-}(\mu_{ij}, \sigma^2)$, where the marginal mean follows a GLM with the logit link; that is, $\log \frac{\mu_{ij}}{1-\mu_{ij}} = \mathbf{x}_{ij}^T \boldsymbol{\beta}$.

From equation (2.25), the modified score residual of the simplex distribution is

$$\begin{aligned} r_{ij} &= \mu_{ij}^3 (1-\mu_{ij})^3 \delta(y_{ij}; \mu_{ij}) \\ &= (y_{ij} - \mu_{ij}) \left\{ d(y_{ij}; \mu_{ij}) \mu_{ij}^2 (1-\mu_{ij})^2 + 1 \right\}. \end{aligned}$$

The working covariance matrix takes the form

$$\Sigma_i = \mathbf{G}_i^{1/2} \mathbf{R}(\gamma) \mathbf{G}_i^{1/2},$$

where by Proposition 2.19, the j-th main diagonal element of $\mathbf{G}_i$ is

$$\text{Var}(r_{ij}) = \sigma^2 \mu_{ij}^3 (1-\mu_{ij})^3 \left\{ 3\sigma^2 \mu_{ij}^2 (1-\mu_{ij})^2 + 1 \right\}.$$

Similarly,

$$\begin{aligned} \mathbf{A}_i &= \text{diag}\left\{ 3\sigma^2 \mu_{i1}^2 (1-\mu_{i1})^2 + 1, \cdots, 3\sigma^2 \mu_{in_i}^2 (1-\mu_{in_i})^2 + 1 \right\}, \\ \mathbf{D}_i^T &= \mathbf{X}_i \text{diag}\{\mu_{i1}(1-\mu_{i1}), \cdots, \mu_{in_i}(1-\mu_{in_i})\}. \end{aligned}$$

Note that in this case, the dispersion parameter σ^2 fully involves in the GEE and cannot be factorized out of the equation. This complicates not only the search for the root of the GEE, but also the performance of the GEE estimator itself. Therefore, the dispersion parameter σ^2 appears much more substantially influential in the simplex GEE than in other GEE cases.

Example 5.3 (von Mises GEE).

Suppose marginally an angular response $Y_{ij} \sim \text{vM}(\mu_{ij}, \sigma^2)$, where similar to equation (2.31) the marginal mean direction follows the model of the form: $\mu_{ij} = \mu_0 + 2\arctan(\mathbf{x}_{ij}^T \boldsymbol{\beta})$, and $\boldsymbol{\beta}^* = (\mu_0, \boldsymbol{\beta})$.

From Section 2.6.4, the unit variance function of the von Mises distribution is $V(\mu_{ij}) = 1$, so the modified score residual is the same as the deviance score:

$$r_{ij} = \delta(y_{ij}; \mu_{ij}) = \sin(y_{ij} - \mu_0 - 2\arctan(\mathbf{x}_{ij}^T \boldsymbol{\beta})).$$

By Proposition 2.17, it is easy to obtain the j-th main diagonal element of $\mathbf{G}_i$ as

$$\text{Var}(r_{ij}) = \text{Var}\{\delta(y_{ij}; \mu_{ij})\} = A_1(\lambda)/\lambda,$$

where $A_1(\lambda) = \frac{I_1(\lambda)}{I_0(\lambda)}$ is the mean resultant length and the same for all j, with $\lambda = 1/\sigma^2$, and $I_0(\lambda)$ and $I_1(\lambda)$ being the modified Bessel function of the first kind given in Section 2.6.4. It is interesting to note that because this term is constant, it can be factorized out the estimating equation with no effect on the estimation of $\boldsymbol{\beta}$. The components in GEE (5.9) are given by

$$\begin{aligned}\mathbf{D}_i^T &= \begin{bmatrix}\mathbf{1}\\ \mathbf{X}_i\end{bmatrix}\left[\text{diag}\{1, 1+\eta_{i1}^2, \ldots, 1+\eta_{in_i}^2\}\right]^{-1}\\ \mathbf{A}_i &= \lambda\text{diag}\left\{\text{Var}(\delta_{ij}), \ldots, \text{Var}(\delta_{in_i})\right\} = A_1(\lambda)I_{n_i}\end{aligned}$$

where $\eta_{ij} = \mathbf{x}_{ij}^T\boldsymbol{\beta}, j = 1, \ldots, n_i$.

5.2.3 Wald Test for Nested Models

Assume the nuisance parameters $\boldsymbol{\gamma}$ and σ^2 are fixed. Suppose the full MGLM (5.2) can be decomposed into two parts,

$$g(\mu_{ij}) = \mathbf{x}_{ij}^T\boldsymbol{\beta} = \mathbf{x}_{1ij}^T\boldsymbol{\beta}_1 + \mathbf{x}_{2ij}^T\boldsymbol{\beta}_2$$

where $\boldsymbol{\beta}_1 \in \mathcal{R}^{p_1}$ and $\boldsymbol{\beta}_2 \in \mathcal{R}^{p_2}$, $p_1 + p_2 = p$. The hypothesis of interest is $H_0 : \boldsymbol{\beta}_2 = \mathbf{0}$. Under the null hypothesis, the model reduces to

$$g(\mu_{ij}) = \mathbf{x}_{1ij}^T\boldsymbol{\beta}_1,$$

nested to the full model. It follows from the asymptotic normality for the full model that $\widehat{\boldsymbol{\beta}}_2$ is asymptotically multivariate normal with mean $\boldsymbol{\beta}_2$ and covariance matrix $[\mathbf{j}^{-1}]_{22}$, the submatrix of the inverse of Godambe matrix $\mathbf{j}^{-1}$ corresponding to subvector $\boldsymbol{\beta}_2$ of $\boldsymbol{\beta}$. Let $\mathbf{j}^{22} = [\mathbf{j}^{-1}]_{22}$. The Wald statistic is then $W^2 = \widehat{\boldsymbol{\beta}}_2^T\left(\mathbf{j}^{22}\right)^{-1}\widehat{\boldsymbol{\beta}}_2$, which asmptotically follows $\chi^2_{p_2}$ distribution.

Some other approaches have been proposed for testing complex hypotheses in the mean and/or the association structure in marginal models for longitudinal data. See, for example, Rotnitzky and Jewell (1990), Liang and Self (1996), and Zeigler et al. (1998).

5.3 GEE2

A crucial step in the application of GEE is to plug in a $\sqrt{K}$-consistent estimator of the correlation parameter $\boldsymbol{\gamma}$, as well as a $\sqrt{K}$-consistent estimator of the dispersion parameter σ^2 if relevant. This section introduces two approaches to handling these nuisance parameters: one is the so-called GEE2 that adds additional unbiased estimating equations to estimate the nuisance parameters to guarantee the required consistency, and the other is the quadratic inference function that avoids estimating the nuisance parameters but just uses some basis matrices from a given working correlation structure. One drawback of

GEE2 is that, unlike GEE1, it is not robust to misspecification of the association structure. Chan et al. (1998) moreover showed that GEE2 may be computationally intensive, as it involves the inversion of matrices of dimension $O(n_i^2) \times O(n_i^2)$.

In the ED GEE or von Mises GEE, the dispersion parameter σ^2 is not involved in the GEE $\Psi(\mathbf{Y};\boldsymbol{\beta}) = 0$, and hence can be separately estimated. In contrast, the simplex GEE involves the σ^2 deeply, which has to be estimated simultaneously with the $\boldsymbol{\beta}$. Note that in some distributions, such as binomial and Poisson, the σ^2 is known, unless over- or under-dispersion needs to be accounted for in the modeling.

5.3.1 Constant Dispersion Parameter

First consider the case where σ^2 is not involved in the GEE. Prentice (1988) suggested to use an unbiased inference function to estimate γ, so the consistency for both estimators $\hat{\boldsymbol{\beta}}$ and $\hat{\gamma}$ can be guaranteed. That is, one may include a second estimating equation, say, $\Psi_*(\mathbf{Y};\boldsymbol{\beta},\boldsymbol{\gamma}) = \mathbf{0}$, based on certain moment properties of residuals. Therefore, estimates of $\boldsymbol{\beta}$ and $\boldsymbol{\gamma}$ can be found by simultaneously solving the joint estimating equation (GEE2):

$$\boldsymbol{\Psi}(\mathbf{Y};\boldsymbol{\beta},\boldsymbol{\gamma}) = \begin{bmatrix} \Psi(\mathbf{Y};\boldsymbol{\beta},\boldsymbol{\gamma}) \\ \Psi_*(\mathbf{Y};\boldsymbol{\beta},\boldsymbol{\gamma}) \end{bmatrix} = \mathbf{0}.$$

To proceed, define residuals as follows:

$$e_{ij} = \frac{\delta_{ij}}{\sqrt{\mathrm{E}\{-\dot{\delta}(Y_{ij};\mu_{ij})\}}}, j = 1,\ldots,n_i, i = 1,\ldots,K, \tag{5.14}$$

which may be regarded as an extended version of the Pearson residual in the classical GLM theory with the ED model margins. This is because, by Proposition 2.18, when the marginal distribution is an ED model, $\mathrm{E}\{-\dot{\delta}(Y_{ij};\mu_{ij})\} = 1/V(\mu_{ij})$, and thus $\delta_{ij} = (y_{ij} - \mu_{ij})/V(\mu_{ij})$. This leads to $e_{ij} = (y_{ij} - \mu_{ij})/\sqrt{V(\mu_{ij})}$, which is the Pearson residual given in Table 2.3.

Moreover, by Proposition 2.17, it is easy to see that $\mathrm{E}(e_{ij}) = 0$, $\mathrm{Var}(e_{ij}) = \sigma^2$ and

$$\mathrm{E}(e_{ij}e_{ij'}) = \sigma^2\mathrm{corr}(\delta_{ij},\delta_{ij'}) = \sigma^2\mathrm{corr}(r_{ij},r_{ij'}),\ j \neq j'.$$

Thus, the second estimating equation may be formulated as follows,

$$\Psi_*(\mathbf{Y};\boldsymbol{\beta},\boldsymbol{\gamma}) = \sum_{i=1}^{K} \left(\frac{\partial \boldsymbol{\xi}_i}{\partial \boldsymbol{\gamma}}\right)^T \mathbf{H}_i^{-1}(\mathbf{e}_i - \boldsymbol{\xi}_i) \tag{5.15}$$

where $\mathbf{e}_i$ is a vector that contains all possible distinct pairwise cross-products of the residuals, namely,

$$\mathbf{e}_i = (e_{i1}e_{i2}, e_{i1}e_{i3}, \ldots, e_{in_i-1}e_{in_i})^T.$$

$\mathbf{H}_i$ is a working covariance matrix and $\boldsymbol{\xi}_i = \mathrm{E}(\mathbf{e}_i)$. Clearly, inference function Ψ_* attains the Crowder optimality (Theorem 3.10), when $\mathbf{H}_i = \mathrm{Var}(\mathbf{e}_i)$. However, it is consistency that is more crucial than efficiency in the estimation of correlation parameter $\boldsymbol{\gamma}$.

Example 5.4 (Interchangeable Structure). The interchangeable or compound symmetry structure is a working correlation that defines $\mathrm{corr}(r_{ij}, r_{ij'}) = \gamma$ for all i and all $j \neq j'$. In such a case,

$$\frac{\partial \boldsymbol{\xi}_i}{\partial \gamma} = \mathbf{1}^T,$$

where $\mathbf{1}$ is an $n_i(n_i-1)/2$ dimensional vector with all elements being 1. If $\mathbf{H}_i$ is taken to be the identity matrix, then

$$\Psi_*(\mathbf{Y}; \boldsymbol{\beta}, \gamma) = \sum_{i=1}^K \mathbf{1}^T(\mathbf{e}_i - \sigma^2\gamma\mathbf{1}) = \sum_{i=1}^K \mathbf{1}^T\mathbf{e}_i - \sigma^2\gamma \sum_{i=1}^K \frac{1}{2} n_i(n_i-1).$$

It follows immediately that the solution to this equation is

$$\sigma^{-2} \sum_{i=1}^K \sum_{j>j'} e_{ij} e_{ij'} \Big/ \left\{ \sum_{i=1}^K \frac{1}{2} n_i(n_i-1) \right\}.$$

When the $\widehat{\boldsymbol{\beta}}$ is plugged in the calculation of the residuals, with the degrees of freedom being adjusted, the resulting estimator of γ is given by

$$\hat{\gamma} = \sigma^{-2} \sum_{i=1}^K \sum_{j>j'} \hat{e}_{ij} \hat{e}_{ij'} \Big/ \left\{ \sum_{i=1}^K \frac{1}{2} n_i(n_i-1) - (p+1) \right\},$$

which is identical to the the method of moments estimate for γ.

Example 5.5 (AR-1 Structure). The AR-1 correlation structure is useful to incorporate the serial correlation,

$$\mathrm{corr}(r_{ij}, r_{ij'}) = \gamma^{|t_{ij} - t_{ij'}|}, \ |\gamma| < 1.$$

To relax the constraint of the γ parameter, namely $|\gamma| < 1$, the exponential correlation (EC) structure is used,

$$\mathrm{corr}(z_{ij}, z_{ij'}) = \exp(-\tilde{\gamma}|t_{ij} - t_{ij'}|), \tilde{\gamma} > 0,$$

which however only suits for positive serial correlation. The EC structure appears more stable numerically.

Take the EC structure for illustration. When the matrix $\mathbf{H}_i$ is set to be the identity matrix, the resulting estimating equation for parameter $\tilde{\gamma}$ is given by

$$\Psi_*(\mathbf{Y};\boldsymbol{\beta},\gamma) = \sum_{i=1}^{K} \mathbf{c}_i^T\{\mathbf{e}_i - \boldsymbol{\xi}_i\} = 0, \tag{5.16}$$

with

$$\mathbf{c}_i = \left[|t_{i1} - t_{i2}|\exp(-\tilde{\gamma}|t_{i1} - t_{i2}|), \ldots, |t_{in_i-1} - t_{in_i}|\exp(-\tilde{\gamma}|t_{in_i-1} - t_{in_i}|)\right]^T.$$

The solution of this equation does not have a closed form expression, and a numerical algorithm is required to solve this equation jointly with the other estimating equations with respect to the parameter $\boldsymbol{\beta}$.

Note that in the two examples above, for simplicity, matrix $\mathbf{H}_i$ has been taken to be the identity matrix, which will lead to a potential loss of efficiency in estimation for γ. However, as pointed out by Diggle et al. (2002), this loss of efficiency does not affect the $\boldsymbol{\beta}$ estimation much.

In summary, the GEE2 for parameter $\boldsymbol{\theta} = (\boldsymbol{\beta}, \gamma)$ takes a form of

$$\begin{aligned}\boldsymbol{\Psi}(\mathbf{Y};\boldsymbol{\theta}) &= \begin{bmatrix} \Psi(\mathbf{Y};\boldsymbol{\beta},\gamma) \\ \Psi_*(\mathbf{Y};\boldsymbol{\beta},\gamma) \end{bmatrix} \\ &= \begin{bmatrix} \mathbf{D}_i^T & \mathbf{0} \\ \left(\frac{\partial \boldsymbol{\xi}_i}{\partial \boldsymbol{\beta}}\right)^T & \left(\frac{\partial \boldsymbol{\xi}_i}{\partial \gamma}\right)^T \end{bmatrix} \begin{bmatrix} \mathbf{A}_i & \mathbf{0} \\ \mathbf{0} & \mathbf{I}_q \end{bmatrix} \begin{bmatrix} \Sigma_i & \mathbf{0} \\ \mathbf{0} & \mathbf{H}_i \end{bmatrix}^{-1} \begin{bmatrix} \mathbf{r}_i \\ \mathbf{e}_i - \boldsymbol{\xi}_i \end{bmatrix} = \mathbf{0}.\end{aligned}$$

Again, the joint inference function $\boldsymbol{\Psi}(\mathbf{Y};\boldsymbol{\theta})$ is unbiased as $\mathrm{E}\{\boldsymbol{\Psi}(\mathbf{Y};\boldsymbol{\theta})\} = \mathbf{0}$. The asymptotic results in Section 3.5 suggest that the estimator $\widehat{\boldsymbol{\theta}} = (\widehat{\boldsymbol{\beta}}, \widehat{\gamma})$ is consistent. Moreover, under some mild regularity conditions, $K^{1/2}(\widehat{\boldsymbol{\theta}} - \boldsymbol{\theta})$ is asymptotically multivariate normal with zero mean and covariance matrix of the form $\lim_K K\mathbf{j}^{-1}(\boldsymbol{\theta})$ where $\mathbf{j}(\boldsymbol{\theta})$ is the Godambe information matrix given by

$$\mathbf{j}(\boldsymbol{\theta}) = \mathbf{S}^T(\boldsymbol{\theta})\mathbf{V}^{-1}(\boldsymbol{\theta})\mathbf{S}(\boldsymbol{\theta}).$$

The sensitivity matrix $\mathbf{S}$ and variability matrix $\mathbf{V}$ are given as follows. First, $\mathbf{S}$ appears to be a block diagonal matrix, $\mathrm{diag}\{\mathbf{S}_1(\boldsymbol{\theta}), \mathbf{S}_2(\boldsymbol{\theta})\}$, where $\mathbf{S}_1(\boldsymbol{\theta})$ is given in equation (5.10) and

$$\mathbf{S}_2(\boldsymbol{\theta}) = \mathrm{E}\nabla_{\boldsymbol{\gamma}}\Psi_*(\mathbf{Y};\boldsymbol{\theta}) = -\sum_{i=1}^{K} \left(\frac{\partial \boldsymbol{\xi}_i}{\partial \gamma}\right)^T \mathbf{H}_i^{-1} \left(\frac{\partial \boldsymbol{\xi}_i}{\partial \gamma}\right). \tag{5.17}$$

The variability matrix can be written as a 2×2 block matrix as follows,

$$\mathbf{V}(\boldsymbol{\theta}) = \mathrm{E}\{\boldsymbol{\Psi}(\mathbf{Y};\boldsymbol{\theta})\boldsymbol{\Psi}^T(\mathbf{Y};\boldsymbol{\theta})\} = \begin{pmatrix} \mathbf{V}_{11}(\boldsymbol{\theta}) & \mathbf{V}_{12}(\boldsymbol{\theta}) \\ \mathbf{V}_{21}(\boldsymbol{\theta}) & \mathbf{V}_{22}(\boldsymbol{\theta}) \end{pmatrix},$$

where $\mathbf{V}_{11}(\boldsymbol{\theta})$ is given by equation (5.11), and the other three terms are

$$\mathbf{V}_{22}(\boldsymbol{\theta}) = \sum_{i=1}^{K} \left(\frac{\partial \boldsymbol{\xi}_i}{\partial \gamma}\right)^T \mathbf{H}_i^{-1} \mathrm{Var}(\mathbf{e}_i) \mathbf{H}_i^{-1} \left(\frac{\partial \boldsymbol{\xi}_i}{\partial \gamma}\right),$$

$$\mathbf{V}_{12}(\boldsymbol{\theta}) = \sum_{i=1}^{K} \mathbf{D}_i^T \mathbf{A}_i \Sigma_i^{-1} \mathrm{cov}(\mathbf{r}_i, \mathbf{e}_i) \mathbf{H}_i^{-1} \left(\frac{\partial \boldsymbol{\xi}_i}{\partial \gamma}\right),$$

$$\mathbf{V}_{21}(\boldsymbol{\theta}) = \mathbf{V}_{12}^T.$$

In the above, the dispersion parameter σ^2 is assumed known (e.g., binomial and Poisson); otherwise, it would be replaced by a $\sqrt{K}$-consistent estimator $\widehat{\sigma}^2$. Similarly, $\mathrm{Var}(\mathbf{e}_i)$ and $\mathrm{cov}(\mathbf{r}_i, \mathbf{e}_i)$ may replaced, respectively, by $\mathbf{e}_i \mathbf{e}_i^T$ in $\mathbf{V}_{22}$ and $\mathrm{diag}\{V(\mu_{i1}), \cdots, V(\mu_{in_i})\} \boldsymbol{\delta}_i \mathbf{e}_i^T$ in $\mathbf{V}_{12}$.

Several versions of the method of moments estimation for σ^2 are presented below. The first is to utilize the fact, $\mathrm{Var}(e_{ij}) = \sigma^2$, which turns out to be the the mean-variance relation in the case of the ED GEE. This estimate take the form

$$\begin{aligned} \widehat{\sigma}^2 &= \frac{1}{\sum_{i=1}^K n_i - (p+1)} \sum_{i=1}^K \sum_{j=1}^{n_i} \widehat{e}_{ij}^2 \\ &\overset{\text{ED}}{=} \frac{1}{\sum_{i=1}^K n_i - (p+1)} \sum_{i=1}^K \sum_{j=1}^{n_i} \frac{(y_{ij} - \widehat{\mu}_{ij})^2}{V(\widehat{\mu}_{ij})}, \end{aligned} \tag{5.18}$$

where the last equality holds for the family of ED marginal models.

The second one is the Jørgensen estimator derived from the application of the small-dispersion asymptotics. Similar to equation (2.21),

$$\widehat{\sigma}^2 = \frac{1}{\sum_{i=1}^K n_i - (p+1)} \sum_{i=1}^K \sum_{j=1}^{n_i} d(y_{ij}; \widehat{\mu}_{ij}), \tag{5.19}$$

which is the average of the marginal squared deviance residuals.

The third one uses the moment property given by Proposition 2.17 and similar to equation (2.22),

$$\widehat{\sigma}^2 = \frac{\sum_{i=1}^K \sum_{j=1}^{n_i} (\hat{\delta}_{ij} - \bar{\hat{\delta}})^2}{\sum_{i=1}^K \sum_{j=1}^{n_i} (-\hat{\dot{\delta}}_{ij})}, \tag{5.20}$$

where $\hat{\delta}_{ij} = \delta_{ij}(y_{ij}; \widehat{\mu}_{ij})$, $\hat{\dot{\delta}}_{ij} = \dot{\delta}_{ij}(y_{ij}; \widehat{\mu}_{ij})$ and $\bar{\hat{\delta}} = \frac{1}{\sum_i n_i} \sum_{i,j} \hat{\delta}_{ij}$.

Our own experience suggests that the first one (5.18) usually works well for the MGLM with any DM margin. Although the second one (5.19) is suitable for the MGLM with any DM margin, it is only recommended when the dispersion parameter is not large, say $\sigma^2 \le 5$. The third one (5.20) is an alternative to the first one.

5.3.2 Varying Dispersion Parameter

Consider the MGLM in that the dispersion parameter σ^2 requires special attention. The dispersion parameter σ^2 describes the distributional shape, which is beyond what the location or mean parameter alone can describe. A few studies have investigated the impact of the dispersion parameter on the GEE approach, when it is incorrectly modeled and estimated. See, for example, Paik (1992) and Song et al. (2004). A technical advantage by setting a constant dispersion parameter is that the regression coefficients can be separately estimated from the dispersion parameter.

However, in practice the assumption of a constant dispersion may be questionable. For example, the magnitude of dispersion may vary across drug treatment cohorts due to different rates of disease progression or over different follow-up times due to different environmental exposures. It is clear that the marginal pattern of a population depends not only on its averaged trend but also on its dispersion characteristics, as described by the dispersion models. Therefore, incorporating varying dispersion in the modeling process allows one to assess the heterogeneity of dispersion and to develop a simultaneous inference for the entire marginal models concerning both trend and dispersion components. Such an access to the profile of the dispersion parameter is important, and mistakenly assuming a varying dispersion to be constant in the application of the GEE method could cause some serious problems in statistical inference. For example, the asymptotic normality theory for the estimators may no longer be valid, and this theory is crucial to test for statistical significance for the effects of some covariates of interest. In addition, a proper estimation for the dispersion parameter is appealing, for example, in residual analysis, where a standardization for residuals is usually taken to stablize their variances. The computation of standardized residuals always asks for an appropriate estimate of the dispersion parameter.

Model $\log(\sigma_{ij}^2) = \mathbf{z}_{ij}^T \boldsymbol{\alpha}$ given in (5.3) may be assumed to study the profile of the dispersion in data analysis. When the covariate vector $\mathbf{z}_{ij}$ contains only the intercept, this model reduces to the constant dispersion case. In other words, this dispersion model allows us to perform a formal statistical test to determine whether the constant dispersion assumption is appropriate. Given an appropriate moment property, one may set up a third unbiased inference function, $\Psi_{**}(\mathbf{Y};\boldsymbol{\theta})$, and then the GEE2 for parameter $\boldsymbol{\theta} = (\boldsymbol{\beta}, \boldsymbol{\gamma}, \boldsymbol{\alpha})$ becomes

$$\boldsymbol{\Psi}(\mathbf{Y};\boldsymbol{\theta}) = \begin{bmatrix} \Psi(\mathbf{Y};\boldsymbol{\theta}) \\ \Psi_{*}(\mathbf{Y};\boldsymbol{\theta}) \\ \Psi_{**}(\mathbf{Y};\boldsymbol{\theta}) \end{bmatrix} = \mathbf{0}. \tag{5.21}$$

An version of inference function $\Psi_{**}(\mathbf{Y};\boldsymbol{\theta})$ may be formed on the basis of the the moment property, $\mathrm{Var}(e_{ij}) = \sigma_{ij}^2$, where e_{ij} is defined in (5.14). That is,

$$\Psi_{**}(\mathbf{Y};\boldsymbol{\theta}) = \sum_{i=1}^{K} \left(\frac{\partial \sigma_{ij}^2}{\partial \boldsymbol{\alpha}} \right)^T \mathbf{M}_i^{-1} (\mathbf{e}_i^2 - \boldsymbol{\sigma}_i^2(\boldsymbol{\alpha})) = 0,$$

where $\mathbf{e}_i^2 = (e_{i1}^2, \cdots, e_{in_i}^2)^T$, $\boldsymbol{\sigma}_i^2(\boldsymbol{\alpha}) = \mathrm{E}(\mathbf{e}_i^2)$, and $\mathbf{M}_i$ is a certain weighting matrix. Note that the e_{ij} reduces to the Pearson residual when the response y_{ij} follows an ED model, and hence this moment property is effectively equivalent to the mean-variance relation, $\mathrm{Var}(Y_{ij}) = \sigma_{ij}^2 V(\mu_{ij})$. For non-ED models, other kinds of moment properties can be also considered. For instance, in the simplex GEE for continuous proportional tdata, Song et al. (2004) suggested using a special moment property, $\mathrm{E}d(Y_{ij}; \mu_{ij}) = \sigma_{ij}^2$ to construct the Ψ_{**}.

Since the joint inference function $\boldsymbol{\Psi}(\mathbf{Y}; \boldsymbol{\theta})$ preserves the unbiasedness, similar asymptotics, including the Godambe information matrix, hold in this case.

5.4 Residual Analysis

This section concerns residual-based model diagnostics. As a semi-parametric modeling approach, the MGLM is dependent on the appropriateness of some parametric model assumptions. Three key parametric assumptions are: the assumption for marginal distribution, the assumption for link function, and the assumption for working correlation structure. Like every residual analysis, the proposed diagnostic tools are only able to provide a graphic check, which can only detect strong signals of model assumption violation. It is worth pointing out that the GEE method cannot provide a rigorous goodness-of-fit test for the first moment assumption, i.e., $\mathrm{E}\{\delta(Y_{ij}; \mu_{ij})\} = 0$ or the condition $\mathrm{E}(Y_{ij} - \mu_{ij}) = 0$ in the ED GEE. This is the most crucial assumption for a valid MGLM analysis in order to obtain consistent estimation. Most of the existing solutions are *ad hoc*, including Barnhart and Williamson (1998) and Pan (2001; 2002). This weakness of the GEE can be overcome by the QIF approach discussed in Section 5.5.

After an MGLM is fitted, the residuals can be calculated according to

$$e_{ij} \overset{\mathrm{DM}}{=} \frac{\delta_{ij}}{\sqrt{\mathrm{E}\{-\dot{\delta}(Y_{ij}; \mu_{ij})\}}} \overset{\mathrm{ED}}{=} \frac{y_{ij} - \mu_{ij}}{\sqrt{V(\mu_{ij})}}, \; j = 1, \ldots, n_i; i = 1, \ldots, K.$$

It is known that $\mathrm{E}(e_{ij}) = 0$ and $\mathrm{Var}(e_{ij}) = \sigma^2$. The standardized residuals are $\bar{e}_{ij} = e_{ij}/\sigma$, which consequently have mean zero and variance (or standard deviation) 1. It is also known that

$$\mathrm{corr}(\bar{e}_{ij}, \bar{e}_{ij'}) = \mathrm{corr}(\delta_{ij}, \delta_{ij'}).$$

The sample counterpart of $\bar{e}_{ij}$ is

$$\widehat{\bar{e}}_{ij} \overset{\mathrm{DM}}{=} \frac{\widehat{\delta}_{ij}}{\hat{\sigma}\sqrt{\widehat{\mathrm{E}}\{-\dot{\delta}(Y_{ij}; \widehat{\mu}_{ij})\}}} \overset{\mathrm{ED}}{=} \frac{y_{ij} - \widehat{\mu}_{ij}}{\hat{\sigma}\sqrt{V(\widehat{\mu}_{ij})}}.$$

5.4.1 Checking Distributional Assumption

Take a representative moment property as the target for diagnosis. One proposal is to check the property $\mathrm{Var}(e_{ij}) = \sigma^2$, which is equivalent to checking the mean-variance relation, $\mathrm{Var}(y_{ij}) = \sigma^2 V(\mu_{ij})$ in the case of the ED margin. If this relation is true, then the variance of the Pearson-type residual is $\mathrm{Var}(\bar{e}_{ij}) = 1$ and independent of mean μ_{ij}. Therefore the plot of $\widehat{\bar{e}}_{ij}$ against $\widehat{\mu}_{ij}$ may be invoked for checking the mean-variance relation, and hence the assumption of marginal distribution. The ideal appearance of the plot would be that all points randomly scatter around the horizontal line at zero, with approximately 95% points in the band $(-2, 2)$. Any departure from this suggests that the assumed distribution may not be appropriate for the data.

5.4.2 Checking Constant Dispersion Assumption

The plot of $\widehat{\bar{e}}_{ij}$ against $\widehat{\mu}_{ij}$ can also be used to check if the dispersion parameter is constant over the mean value of the response, if the residuals are calculated from the GEE under the assumption of a constant dispersion. In the mean time, plots of $\widehat{\bar{e}}_{ij}$ against individual covariates $x_{ijl}, l = 1, \ldots, p$, can indicate whether the dispersion parameter depends any specific covariates, such as the covariate of time.

5.4.3 Checking Link Functions

An informal check for the link function assumption could be done by following the McCullagh and Nelder (1989) plot of the adjusted dependent variable w against the linear predictor $\hat{\eta}$. In our setting, define

$$w_{ij} = g(\mu_{ij}) + \dot{g}(\mu_{ij})(y_{ij} - \mu_{ij}),\ j = 1, \ldots, n_i; j = 1, \ldots, K.$$

Clearly $\mathrm{E}(w_{ij}) = g(\mu_{ij})$ since $\mu_{ij} = \mathrm{E}(y_{ij})$. If the link function is appropriate, the plot of $\widehat{w}_{ij} = g(\widehat{\mu}_{ij}) + \dot{g}(\widehat{\mu}_{ij})(y_{ij} - \widehat{\mu}_{ij})$ against $\hat{\eta}_{ij} = \mathbf{x}_{ij}^T\widehat{\boldsymbol{\beta}}$ should show a straight line. Practically, drawing a least-squares fitted line of $\widehat{w}_{ij}$ on $\hat{\eta}_{ij}$ helps in visualizing possible deviations in the diagnosis.

5.4.4 Checking Working Correlation

Although it is difficult to model the true correlation structure of correlated data, approximate correlation structures would be appealing to achieve high efficiency of estimation for regression coefficients. It is important to assess the appropriateness of working correlation used in GEE via residual analysis. The idea behind is that the resulting residuals should not show strong and systematic patterns in the serial correlation if the working correlation structure is a good approximation to the true one. Note that

$$\text{corr}(\bar{e}_{ij}, \bar{e}_{ij'}) = \text{corr}(r_{ij}, r_{ij'}),$$

which means that the true correlation of variable z_{ij} is equal to that of the standardized residuals $\bar{e}_{ij}$. Now form cross-sectionally a vector of residuals at the jth time, denoted by

$$\bar{\mathbf{e}}_j = (\bar{e}_{1j}, \ldots, \bar{e}_{Kj})^T, \ j = 1, \ldots, n^* = \max(n_1, \ldots, n_K).$$

In general, the dimension of the residual vector is variable over time. Then, calculate the $n^* \times n^*$ sample correlation matrix based on these cross-sectional residual vectors $\bar{\mathbf{e}}_j$'s. All off-diagonal values should be modest (less than 0.4, say) if the working correlation is appropriate. Alternatively, one may construct a scatterplot matrix for all possible pairs of $\bar{\mathbf{e}}_j$'s, and if the working correlation is an appropriate approximation to the true one, all plots should show no strong dependence and patterns over time.

5.5 Quadratic Inference Functions

Quadratic inference functions (QIF) provide another quasi-likelihood inference in the MGLM. In comparison to the GEE approach, QIF has the following advantages:

(a) The application of QIF does not require more model assumptions than the GEE.
(b) It constructs more estimating functions than the number of parameters, so extra degrees of freedom are available to perform the goodness-of-fit test. Moreover, some model selection criteria such as AIC and BIC can be established in QIF. Note that such types of procedures are unavailable in the GEE.
(c) Qu et al. (2000) showed that the QIF estimator of $\boldsymbol{\beta}$ in the MGLM is more efficient than the GEE estimator, when the working correlation is misspecified, but equally efficient when the working correlation is correctly specified. This efficiency gain is due to the fact that QIF does not need to estimate the parameters in a given correlation structure.
(d) The QIF estimators are robust with a bounded influence function against unduly large outliers or contaminated data points, whereas the GEE is not robust and very sensitive to influential data cases. The current effort in the literature is on the detection and removal of outliers (Preisser and Qaqish, 1996), rather than dealing with outliers via robust estimating functions. Refer to Qu and Song (2004) for more details.

The formulation of QIF is rooted in the availability that the inverse of the working correlation $\mathbf{R}(\gamma)$ can be expressed by a linear combination of basis matrices; that is,

$$\mathbf{R}^{-1}(\gamma) = \sum_{l=1}^{m} \gamma_l M_l, \tag{5.22}$$

where $M_1, \ldots, M_m$ are known matrices and $\gamma_1, \ldots, \gamma_m$ are unknown coefficients.

Example 5.6 (Interchangeable Structure). The interchangeable (or compound symmetry) working correlation matrix $\mathbf{R}$ gives rise to $\mathbf{R}^{-1}(\gamma) = \gamma_1 M_1 + \gamma_2 M_2$, where $\gamma_l, l = 1, 2$ are both functions of the equi-correlation parameter γ. The two basis matrices are $M_1 = \mathbf{I}_{n_i}$, the n_i-dimensional identity matrix, and M_2, a matrix with 0 on the diagonal and 1 off the diagonal.

Example 5.7 (AR-1 Structure). The inverse of the AR-1 working correlation $\mathbf{R}$ can be written as $\mathbf{R}^{-1}(\gamma) = \gamma_1 M_1 + \gamma_2 M_2 + \gamma_3 M_3$, where $\gamma_j, j = 1, 2, 3$ are functions of the auto-correlation parameter γ. These three basis matrices can be found as $M_1 = \mathbf{I}_{n_i}$, M_2 with 1 on the two main off-diagonals and 0 elsewhere, and M_3 with 1 on the corners $(1, 1)$ and (n_i, n_i), and 0 elsewhere.

Note that this decomposition (5.22) is in general not unique, and the basis matrices given in the above examples are suggested in Qu et al. (2000). What really matters in (5.22) is that it gives rise to different moment conditions, which is essential to employ the generalized method of moments (GMM) introduced in Section 3.6.

Plugging the form (5.22) into the GEE (5.9) results in

$$\sum_{i=1}^{K} \mathbf{D}_i^T \mathbf{A}_i \mathbf{G}_i^{-1/2} (\gamma_1 M_1 + \ldots + \gamma_m M_m) \mathbf{G}_i^{-1/2} \mathbf{r}_i = \mathbf{0}. \tag{5.23}$$

This is in fact a linear combination of elements of the following inference function vector, each being related to one basis matrix,

$$\begin{aligned} \boldsymbol{\varphi}(\mathbf{Y}; \boldsymbol{\beta}) &= \frac{1}{K} \sum_{i=1}^{K} \varphi_i(\mathbf{Y}_i; \boldsymbol{\beta}) \\ &= \frac{1}{K} \begin{bmatrix} \sum_{i=1}^{K} \mathbf{D}_i^T \mathbf{A}_i \mathbf{G}_i^{-1/2} M_1 \mathbf{G}_i^{-1/2} \mathbf{r}_i \\ \vdots \\ \sum_{i=1}^{K} \mathbf{D}_i^T \mathbf{A}_i \mathbf{G}_i^{-1/2} M_m \mathbf{G}_i^{-1/2} \mathbf{r}_i \end{bmatrix}, \end{aligned} \tag{5.24}$$

where the coefficients γ_l's are not involved. If an independence working correlation is assumed, the $\boldsymbol{\varphi}$ reduces to the GEE, with $m = 1$ and $M_1 = I_{n_i}$.

Since the number of components in inference function $\boldsymbol{\varphi}$ is greater than the dimension of $\boldsymbol{\beta}$, it is impossible to directly solve $\boldsymbol{\varphi}(\boldsymbol{\beta}) = 0$ for $\boldsymbol{\beta}$. This is because the $\boldsymbol{\beta}$ is overidentified. Following the GMM in Section 3.6, one may utilize the optimal quadratic distance function of the $\boldsymbol{\varphi}$

$$Q(\boldsymbol{\beta}) = \boldsymbol{\varphi}^T(\mathbf{Y}; \boldsymbol{\beta}) \mathbf{C}(\boldsymbol{\beta})^{-1} \boldsymbol{\varphi}(\mathbf{Y}; \boldsymbol{\beta}), \tag{5.25}$$

and take its minimizer as the estimator of $\boldsymbol{\beta}$,

$$\widehat{\boldsymbol{\beta}} = \arg\min_{\boldsymbol{\beta}} Q(\boldsymbol{\beta}).$$

Here matrix $\mathbf{C}(\boldsymbol{\beta}) = (1/K^2) \sum_{i=1}^{K} \boldsymbol{\varphi}_i(\boldsymbol{\beta}) \boldsymbol{\varphi}_i(\boldsymbol{\beta})^T$ is a consistent estimate of the variance of $\boldsymbol{\varphi}(\mathbf{Y}; \boldsymbol{\beta})$. Under some mild regularity conditions, this minimizer is unique. This objective function Q is referred to as the *quadratic inference function* (QIF).

Qu et al. (2000) showed that the QIF estimator $\widehat{\boldsymbol{\beta}}$ is consistent and asymptotically normal, and that the asymptotic variance is equal to the inverse of the Godambe information matrix, $\mathbf{j}(\boldsymbol{\beta})^{-1}$, where $\mathbf{j}(\boldsymbol{\beta}) = \mathbf{S}_{\boldsymbol{\varphi}}^T(\boldsymbol{\beta}) \mathbf{V}_{\boldsymbol{\varphi}}(\boldsymbol{\beta})^{-1} \mathbf{S}_{\boldsymbol{\varphi}}(\boldsymbol{\beta})$. Also, the Godambe information matrix may be consistently estimated by

$$\hat{\mathbf{j}}(\widehat{\boldsymbol{\beta}}) = \dot{\boldsymbol{\varphi}}(\widehat{\boldsymbol{\beta}})^T \mathbf{C}(\widehat{\boldsymbol{\beta}})^{-1} \dot{\boldsymbol{\varphi}}(\widehat{\boldsymbol{\beta}}).$$

An important contribution of the QIF is that it provides a goodness-of-fit test for the first moment assumption of the MGLM, i.e., $\mathrm{E}\{\delta_{ij}(Y_{ij}; \mu_{ij})\} = 0$ or $\mathrm{E}(y_{ij} - \mu_{ij}) = 0$ under ED margins. This test not only closely mimics an extension of the minimum χ^2 method of generating the best asymptotically normal estimates as originally introduced by Neyman (1949) and later expanded by Ferguson (1958), but is also analogous to the likelihood ratio test in the framework of likelihood-based inferences. Theorem 3.14 implies that under the null hypothesis of the first moment assumption being valid, the asymptotic distribution of $Q(\widehat{\boldsymbol{\beta}})$ is χ^2 with degrees of freedom equal to $\{\dim(\boldsymbol{\varphi}) - \dim(\boldsymbol{\beta})\}$. In the QIF (5.25), where each component is p-dimensional, it is easy to see that $df = mp - p = (m-1)p$, where $p = \dim(\boldsymbol{\beta})$. Therefore, the null will be rejected if the corresponding p-value of the observed statistic is smaller than 0.05, say. Note that this test holds whether or not the working correlation is correctly specified.

Example 5.8 (Unstructured Correlation).

The construction of QIF for the case of unstructured correlation is done in a very different fashion. Qu and Lindsay (2003) found that an approximately optimal inference function can be obtained through a sequence of basis matrices $I, V, V^2, \ldots$, where $V = \mathrm{Var}(\mathbf{Y})$. Therefore, they suggested using $M_1 = I, M_2 = V, \ldots, M_m = V^m$ to form the QIF. In most applications, $m = 2$ or $m = 3$ would give a satisfactory efficiency gain, and adding more basis matrices will not generally improve the efficiency much but, on the other hand, make related numerical computations much more difficult.

This variance matrix can be estimated simply by the sample covariance matrix $(\mathbf{y} - \boldsymbol{\mu})(\mathbf{y} - \boldsymbol{\mu})^T$, where $\boldsymbol{\mu} = \boldsymbol{\mu}(\boldsymbol{\beta})$ will be updated along the updating of regression coefficient $\boldsymbol{\beta}$ in the iterative algorithm. An advantage of this method is that it avoids the need of inverting the correlation matrix, which may be difficult in some model settings.

Finally, the objective function Q can be used to define certain model selection criteria such as Akaike's information criterion (AIC) and the traditional Bayes information criterion (BIC), respectively, as follows,

$$\text{AIC} = Q(\widehat{\boldsymbol{\beta}}) + 2(m-1)p,$$
$$\text{BIC} = Q(\widehat{\boldsymbol{\beta}}) + \{(m-1)p\}\ln(K).$$

Similar to Section 5.2.3, a score-type test for a nested model can be derived. Suppose the hypothesis of interest is $H_0 : \boldsymbol{\beta}_2 = \mathbf{0}$, under a partition of $\boldsymbol{\beta} = (\boldsymbol{\beta}_1, \boldsymbol{\beta}_2)$ in the full model,

$$g(\mu_{ij}) = \mathbf{x}_{ij}^T\boldsymbol{\beta} = \mathbf{x}_{1ij}^T\boldsymbol{\beta}_1 + \mathbf{x}_{2ij}^T\boldsymbol{\beta}_2.$$

Here $\dim(\boldsymbol{\beta}_1) = p_1, \dim(\boldsymbol{\beta}_2) = p_2$, and $p_1 + p_2 = p$. Then under the H_0, the difference of QIF,

$$\text{DQIF} = Q(\widehat{\boldsymbol{\beta}_1}) - Q(\widehat{\boldsymbol{\beta}}) \overset{asy.}{\sim} \chi^2_{p_1}.$$

Thus, reject H_0 when the p-value of the observed DQIF is smaller than 0.05, say.

5.6 Implementation and Softwares

5.6.1 Newton-Scoring Algorithm

The Newton-scoring algorithm is the key numerical recipe in the search for the root of a system of nonlinear equations. Take the GEE2 with a known dispersion parameter as an example for illustration, and the implementation in the other versions of the GEEs can be done similarly.

The numerical task is to solve the following GEE2 for $\widehat{\boldsymbol{\beta}}$ and $\widehat{\gamma}$, with given σ^2,

$$\boldsymbol{\Psi}(\mathbf{Y};\boldsymbol{\theta}) = \begin{bmatrix} \Psi(\mathbf{Y};\boldsymbol{\beta},\gamma) \\ \Psi_*(\mathbf{Y};\boldsymbol{\beta},\gamma) \end{bmatrix} = \mathbf{0}. \tag{5.26}$$

Because the sensitivity matrix $\mathbf{S}$ of the $\boldsymbol{\Psi}$ is block-diagonal, $\text{diag}(\mathbf{S}_1, \mathbf{S}_2)$, solving $\boldsymbol{\Psi}(\mathbf{Y};\boldsymbol{\theta}) = \mathbf{0}$ can be done iteratively between the following two Newton-scoring updates:

$$\boldsymbol{\beta}^{(l+1)} = \boldsymbol{\beta}^{(l)} - \mathbf{S}_{11}^{-1}\left(\boldsymbol{\theta}^{(l)}\right)\Psi\left(\boldsymbol{\theta}^{(l)}\right),$$
$$\gamma^{(l+1)} = \gamma^{(l)} - \mathbf{S}_2^{-1}\left(\boldsymbol{\theta}^{(l)}\right)\Psi_*\left(\boldsymbol{\theta}^{(l)}\right),$$

where $\mathbf{S}_1$ and $\mathbf{S}_2$ are two sensitivity matrices given respectively by (5.10) and (5.17) .

It is worth pointing out that at each iteration, the Newton-scoring update for the $\boldsymbol{\beta}$ effectively performs a weighted least squares calculation. Rewrite the iterative step as

$$\boldsymbol{\beta}^{(l+1)} = \left\{\sum_{i=1}^K \mathbf{X}_i\mathbf{W}_i^{-1}\left(\boldsymbol{\theta}^{(l)}\right)\mathbf{X}_i^T\right\}^{-1}\sum_{i=1}^K \mathbf{X}_i\mathbf{W}_i^{-1}\left(\boldsymbol{\theta}^{(l)}\right)\widetilde{\mathbf{r}}_i\left(\boldsymbol{\theta}^{(l)}\right), \tag{5.27}$$

where, suppressing the $\boldsymbol{\theta}^{(l)}$

$$\mathbf{W}_i = \text{diag}\{\dot{g}(\mu_{i1}), \ldots, \dot{g}(\mu_{in_i})\}\mathbf{A}_i^{-1}\Sigma_i\mathbf{A}_i^{-1}\text{diag}\{\dot{g}(\mu_{i1}), \ldots, \dot{g}(\mu_{in_i})\}$$
$$\widetilde{\mathbf{r}}_i = \mathbf{X}_i^T\boldsymbol{\beta}^{(l)} + \text{diag}\{\dot{g}(\mu_{i1}), \ldots, \dot{g}(\mu_{in_i})\}\mathbf{A}_i^{-1}\mathbf{r}_i.$$

The formula (5.27) is exactly the same as the weighted least squares estimation of $\boldsymbol{\beta}$ in a linear regression model

$$\widetilde{\mathbf{r}}_i = \mathbf{X}_i^T\boldsymbol{\beta} + \widetilde{\varepsilon}_i, \ \widetilde{\varepsilon}_i \sim \text{MVN}_{n_i}(\mathbf{0}, \mathbf{W}_i).$$

The inverse matrix in (5.27) may be obtained by using a generalized inverse. A generalized inverse of a matrix A is any matrix G such that $AGA = A$.

5.6.2 SAS PROC GENMOD

`SAS` PROC GENMOD currently performs the GEE (not GEE2), in which parameters $\boldsymbol{\gamma}$ and σ^2 are separately estimated from the estimation of $\boldsymbol{\beta}$ via (5.12). It is noted that this `SAS` PROC is only applicable for the MGLM with the ED margins, including normal, binomial, Poisson, and gamma, and it does not work for the simplex or von Mises marginal margins.

Take the multiple sclerosis data of Section 1.3.6 as an example, in which the response Y_{ij} is binary variable with $Y_{ij} = 1$ indicating the presence of exacerbation and 0 otherwise. The covariates includes $\texttt{trt}_i$ for the administrated dose level for subject i (independent of time index j), $\texttt{dur}_i$ for the baseline duration of disease since the diagnosis, and two time variables $\texttt{t}_j$ and $\texttt{t}_j^2$, which are independent of subject index i because of the same scheduled visit times during the trial. The marginal logistic marginal model is

$$\log\frac{\pi_{ij}}{1-\pi_{ij}} = \beta_0 + \beta_1\text{trt}_i + \beta_2\text{t}_j + \beta_3\text{t}_j^2 + \beta_4\text{dur}_i,$$

where $\pi_{ij} = \text{prob}(Y_{ij} = 1|\mathbf{x}_{ij})$ is the probability of exacerbation. `SAS` PROC GENMOD allows several working correlation structures in the GEE estimation of the regression coefficients. The following `SAS` codes are based on unstructured, interchangeable, and AR-1, respectively.

```
title "UNSTRUCTURED CORRELATION";
:::::::(DATA IMPORT):::::::::::
proc genmod data=exacerb;
class id;
model rel= trt t1 t2 dur / dist=bin link=logit;
repeated subject=id / type=un corrw covb modelse;
run;

title "INTERCHANGEABLE CORRELATION (type=cs)";
:::::::(DATA IMPORT):::::::::::
```

```
proc genmod data=exacerb;
class id;
model rel= dose t1 t2 dur / dist=bin link=logit;
repeated subject=id / type=exch corrw covb modelse;
run;
/*cs stands for compound symmetry*/

title "AR-1 CORRELATION";
:::::::(DATA IMPORT):::::::::::
proc genmod data=exacerb;
class id;
model rel= dose t1 t2 dur / dist=bin link=logit;
repeated subject=id / type=ar corrw covb modelse;
run;
```

The data imported to SAS has to be formatted in the following panel form:

```
1 x  x  x  x
1 x  x  x  x
1 x  x  x  x
2 x  x  x  x
2 x  x  x  x
2 x  x  x  x
2 x  x  x  x
... ...
```

where subject 1 has three repeated measurements and subject 2 has four repeated measurements, and so on, listed vertically in columns.

5.6.3 SAS MACRO QIF

An α-test version of a SAS MARCO QIF is available for a secured download at the webpage `www.stats.uwaterloo.ca/~song/BOOKLDA.html`. Outpouts of this macro include estimates of the model parameters, asymptotic covariance matrix, standard errors, χ^2 statistic for goodness-of-fit tests, and two model selection criteria AIC and BIC. It is applicable to the following marginal distributions:

Distribution	Canonical link function
Normal	Identity $g(\mu) = \mu$
Poisson	Log $g(\mu) = \log(\mu)$
Binary	Logit $g(\mu) = \log\{\mu/(1-\mu)\}$
Gamma	Reciprocal $g(\mu) = 1/\mu$

In addition, this macro accommodates the following working correlation structures: independent, unstructured, AR-1, and interchangeable. Users can choose from these available options in their data analysis.

This QIF macro works when the sizes of repeated measurement are different for clusters (except for the unstructure correlation case), where the dimension of the individual inference function φ_i in (5.24) is modified accordingly. It implements a Newton-scoring algorithm with the starting values being assigned as the estimates from PROC GENMOD under the same correlation structure.

Additional features of macro QIF are:

(a) The format of dataset for the macro is the same as that for PROC GENMOD.
(b) The α-test version implements only the listwise deletion method for handling missing values. That is, it assumes the mechanism of missingness to be missing completely at random (MCAR). However, this macro does include an argument of `weight`, with the default set at 1, which provides flexibility in possibly incorporating suitable values of weights such as those determined by the inverse probability weighting scheme, in order to handle the missing at random (MAR) situation.
(c) For binary data, this macro specifies the default probability of "outcome being 1" to be modeled.
(d) This macro is developed under `SAS` version 9.1.3.
(e) The main program is coded in PROC IML.

Macro QIF first validates input arguments before the numerical computation begins. All outputs are formatted in the standard `SAS` module, which is easy to read and edit. The input arguments include dataset, response, covariates, cluster identification, type of marginal distribution, type of correlation structure, and flag of displaying results. Outputs include parameter estimates, asymptotic covariance matrix, fitted values and pearson/deviance residuals, goodness-of-fit test statistic, and AIC/BIC. All of these are stored as `SAS` datasets located in `SAS` Work library.

Based on the same data example as in the illustration of PROC GENMOD, the macro code is

```
\%qif(data=exacerb, yvar=rel, xvar=dose dur t1 t2, id=id,
       dist=bin, corr=exch, print=Y, outpar=par2, outqif=qif2,
       outcov=cov2,outres=binres);
run;
```

5.7 Examples

This section presents three data examples. The first is the analysis of the multiple sclerosis data introduced in Section 1.3.6, where a binary response, `exacerbation`, is of interest. The second example is the analysis of the epileptic seizures data discussed in Section 1.3.2, in which the response of interest is the number of seizures observed during a period of two weeks. The third one is the analysis of the retinal surgery data described in Section 1.3.3, where the response is a percentage of gas absorption confined between 0 and 1.

Each example emphasizes different aspects of the quasi-likelihood methods discussed in this chapter, including estimation, residual analysis, goodness-of-fit, model selection, and robustness.

5.7.1 Longitudinal Binary Data

Refer to Section 1.3.6 for a detailed data description. Both the GEE and QIF are applied to fit the data, and the QIF is applied to perform a goodness-of-fit test and to select the correlation structures. Among the three response variables, only one binary response, `exacerbation`, is analyzed in this example, which refers to whether an exacerbation began since the previous MRI scan, and 1 for yes and 0 for no.

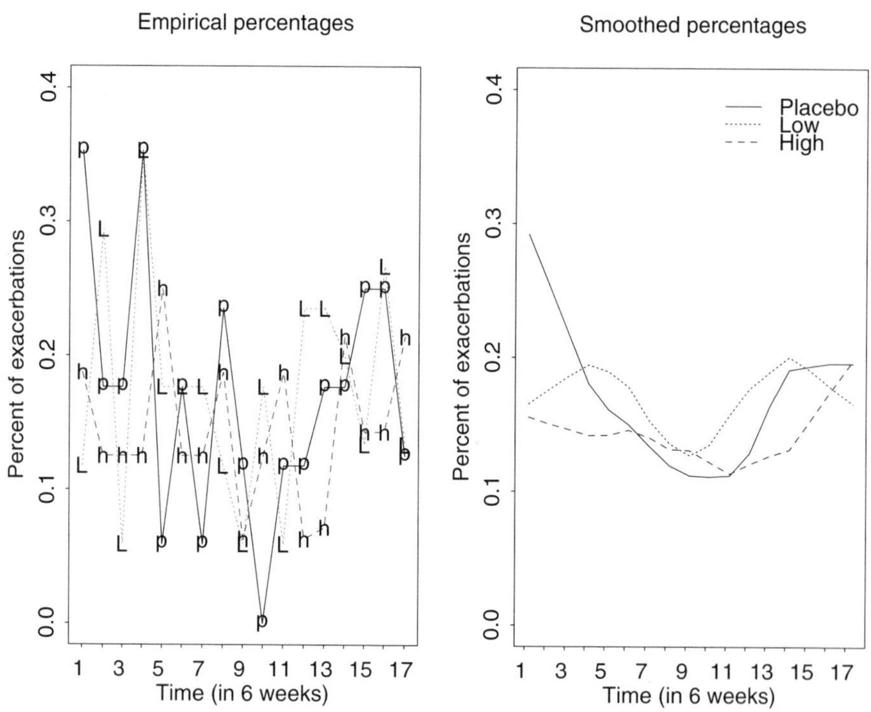

Fig. 5.2. Empirical percentage and smoothed empirical percentage of exacerbations over 17 weeks across three treatment arms.

Figure 5.2 displays the empirical percentage of exacerbations (on the left panel) over the study period of 17 weeks across three treatment arms (placebo, low dose, and high dose). To better visualize patterns regarding the changes on the percentage of exacerbations over time, the right panel shows a smoothed version of the left panel using a local smoothing technique, LOWESS. This figure clearly unveils that the time effect is not linear, so a quadratic polynomial in time is imposed in the fit. The central objective is to examine whether the drug helps to reduce the risk of exacerbation for multiple sclerosis patients.

Several baseline covariates are included in the model. They are, treatment (`trt`), time (`t`) in weeks, and squared time ($\mathtt{t}^2$), and duration of disease (`dur`) in years. Here `trt` is treated as an ordinal covariate with scales $0, 1, 2$ representing zero (placebo), low, and high dosage of the drug treatment. This leads to the marginal logistic model for the data:

$$\text{logit}(\mu_{ij}) = \beta_0 + \beta_1 \, \text{trt}_i + \beta_2 \, \text{t}_j + \beta_3 \, \text{t}_j^2 + \beta_4 \, \text{dur}_i, \tag{5.28}$$

where μ_{ij} is the probability of exacerbation at visit j for subject i.

Three correlation structures (independence, interchangeable, and AR-1) are considered in both GEE and QIF analyses. Since under the independence structure both methods give the same results, it is excluded in the comparison. The estimates produced from PROC GENMOD and macro QIF are listed in Table 5.1.

Table 5.1. GEE estimates (standard errors) from PROC GENMOD and MACRO QIF.

	AR-1		Interchangeable	
	GEE	QIF	GEE	QIF
Par.	Est(Std Err)	Est(Std Err)	Est(Std Err)	Est(Std Err)
intcpt	−0.6793(0.3490)	−0.4955(0.3443)	−0.6847(0.3502)	−0.5419(0.3169)
trt	−0.0151(0.1501)	−0.0222(0.1491)	−0.0175(0.1497)	−0.0650(0.1448)
time	−0.0259(0.0128)	−0.0269(0.0128)	−0.0251(0.0129)	−0.0267(0.0127)
time^2	0.0002(0.0001)	0.0002(0.0001)	0.0002(0.0001)	0.0002(0.0001)
dur	−0.0449(0.0229)	−0.0715(0.0242)	−0.0458(0.0228)	−0.0586(0.0236)

This analysis did not find strong evidence that the population average effect of the drug treatment is significant in reducing the risk of exacerbation. The baseline disease severity measured as the duration of disease before the trial is an important explanatory variable associated with the risk of exacerbation. For the time-course effect, both linear and quadratic time covariates are significant. This is due partly to the fact that periodic recurrences of the disease behave in a more complicated fashion than a linear function. To better understand the treatment effect along with the cycle of disease recurrences, it may be appealing to invoke a varying coefficient model in that the treatment effectiveness would be modeled as a time-dependent function. Interested readers may refer to Wu and Zhang (2006) for fitting a time-varying coefficient model.

Table 5.2. The goodness-of-fit test statistic Q and AIC/BIC model selection criteria given by macro QIF.

Statistic	AR-1	Interchangeable
Q	4.3	2.5
df	5.0	5.0
AIC	14.3	12.5
BIC	23.3	21.5

Macro QIF also reports the goodness-of-fit test statistic Q and AIC/BIC model selection criteria, which are listed in Table 5.2. According to the χ^2 distribution with 5 degrees of freedom, the p-value of the goodness-of-fit test statistic is 0.507 under AR-1 structure and 0.776 under interchangeable structure, both suggesting that the specification of the first moment structure, namely model (5.28), is appropriate. The comparison of AIC/BIC implies that the interchangeable structure is slightly better than the AR-1 structure. The lorelograms of the observed exacerbation incidences across the three groups, shown in Figure 5.3, confirms the result of model selection, because there are no obvious patterns suggested in these lorelograms.

Since both correlation structures give the same conclusions regarding the statistical significance for the covariates as well as the goodness-of-fit, it is not very crucial to determine which structure to be used in the analysis.

5.7.2 Longitudinal Count Data

This section presents an analysis of the epileptics data introduced in Section 1.3.2 using both GEE and QIF methods. The emphasis of this analysis is on the robustness of QIF. As pointed in Section 1.3.2, patient ID 207 is a noticeable outlier. To assess how outlier patient 207 influences the performance of GEE or QIF, the same analysis is done twice, once for the full data including patient 207 and then for the subset of the data excluding this patient. Then the amount of change occurred in the estimation between the two settings is

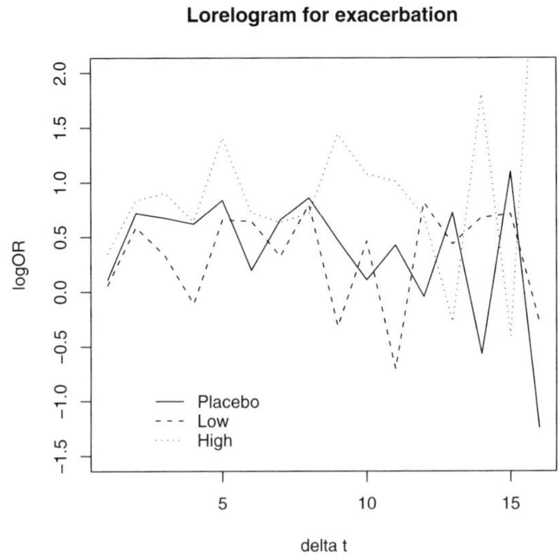

Fig. 5.3. Lorelograms of repeated exacerbation incidences over 17 weeks across three treatment arms.

measured by the well-known criterion DFBETAS. That is, for each parameter, a ratio of relative change is defined as:

$$RC(\theta_j) = \frac{|\theta_{j,gee}^{with} - \theta_{j,gee}^{without}|}{s.e.(\theta_{j,gee}^{without})} / \frac{|\theta_{j,qif}^{with} - \theta_{j,qif}^{without}|}{s.e.(\theta_{j,qif}^{without})}.$$

If RC is larger than 1, then the outlier affects the GEE more severely than the QIF. The larger the RC is, the more robust the QIF method is relative to the GEE.

The response variable is the number of seizures in a two-week period, and covariates include a logarithm of a quarter of baseline seizure counts (`bsln`), logarithm of age (`logage`), treatment (`trt`, 1 for progabide and 0 for placebo), and visit (`vst` $= 1, 2, 3, 4$). The marginal log-linear model takes the form

$$\log(\mu_{ij}) = \beta_0 + \beta_1 \,\text{bsln}_i + \beta_2 \,\text{trt}_i + \beta_3 \,\text{logage}_i + \beta_4 \,\text{vst}_j, \tag{5.29}$$

where μ_{ij} is the expected number of seizures during the two-week period j for subject i. Since the GEE and QIF give the same estimates and standard errors under the independence correlation structure, the comparison for the two methods is conducted for the AR-1 and interchangeable structures.

Table 5.3 lists all the estimates and standard errors, with or without patient 207, under AR-1 correlation structure.

For this analysis, the mean zero assumption for the first moment (i.e., the log-linear model specification) in model (5.29) is confirmed, since the

Table 5.3. GEE and QIF estimates and standard errors with AR-1 correlation structure for the complete data and the data without patient 207.

	Complete data				Without 207			
	Estimate		Std Err		Estimate		Std Err	
Par	GEE	QIF	GEE	QIF	GEE	QIF	GEE	QIF
intcpt	−2.522	−2.233	1.034	1.006	−2.380	−2.017	0.863	0.892
bsln	1.247	1.193	0.163	0.099	0.987	0.960	0.080	0.066
trt	−0.020	−0.046	0.190	0.141	−0.255	−0.281	0.152	0.146
logage	0.653	0.581	0.287	0.270	0.783	0.680	0.247	0.261
vst	−0.064	−0.052	0.034	0.026	−0.045	−0.047	0.035	0.031
Q-stat	–	3.7	–	–	–	5.9	–	–
AIC	–	13.7	–	–	–	16.9	–	–
BIC	–	24.1	–	–	–	26.2	–	–

p-value is 0.5934 based on the χ^2 distribution with 5 degrees of freedom for the complete data, and the p-value is 0.3161, based on the same distribution for the data without patient 207. Under the AR-1 correlation structure, the RC values are given in Table 5.4.

Table 5.4. Relative changes of estimates with respect to data point of patient 207 between GEE and QIF under AR-1 correlation.

	Parameter				
Covariates	intcpt	bsln	trt	logage	vst
RC	0.68	0.92	0.96	1.39	3.37

Table 5.4 indicates that the performances of GEE and QIF are close, and the only noticeable difference is at the covariate of `visit`, where the GEE is about three times more unstable than the QIF caused by patient 207.

Table 5.5 reports all the estimates and standard errors, with or without patient 207, under interchangeable correlation structure.

By comparison, the model selection criteria (both AIC and BIC) suggest that the interchangeable correlation is preferred over the AR-1 correlation. The p-values for the goodness-of-fit test, with and without patient 207, are 0.8628 and 0.7308 based on the χ^2 distribution with 5 degrees of freedom, respectively. Both imply that the mean model given in (5.29) is appropriate. Based on Table 5.6, both `basline seizure` and `logage` are statistically

Table 5.5. GEE and QIF estimates and standard errors with interchangeable correlation structure for the complete data and the data without patient 207.

	Complete data				Without 207			
	Estimate		Std Err		Estimate		Std Err	
Par	GEE	QIF	GEE	QIF	GEE	QIF	GEE	QIF
intcpt	−2.323	−1.870	1.045	0.991	−2.176	−1.793	0.888	0.963
bsln	1.227	1.181	0.156	0.115	0.986	0.961	0.084	0.082
trt	−0.010	−0.003	0.190	0.140	−0.223	−0.157	0.160	0.152
logage	0.604	0.497	0.288	0.277	0.721	0.592	0.250	0.269
vst	−0.059	−0.070	0.035	0.024	−0.043	−0.046	0.038	0.026
Q-stat	–	1.9	–	–	–	2.8	–	–
AIC	–	11.9	–	–	–	12.8	–	–
BIC	–	22.3	–	–	–	23.1	–	–

significant. This analysis does not find evidence that the treatment is helping to lessen the disease symptom.

To see the influence of patient 207 on the GEE and QIF, Table 5.6 gives the RC values. This table indicates that the performances of GEE and QIF are close, and the QIF is always slightly better except for the covariate of `visit`. As a result of the robust analysis, patient 207 is not very influential and does not cause much difference between the GEE and QIF analyses in terms of the influence measure DFBETAS. If a yardstick of ± 2 is used as a cutoff, patient ID 207 affects the GEE's estimation on the effect of `vst` in the AR-1 case and on the effect of `intercept` in the interchangeable case, more than it does for QIF. In addition, the robustness behavior may vary from one working correlation structure to another, due to different basis matrices used in the formulation of QIF.

Table 5.6. Relative changes of estimates with respect to data point of patient 207 between GEE and QIF under interchangeable correlation.

	Parameter				
Covariates	intcpt	bsln	trt	logage	vst
RC	2.07	1.07	1.31	1.33	0.64

5.7.3 Longitudinal Proportional Data

This section presents an analysis of the eye surgery data discussed previously in Section 1.2.3, using the GEE2 method. Let Y_{ij} be the j-th gas (C_3F_8) volume for the i-th individual at day t_{ij}. Recall they are all percentages and thus are assumed to follow marginally the simplex distribution $S^{-}(\mu_{ij}, \sigma^2)$. It begins with the case of a constant dispersion, and this assumption will be confirmed later via residual analysis. Figure 1.2 shows the longitudinal plot of observed y_{ij} versus t_{ij} for all 31 subjects. The LOWESS curve of the data with the fraction parameter set to 1/3 was plotted in Figure 5.4, which indicates that the volume (in percent) of the intraocular expansile gas decreases slowly in the first few days after maximal expansion, and then it decreases more rapidly and finally, more slowly.

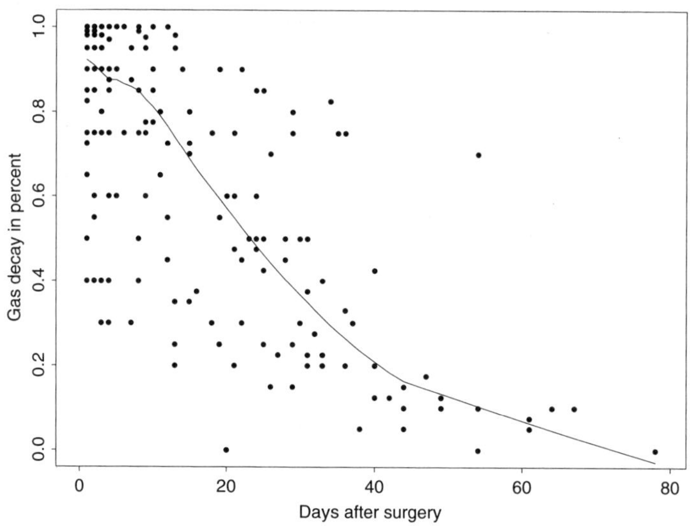

Fig. 5.4. Scatterplots of the smooth LOWESS curve for the raw eye surgery data.

The marginal model for the mean gas volume takes the form of

$$\text{logit}(\mu_{ij}) = \beta_0 + \beta_1 \log(\text{time}_{ij}) + \beta_2 \log^2(\text{time}_{ij}) + \beta_3 \text{gas}_i, \tag{5.30}$$

where the covariate of gas concentration level is defined as

$$\text{gas}_i = \frac{\text{gas}_i - 20}{5} = \begin{cases} -1, & \text{gas concentration level is 15} \\ 0, & \text{gas concentration level is 20} \\ 1, & \text{gas concentration level is 25.} \end{cases}$$

The GEE2 (5.26) is invoked to fit the model with three different correlation structures: independence, interchangeability, and AR-1. The Newton-scoring algorithm is used to solve the equation, where the dispersion parameter σ^2 is estimated by using the Jørgensen estimator:

$$\hat{\sigma}^2 = \frac{1}{\sum_i n_i - 4} \sum_i \sum_j d(y_{ij}; \hat{\mu}_{ij}).$$

Table 5.7. Results of the ophthalmology study under homogeneous dispersion.

	Independence		Interchangeability		AR-1	
Variable	Estimate	Std Err	Estimate	Std Err	Estimate	Std Err
Intercept	2.6850	0.3002	2.6534	0.2958	2.7330	0.2729
Log.time	0.0648	0.2491	0.1604	0.1991	0.0962	0.1991
Log.time^2	−0.3354	0.0662	−0.3790	0.0493	−0.3496	0.0470
Gas	0.3250	0.1945	0.2496	0.1778	0.3034	0.1716
$\rho = \exp(-\gamma)$	–	–	0.2515	0.0388	0.4877	0.1418

Table 5.7 summarizes the results obtained from the GEE2 with a constant dispersion. The estimate of dispersion parameter σ^2 is equal to 201.64, almost identical in the three models, indicating that the data are very strongly dispersed. Hence it is not reasonable to assume the data are from a normal distribution that corresponds to the case of dispersion close to 0. Thus, treating the response as to be normally distributed is worrisome. To confirm the appropriateness of the constant dispersion assumption, Figure 5.5 shows the residual plot, where panel (a) plots the standardized residuals $\widehat{\bar{e}}_{ij}$ versus covariate `log.time` and panel (b) plots the same residuals against covariate `gas` concentration level. Both plots indicate that the variation in the residuals varies in time and in the gas level. In fact, the residuals appears substantially volatile at the beginning of the observation period but gets stablized in later time. Similarly, the low concentration levels seem to be associated with bigger variability in the residuals. This requires investigation to see if the modeling of the dispersion parameter would lead to better behavior of the residuals.

Therefore, a model that can address the varying dispersion in two covariates of time and gas concentration level is proposed:

$$\log\left(\sigma_{ij}^2\right) = \alpha_0 + \alpha_1 \log(\text{time}_{ij}) + \alpha_2 \text{gas}_i. \tag{5.31}$$

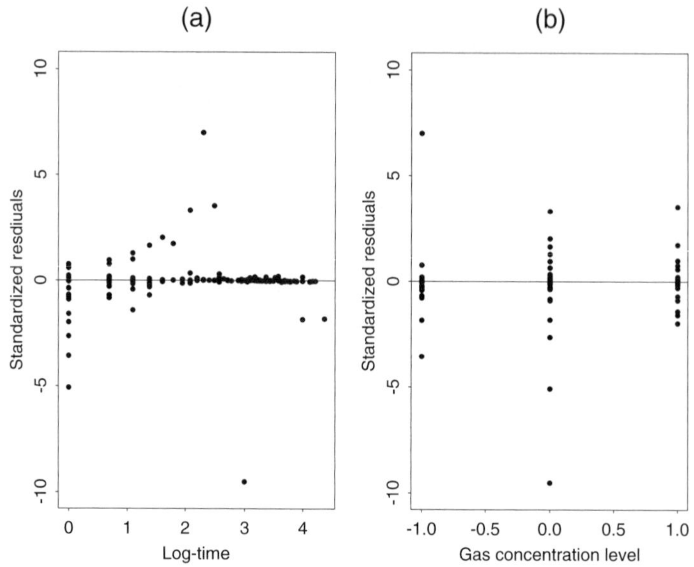

Fig. 5.5. Plots of the standardized residuals against two covariates, respectively, the log-time in panel (a) and the concentration level in panel (b).

The third estimating equation of the GEE2 (5.21) is formed on the basis of unit deviance $d_{ij}(y_{ij};\mu_{ij})$ since for the simplex distribution $\mathrm{E}\{d(Y_{ij};\mu_{ij})\} = \sigma_{ij}^2$, by Proposition 2.19 (a). The estimating equation takes the form

$$\Psi_{**}(\boldsymbol{\beta},\boldsymbol{\alpha},\boldsymbol{\gamma}) = \sum_{i=1}^{K} \left(\frac{\partial \boldsymbol{\sigma}_i^2}{\partial \boldsymbol{\alpha}}\right)^T \mathbf{M}_i^{-1}(\mathbf{d}_i - \boldsymbol{\sigma}_i^2) = \mathbf{0}, \tag{5.32}$$

where $\mathbf{d}_i = (d(y_{i1};\mu_{i1}),\ldots,d(y_{in_i};\mu_{in_i}))^T$, $\mathbf{M}_i = \mathrm{diag}\{\mathrm{Var}(d_{i1}),\ldots,\mathrm{Var}(d_{in_i})\}$ is a working covariance matrix with the independence correlation structure, and $\boldsymbol{\sigma}_i^2 = \mathrm{E}(\mathbf{d}_i) = (\sigma_{i1}^2,\ldots,\sigma_{in_i}^2)^T$. The first component Ψ in the GEE2 (5.21) uses AR-1 working correlation structure. Estimates and standard errors are listed in Table 5.8.

Clearly, both covariates of time and treatment are significant factors attributed to the varying dispersion in model (5.31). Figure 5.6 displays the fitted curves for the pattern of dispersion profile over time across three different gas concentration levels.

Based on the model with the varying dispersion, the findings for the other parameters are very similar to those given in Table 5.7. The squared log-time term is found significant, the linear log-time term is found not significant, and the gas concentration covariate is found marginally insignificant, at the significance level 0.05. Also, the estimated lag-1 autocorrelation $\widehat{\rho} = e^{-\widehat{\gamma}} =$

Table 5.8. Estimates, standard errors, and robust z-statistics from the heterogeneous dispersion model for the eye surgery data.

Parameter	β_0 (Int.)	β_1 (Log.T)	β_2 (Log^2.T)	β_3 (Gas)	α_0 (Int.)	α_1 (Log.T)	α_2 (Gas)	γ $(-\log(\rho))$
Estimate	2.7445	−0.0223	−0.3144	0.4114	6.1551	−0.4583	−0.4938	1.8484
Std Err	0.2107	0.3367	0.0855	0.2122	0.1988	0.0803	0.1427	0.3881
Z	13.0256	−0.0663	−3.6771	1.9393	30.9613	−5.7073	−3.4604	4.7627

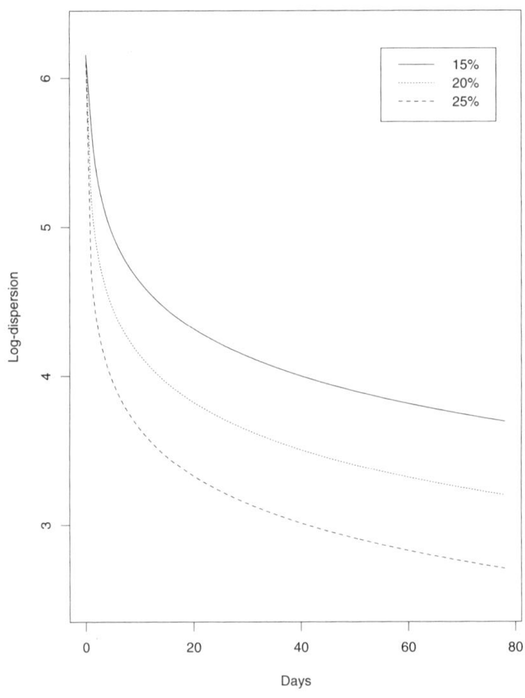

Fig. 5.6. Fitted curves for the pattern of heterogeneous dispersion over time across three treatment levels.

0.1575(0.0611) and its Z-statistic is 2.5769, suggesting that ρ is significantly different from zero.

A further residual analysis for the above model is given as follows. The left panel in Figure 5.7 shows the scatterplot of the estimated standardized

Pearson residuals $\widehat{e}_{ij}$ against the fitted mean values $\widehat{\mu}_{ij}$, to check the distributional assumption. The dashed lines at 2 and -2 represent the asymptotic 95% upper and lower limits, respectively. The residuals seem to behave reasonably well as expected, with only three of them lying outside of the region. The plot seems to be in agreement with the simplex marginal distribution.

The right panel in Figure 5.7 provides a rough check of the logit link function used in the proposed model, showing the scatterplot of the estimated adjusted dependent variables $\widehat{w}_{ij}$ against the estimated logit linear predictor $\widehat{\eta}_{ij}$. The two solid lines stand for the asymptotic 95% confident bands within which almost 96% points are contained. This clearly supports the logit link function assumption.

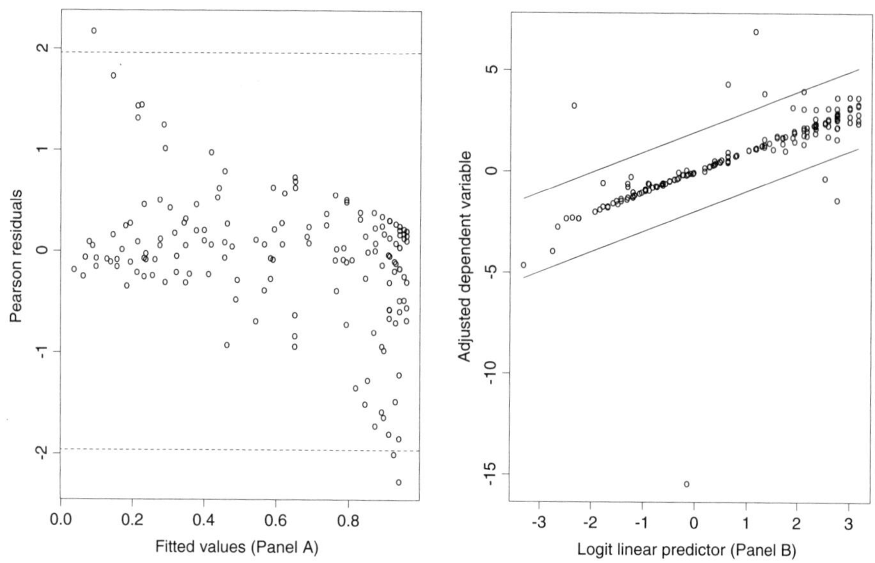

Fig. 5.7. Diagnostic plots in the eye surgery data analysis. Panel A checks the distribution assumption, and panel B checks the link function.

Checking the working correlation based on the regular autocorrelation matrix seems to be hard in this data analysis, since the data are measured at irregular time points and the residuals available at a given time are sparse. In this particular application, since the correlation is low, it is hard to observe any clear patterns for correlations over time, and moreover it does not appear very crucial to specify a "right" correlation matrix. Alternatively, readers may consider applying Diggle's variogram plot (Diggle, 1990) to reach an appropriate conclusion.

6

Vector Generalized Linear Models

6.1 Introduction

This chapter is devoted to the development of multi-dimensional generalized linear models (GLMs) that are useful to analyze correlated data of equal size. The GEE or QIF method has enjoyed its robustness against the misspecification on the correlation structure and its simplicity in modeling where only the first two moments need to be specified. However, these quasi-likelihood methods may suffer from the loss of estimation efficiency and the lack of procedures for model assessment and selection. In some situations where a good deal of information regarding the study design and data collection has been made available, it is of interest to make the best use of data and undertake a powerful statistical inference. This requires a fully parametric model that allows us to execute the maximum likelihood inference in regression analysis.

The key ingredient required for the extension of the univariate GLM to a general multivariate framework for vector outcomes is the multivariate analogue of the dispersion model (DM) family distributions in (2.3). Suppose that for each subject an n-element response vector $\mathbf{Y} = (Y_1, \ldots, Y_n)^T$ and an associated p-element covariate vector $\mathbf{x}$ are observed. For example, the vector $\mathbf{Y}$ is comprised of measurements from multiple response variables, such as blood pressure, heart-rate, weight, and temperature for a subject. Other examples of such data include clustered data with an equal cluster size, longitudinal data with a fixed number of repeated measurements, and spatial data collected from a fixed number of spatial locations. To analyze such data by the GLM approach, vector GLMs (VGLM) can be formulated as a model for which the conditional distribution of $\mathbf{Y}$ given $\mathbf{x}$ is of the form

$$f(\mathbf{y}|\mathbf{x}; \boldsymbol{\mu}, \boldsymbol{\sigma}^2, \Gamma) = \mathcal{F}(\mathbf{y}, \eta_1, \ldots, \eta_n; \boldsymbol{\sigma}^2, \Gamma), \tag{6.1}$$

where $\boldsymbol{\beta} = (\boldsymbol{\beta}_1^T, \ldots, \boldsymbol{\beta}_m^T)^T$ are the regression coefficients, the j-th linear predictor is $\eta_j = \eta_j(\mathbf{x}) = \mathbf{x}^T \boldsymbol{\beta}_j$, and $\mathcal{F}(\cdot; \boldsymbol{\sigma}^2, \Gamma)$ is a certain joint density function that is parametrized by the vector of dispersion parameters

$\boldsymbol{\sigma}^2 = (\sigma_1^2, \ldots, \sigma_n^2)^T$ and the association matrix Γ. Here $\Gamma = (\gamma_{ij})$ characterizes the dependence among the components of $\mathbf{Y}$. Note that m may not be equal to n, the dimension of $\mathbf{Y}$.

To complete the specification of a VGLM in (6.1), it is necessary to specify the $\mathcal{F}(\cdot)$ function and the parameter set Γ. In the literature, many proposals have been made for the $\mathcal{F}(\cdot)$, some of which will be introduced in this chapter. A desired density function $\mathcal{F}(\cdot;\Gamma)$ should satisfy the following two basic properties:

(i) The VGLM resulting from a chosen $\mathcal{F}(\cdot)$ should be reproducible or marginally closed, namely the lower-dimensional regression models should have the same error distribution type as the joint model. This is because in most practical problems, the data types for the individual components of $\mathbf{Y}$ are relatively easy to recognize, so that the corresponding marginal error distributions can be readily assumed, as is the practice for univariate GLMs. In addition, this marginal closure allows the development of statistical inferences based on lower-dimensional margins, such as the composite likelihood method (Lindsay, 1988).

(ii) The association parameters in Γ are able to characterize both positive and negative associations for vector $\mathbf{Y}$. In practice, positive association is often seen in biomedical studies, while negative association is frequently present in economic or financial data. For example, in the analysis of insurance data, the amount of claims and the number of claims over a given period of time are usually negatively correlated. A model that allows a full range of association certainly provides flexibility to the analysis of a broad range of data types.

Section 6.2 presents the log-linear model representation (Bishop et al., 1975) or Bahadur's representation (1961), a multivariate distribution for correlated binary data. This distribution has been employed by several authors (e.g., Zhao and Prentice, 1990; Fitzmaurice et al., 1993) to specify the $\mathcal{F}(\cdot)$ for the analysis of binary longitudinal data, where the common regression parameter (i.e., $\boldsymbol{\beta}_j = \boldsymbol{\beta}$) is used. The conditional modeling approach discussed in Section 4.3 is another way of specifying the $\mathcal{F}(\cdot)$, resulting in generalized linear mixed models (see Chapter 7). This chapter will focus on the joint modeling approach. The emphasis will be given to the VGLMs based on Gaussian copulas, a general and unified framework suitable for the regression analysis of a broad range of multi-dimensional data types.

6.2 Log-Linear Model for Correlated Binary Data

Let us begin with the binary random variable, namely where a random variable takes only two values, either 1 (success) or 0 (failure). A widely used probability model for multi-dimensional binary data is the *log-linear model* (Bishop et al., 1975) whose probability mass function is defined by

$$p(\mathbf{y}) = c(\boldsymbol{\theta}) \exp\left\{\sum_{j=1}^{n} \theta_j y_j + \sum_{j_1<j_2} \theta_{j_1 j_2} y_{j_1} y_{j_2} + \cdots + \theta_{1\cdots n} y_1 \cdots y_n\right\} \tag{6.2}$$

where $\boldsymbol{\theta} = (\theta_1, \ldots, \theta_n, \theta_{12}, \ldots, \theta_{n-1,n}, \ldots, \theta_{1\cdots n})^T$ is a $(2^n - 1)$-element vector of canonical parameters, and $c(\boldsymbol{\theta})$ is the normalizing term. The representation (6.2) actually gives a saturated model in which there is only one constraint on the 2^n cell probabilities, $\sum_{\mathbf{y}} p(\mathbf{y}) = 1$.

These θ parameters describe the association among the components of $\mathbf{y}$. Consider a special log-linear model with the third and higher order terms equal to 0, which is also known as the *quadratic exponential model* (QEM) considered in Zhao and Prentice (1990). The resulting probability mass function takes the form:

$$\begin{aligned} p(\mathbf{y}) &= c(\boldsymbol{\theta}) \exp\left\{\sum_{j=1}^{n} \theta_j y_j + \sum_{j<k} \theta_{jk} y_j y_k\right\} \\ &= c(\boldsymbol{\theta}_1, \boldsymbol{\theta}_2) \exp\left\{\mathbf{y}^T \boldsymbol{\theta}_1 + \mathbf{w}^T \boldsymbol{\theta}_2\right\} \end{aligned} \tag{6.3}$$

where

$$\boldsymbol{\theta}_1 = (\theta_1, \ldots, \theta_n)^T, \quad \boldsymbol{\theta}_2 = (\theta_{12}, \ldots, \theta_{n-1,n})^T,$$

and

$$\mathbf{y} = (y_1, \ldots, y_n)^T, \quad \mathbf{w} = (y_1 y_2, \ldots, y_{n-1} y_n)^T.$$

Denote the model by $\mathbf{y} \sim QEM(\boldsymbol{\theta})$. For model (6.3), it can be shown that

$$\log\left\{\frac{\mathrm{P}(Y_j = 1 | Y_k = 0, k \neq j)}{\mathrm{P}(Y_j = 0 | Y_k = 0, k \neq j)}\right\} = \theta_j.$$

This implies that the θ_j is equal to the log odds for $Y_j = 1$ given that the remaining responses Y_k, $k \neq j$ are all zero. Similarly,

$$\log\left\{\frac{\mathrm{P}(Y_j = 1, Y_k = 1 | Y_l = y_l, l \neq j, k)\mathrm{P}(Y_j = 0, Y_k = 0 | Y_l = y_l, l \neq j, k)}{\mathrm{P}(Y_j = 1, Y_k = 0 | Y_l = y_l, l \neq j, k)\mathrm{P}(Y_j = 0, Y_k = 1 | Y_l = y_l, l \neq j, k)}\right\} = \theta_{jk};$$

that is, the θ_{jk} is again equal to a log odds ratio, which describes the association between Y_j and Y_k, conditional on all the other responses being withheld. The interpretation of $\boldsymbol{\theta}_2$ as a conditional odds ratio is restrictive, because it depends upon the number of other responses in a cluster. Hence, this joint distribution is most useful when the clusters are of the same size ($n_i = n$).

The above models can be applied to specify the $\mathcal{F}(\cdot)$ function to form VGLMs for correlated binary data. Take the instance of the QEM (6.3). To formulate a VGLM, first define marginal expectations

$$\mu_j = \mathrm{P}(Y_j = 1), j = 1, \ldots, n,$$

which are the parameters of interest because they relate to covariates via the logit model. So, from the interpretation point of view, it is more appealing to reparametrize model (6.3) on the basis of parameters $(\boldsymbol{\mu}, \boldsymbol{\theta}_2)$. Note that under (6.3),

$$\boldsymbol{\mu} = (\mu_1, \ldots, \mu_n)^T = (\mu_1(\boldsymbol{\theta}_1, \boldsymbol{\theta}_2), \ldots, \mu_n(\boldsymbol{\theta}_1, \boldsymbol{\theta}_2))^T,$$

with $\dim(\boldsymbol{\mu}) = \dim(\boldsymbol{\theta}_1)$. With $\boldsymbol{\theta}_2$ remaining the same, there exists a one-to-one correspondence between the canonical parameter $(\boldsymbol{\theta}_1, \boldsymbol{\theta}_2)$ and $(\boldsymbol{\mu}, \boldsymbol{\theta}_2)$.

Second, the marginal mean parameters are modeled with a common $\boldsymbol{\beta}$ as follows:

$$\text{logit}(\mu_j) = \mathbf{x}_j^T \boldsymbol{\beta}, \ j = 1, \ldots, n.$$

Given data $\mathbf{Y}_i | \mathbf{X}_i \stackrel{ind.}{\sim} QEM(\boldsymbol{\theta}_i), i = 1, \ldots, K$, with $\boldsymbol{\theta}_i = (\boldsymbol{\beta}, \boldsymbol{\theta}_{i2})$, the resulting score equation for $\boldsymbol{\beta}$ under the above reparameterization is given by

$$\sum_{i=1}^{K} \left(\frac{\partial \boldsymbol{\mu}_i^T}{\partial \boldsymbol{\beta}} \right) \text{Var}^{-1}(\mathbf{Y}_i)(\mathbf{y}_i - \boldsymbol{\mu}_i) = \mathbf{0},$$

which is exactly of the same form as the optimal GEE Ψ_{op} in (5.7). Note that the variance matrix of $\mathbf{Y}_i$ is a function of both $\boldsymbol{\beta}$ and $\boldsymbol{\theta}_{2i}$. This indicates the solution of the optimal GEE is identical to the MLE under the QEM when the covariance (equivalently, the correlation structure) is specified by the model (6.3). One aspect appearing more complicated than the GEE is the need of estimating $\boldsymbol{\theta}_{i2}$ together with $\boldsymbol{\beta}$, rather than treating it as a nuisance parameter, similar to the GEE setting where correlation parameters are estimated separately.

An alternative parameterization of the log-linear model that directly uses marginal means was proposed by Bahadur (1961), known as the so-called Bahadur's representation. Let

$$R_j = \frac{Y_j - \mu_j}{\{\mu_j(1-\mu_j)\}^{1/2}},$$

$$\rho_{jk} = \text{corr}(Y_j, Y_k) = \text{E}(R_j R_k), \rho_{jkl} = \text{E}(R_j R_k R_l),$$

and so on, up to

$$\rho_{1\ldots n} = \text{E}(R_1 \cdots R_n).$$

The probability mass function of $\mathbf{y}$ can be rewritten as follows,

$$p(\mathbf{y}) = \prod_{j=1}^{n} \mu_j^{y_j} (1-\mu_j)^{1-y_j} \times \left(1 + \sum_{j<k} \rho_{jk} r_j r_k + \sum_{j<k<l} \rho_{jkl} r_j r_k r_l + \cdots + \rho_{1\ldots n} r_1 r_2 \cdots r_n \right). \tag{6.4}$$

Therefore, the joint distribution is now expressed in terms of the marginal means, pairwise Pearson correlations, and higher order moments of the standardized residuals R_j's.

The merit of the Bahadur's representation is that this model uses marginal probabilities and Pearson correlations, which apparently gives a more direct interpretation of association than (6.2). However, this representation has two serious drawbacks. One is that correlations are constrained in a complicated fashion with the marginal means (Carey et al., 1993). The constraint usually causes a substantial shrinkage on the range of correlation. Hence, if μ_j is modeled on $\mathbf{x}_j$ via a logit model, it may be inappropriate to assume that the correlation and higher order moments are independent of $\mathbf{x}_j$, as would be convenient. The other drawback, as pointed by Fitzmaurice et al. (1993), is that the log-linear model representation is not reproducible, which makes the interpretation of model fitting results difficult. In addition, the above model formulation is *ad hoc*, in the sense that it cannot be extended to establish a general VGLM framework for other types of correlated outcomes.

Alternatively, this chapter presents a new class of $\mathcal{F}(\cdot)$ functions based on the multivariate distributions generated by parametric copulas (see Joe, 1997, chapter 5). In Section 6.3, the class of multivariate exponential dispersion (MED) distributions generated by Gaussian copulas (Song, 2000a) will be discussed in detail. Consequently, this class of multivariate distributions is applied to establish a unified framework of VGLMs for correlated continuous outcomes, correlated discrete outcomes, and correlated mixed outcomes. The rest of this chapter focuses on a joint modeling approach to correlated data analysis based on Gaussian copulas.

6.3 Multivariate ED Family Distributions

This section is devoted to the class of multivariate exponential dispersion distributions generated from Gaussian copulas. It provides a class of multivariate error distributions useful for the development of a general and unified VGLM framework. Among many parametric copulas available in the literature, Gaussian copula is of particular interest owing to its advantages described as follows:

(a) The utility of Gaussian copulas is little affected by the dimension of vector outcomes. That is, either theoretical or numerical complexity of related regression models remains nearly the same in terms of the dimension of outcomes. This, however, is not the case for many other parametric copulas, such as Frank copula, Clayton copula, and Gumbel copula. These parametric copulas are relatively simple when the dimension is low (say, two), and become analytically a lot more complicated as the dimension increases.

(b) The association measure resulted from Gaussian copulas inherits good properties of the correlation in the multivariate normal distribution. For example, the components of a vector response are independent if and only if the association matrix is the identity matrix. Therefore, similarly in the multivariate normal analysis, a residual analysis may be used to infer a certain structure about the association matrix. The freedom of manipulating different candidate structures for the association matrix, as done in the multivariate normal, is very desirable in many practical studies. At the end, the selected parsimonious specification of the dependence structure is beneficial to gain better power in inference and better interpretation in data analysis.
(c) When all margins follow the normal linear regression model, the Gaussian copula based VGLM will reduce to the classical multivariate normal linear model. However, the VGLMs based on other parametric copulas do not have this property.

6.3.1 Copulas

Let $\mathbf{u}_{-S}$ be a subvector of $\mathbf{u} = (u_1, \ldots, u_n)^T$ with those components indicated by the set S being omitted, where S is a subset of the indices $\{1, \ldots, n\}$. According to Sklar (1959), a mapping $C: \ (0,1)^n \longrightarrow (0,1)$ is called a *copula* if

(1) it is a continuous distribution function; and
(2) each margin is a univariate uniform distribution, namely

$$\lim_{\mathbf{u}_{-i} \to \mathbf{1}} C(\mathbf{u}) = u_i, u_i \in (0,1)$$

where the limit is taken under $u_j \to 1, \ \forall j \neq i$.

Clearly, $\lim_{u_j \to 0} C(\mathbf{u}) = 0$, for any $j = 1, \ldots, n$. It is easy to prove that for any subset S, the marginal obtained by $\lim_{\mathbf{u}_{-S} \to \mathbf{1}} C(\mathbf{u})$ is a copula. Copulas are easy to construct from a given multivariate distribution.

If $\mathbf{X} = (X_1, \ldots, X_n)^T \sim G$ where G is an n-dimensional distribution function with margins $G_1, \ldots, G_n$, then the copula is of the form

$$C_G(u_1, \ldots, u_n) = G\left\{G_1^{-1}(u_1), \ldots, G_n^{-1}(u_n)\right\}, \quad u_i \in (0,1), \ i = 1, \ldots, n,$$

provided that the marginal inverse distribution functions G_i^{-1} of G_i exist. Gaussian copula is an important special case, which is obtained when $\mathbf{X} \sim \mathrm{MVN}_n(\mathbf{0}, \Gamma)$ with standardized margins and $G_i \equiv \Phi$. Here Φ denotes the cumulative distribution function (CDF) of the standard normal $N(0,1)$. The n-dimensional Gaussian copula is denoted by $C_\Phi(\mathbf{u}|\Gamma)$, and its density is given by

$$c_\Phi(\mathbf{u}|\Gamma) = |\Gamma|^{-1/2} \exp\left\{\frac{1}{2}\mathbf{q}^T(I_n - \Gamma^{-1})\mathbf{q}\right\} \qquad (6.5)$$

where $\mathbf{q} = (q_1, \ldots, q_n)^T$ with normal scores $q_i = \Phi^{-1}(u_i), i = 1, \ldots, n$, and I_n is the n-dimensional identity matrix. Matrix Γ is called the association matrix.

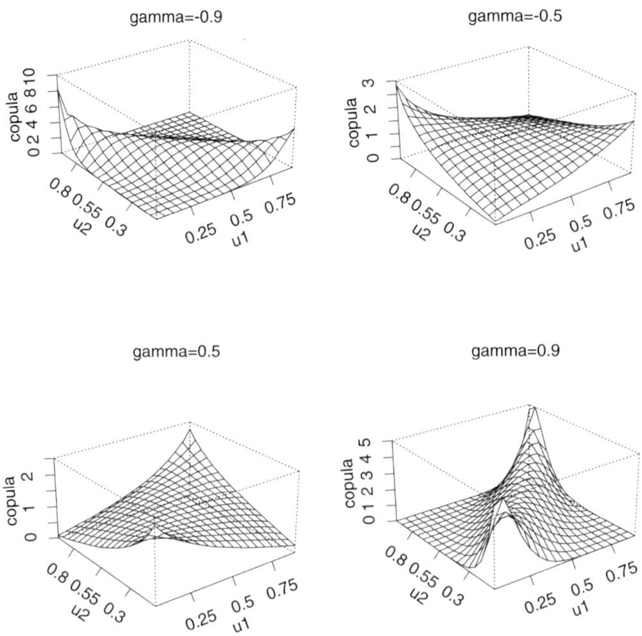

Fig. 6.1. Four bivariate Gaussian copula distributions with different association parameters.

Figure 6.1 displays four bivariate densities of Gaussian copula with different values of the association parameter γ. Clearly, this copula accommodates both positive and negative dependence, indicated by the opposite directions of concentration in the densities. The degree of concentration representing the variation of the distribution increases as the γ parameter tends to ± 1.

It is shown in Joe (1997, Section 5.1) that the bivariate Gaussian copula attains the lower Fréchet bound $\max\{0, u_1 + u_2 - 1\}$, independence, or the upper Fréchet bound $\min\{u_1, u_2\}$, according to the values of the corresponding association parameter γ equal to -1, 0, or 1.

6.3.2 Construction

By complementing the copula C_G with given margins, say $F_1, \ldots, F_n$, a new multivariate distribution can be obtained by

$$F(\mathbf{y}) = C_G \{F_1(y_1), \ldots, F_n(y_n)\}. \tag{6.6}$$

One important property is that the i-th margin of F gives the original F_i, namely, the distribution is marginally closed.

A class of n-variate multivariate dispersion models is obtained by (6.6) when copula $C_G \equiv C_\Phi(\cdot|\Gamma)$ and margins F_i's are dispersion models. The multivariate dispersion models, denoted by $\mathrm{MDM}_n(\boldsymbol{\mu}, \boldsymbol{\sigma}^2, \Gamma)$, are parametrized by three sets of parameters, $\boldsymbol{\mu} = (\mu_1, \ldots, \mu_n)^T$, the vector of position parameters, $\boldsymbol{\sigma}^2 = (\sigma_1^2, \ldots, \sigma_n^2)^T$, the vector of dispersion parameters, and Γ, the association matrix. Refer to Chapter 2 for details of marginal DM distributions, $\mathrm{DM}(\mu_j, \sigma_j^2)$.

Consequently, the multivariate exponential dispersion model is produced by this Gaussian copula construction, denoted by $\mathrm{MED}_n(\boldsymbol{\mu}, \boldsymbol{\sigma}^2, \Gamma)$, when the corresponding margins $\mathrm{ED}(\mu_j, \sigma_j^2)$ are used in construction. Here $\boldsymbol{\mu}$ is the vector of the marginal mean parameters.

When marginal models are continuous, a multivariate dispersion model can be equivalently defined by the density of the following form:

$$f(\mathbf{y}; \boldsymbol{\mu}, \boldsymbol{\sigma}^2, \Gamma) = c_\Phi \{F_1(y_1), \ldots, F_n(y_n)|\Gamma\} \prod_{j=1}^{n} f(y_j; \mu_j, \sigma_j^2). \tag{6.7}$$

Consequently (6.6) gives rise to a large class of continuous multivariate models including multivariate gamma, multivariate inverse Gaussian, multivariate von Mises, and multivariate simplex distribution.

When marginal models are discrete, a multivariate probability mass function is obtained by taking Radon-Nikodym derivative for $F(\mathbf{y})$ in (6.6) with respect to the counting measure,

$$\begin{aligned} f(\mathbf{y}) &= \mathrm{P}(Y_1 = y_1, \ldots, Y_n = y_n) \\ &= \sum_{j_1=1}^{2} \cdots \sum_{j_n=1}^{2} (-1)^{j_1+\ldots+j_n} C_\Phi(u_{1j_1}, \ldots, u_{nj_n}|\Gamma), \end{aligned} \tag{6.8}$$

where $u_{j1} = F_j(y_j)$ and $u_{j2} = F_j(y_j-)$. Here $F_j(y_j-)$ is the left-hand limit of F_j at y_j, which is equal to $F_j(y_j - 1)$ when the support of F_j is an integer set such as the case of Poisson or binomial margins. See more discussions in Examples 6.5 and 6.6 later in Section 6.3.3.

When the n margins appear to be mixed outcomes, say, the first n_1 margins being continuous and the rest $n_2 = n - n_1$ margins being discrete, the joint density function is given as follows. Let $\mathbf{u} = (\mathbf{u}_1^T, \mathbf{u}_2^T)^T$, with $\mathbf{u}_1 = (u_1, \ldots, u_{n_1})^T$ and $\mathbf{u}_2 = (u_{n_1+1}, \ldots, u_n)^T$. The same partition and notation are applied for vectors $\mathbf{x}$ and $\mathbf{q}$. Let

$$C_1^{n_1}(\mathbf{u}_1, \mathbf{u}_2|\Gamma) = \frac{\partial^{n_1}}{\partial u_1 \cdots \partial u_{n_1}} C(u_1, \ldots, u_n|\Gamma)$$
$$= (2\pi)^{-\frac{n_2}{2}} |\Gamma|^{-\frac{1}{2}} \times$$
$$\int_{-\infty}^{\Phi^{-1}(u_{n_1+1})} \cdots \int_{-\infty}^{\Phi^{-1}(u_n)} \exp\left\{ -\frac{1}{2}(\mathbf{q}_1^T, \mathbf{x}_2^T)\Gamma^{-1}(\mathbf{q}_1^T, \mathbf{x}_2^T)^T + \frac{1}{2}\mathbf{q}_1^T \mathbf{q}_1 \right\} d\mathbf{x}_2.$$

Then, the joint density is given by

$$f(\mathbf{y}) = \prod_{j=1}^{n_1} f_j(y_j) \sum_{j_{n_1+1}=1}^{2} \cdots \sum_{j_n=1}^{2} (-1)^{j_{n_1+1}+\ldots+j_n} \times$$
$$C_1^{n_1}(F_1(y_1), \ldots, F_{n_1}(y_{n_1}), u_{n_1+1, j_{n_1+1}}, \ldots, u_{n, j_n}|\Gamma), \tag{6.9}$$

where u_{t,j_t}'s are the same as defined in (6.8).

6.3.3 Interpretation of Association Parameter

An important issue in this copula construction is how to interpret the elements of the matrix Γ in the Gaussian copula (6.5).

First, note from the copula construction that the components $Y_1, \ldots, Y_n$ are mutually independent if matrix Γ is the identity matrix I_n in all these cases, (6.7)–(6.9). In the presence of dependence, three types of dependence measures are employed to explain γ_{ij}, described as follows.

Definition 6.1. *(Kendall's τ)*
Let $F(x_1, x_2)$ be a continuous bivariate distribution function. Then Kendall's dependence measure τ for any independent pairs (X_1, X_2) and (X_1', X_2') with distribution F is defined by,

$$\tau(X_1, X_2) = P\{(X_1 - X_1')(X_2 - X_2') > 0\} - P\{(X_1 - X_1')(X_2 - X_2') < 0\}$$
$$= 4 \int\int F(x_1, x_2) dF(x_1, x_2) - 1.$$

From the definition, it is clear that Kendall's τ is a bivariate measure of monotone dependence for continuous variables and gauges the difference of the probability of two random concordant pairs and the probability of two random discordant pairs.

Definition 6.2. *(Spearman's ρ)*
Let $F(x_1, x_2)$ be a continuous bivariate distribution function with marginal distributions $F_1(x_1)$ and $F_2(x_2)$. Then Spearman's dependence measure ρ is defined as the Pearson correlation of $F_1(X_1)$ and $F_2(X_2)$ for any pair $(X_1, X_2) \sim F$. Therefore,

$$\rho(X_1, X_2) = corr\{F_1(X_1), F_2(X_2)\} = 12 \int\int F_1(x_1) F_2(x_2) dF(x_1, x_2) - 3.$$

Spearman's ρ is also a bivariate measure of monotone dependence for continuous variables, and it reflects the association between two (monotonely increasing) transformed random variables that are non-informative with respect to their original variables. See, for example, Kendall and Gibbons (1990) for more details regarding ρ and τ.

Since Pearson's correlation measures the dependence between two normal random variables, it is natural to compute the Pearson correlation for two non-normal variables under their respective normal scores. This leads to another dependence measure, defined as follows.

Definition 6.3. *(Normal scoring ν)*
Let $F(x_1, x_2)$ be a continuous bivariate distribution function with marginal distributions $F_1(x_1)$ and $F_2(x_2)$. Then the normal scoring dependence measure ν is defined as the Pearson correlation between $q_1(X_1)$ and $q_2(X_2)$ for any pair $(X_1, X_2) \sim F$, namely

$$\nu(X_1, X_2) = corr\{q_1(X_1), q_2(X_2)\},$$

where $q_i(\cdot)$ are two transformations such that $q_i(X_i)$, $i = 1, 2$, follow normal distributions.

Note that the normality transformation $q(\cdot)$ always exists for a continuous random variable.

Proposition 6.4. *Let X be a continuous random variable distributed by $F(x)$. Define*

$$q(x) = \Phi^{-1}\{F(x)\}.$$

Then $q(X)$ follows a normal distribution. The function $q(x)$ is called the normal scoring function *and the values of $q(x)$ are called* normal scores.

Proof. Let $Y = q(X)$ where a one-to-one function $q(\cdot)$ is such that

$$\Phi(y) = P(Y \leq y) = F\left\{q^{-1}(y)\right\},$$

which leads to $q(x) = \Phi^{-1}\{F(x)\}$.

Now we apply the three dependence measures above to interpret the matrix Γ that parametrizes the normal copula given in (6.5). Consider the bivariate marginal normal copula with components (U_i, U_j) obtained from (6.5),

$$\begin{aligned} c_{\Phi_{ij}}(u_i, u_j|\gamma_{ij}) &= |\Gamma_{ij}|^{-1/2} \exp\left\{\frac{1}{2}(x_i, x_j)\left(I_2 - \Gamma_{ij}^{-1}\right)\begin{pmatrix} x_i \\ x_j \end{pmatrix}\right\} \\ &= (1 - \gamma_{ij}^2)^{-1/2} \exp\left\{-\frac{1}{2}\frac{x_i^2 + x_j^2 - 2\gamma_{ij}x_i x_j}{1 - \gamma_{ij}^2} + \frac{1}{2}(x_i^2 + x_j^2)\right\} \end{aligned}$$

where Γ_{ij}, as a 2×2 symmetric submatrix of Γ, has all diagonal elements 1 and all off-diagonal elements γ_{ij} and I_2 is the bivariate identity matrix.

It is easy to see that if marginal distributions F_j are continuous, then

$$\left[\Phi^{-1}\{F_1(Y_1)\}, \ldots, \Phi^{-1}\{F_n(Y_n)\}\right] \sim \mathrm{MVN}_n(\mathbf{0}, \Gamma).$$

It follows obviously that

$$\gamma_{ij} = \mathrm{corr}\left[\Phi^{-1}\{F_i(Y_i)\}, \Phi^{-1}\{F_j(Y_j)\}\right] = \nu(Y_i, Y_j),$$

where γ_{ij} is the Pearson correlation between two normal scores, measuring the association between Y_i and Y_j based on a monotonic nonlinear transformation. This nonlinear transformation is necessary to bring two random variables from two separate probability spaces into a common probability space, in which their association can be properly measured.

Spearman's ρ of (U_i, U_j) is

$$\rho_{ij} = \rho(U_i, U_j) = 12 \int\int_{(0,1)^2} c_{\Phi_{ij}}(u_i, u_j|\gamma_{ij}) du_i du_j - 3.$$

Note that both $\Phi(X_i)$ and $\Phi(X_j)$ are uniformly distributed as Unif$(0,1)$ so it is not hard to prove that

$$\rho_{ij} = 12\mathrm{E}\{\Phi(X_i)\Phi(X_j)\} - 3 = 12\mathrm{cov}\,\{\Phi(X_i), \Phi(X_j)\}$$

where the expectation is taken under bivariate normal $(X_i, X_j) \sim \mathrm{MVN}_2(\mathbf{0}, \Gamma_{ij})$.

Kendall's τ is given by

$$\tau_{ij} = \tau(U_i, U_j) = 4 \int\int_{(0,1)^2} C_{\Phi_{ij}}(u_i, u_j|\gamma_{ij}) dC_{\Phi_{ij}}(u_i, u_j|\gamma_{ij}) - 1.$$

Or equivalently,

$$\tau_{ij} = 4\mathrm{E}\{\Phi_{ij}(X_i, X_j|\gamma_{ij})\} - 1,$$

where the expectation is taken under the bivariate normal distribution $(X_i, X_j) \sim \mathrm{MVN}_2(\mathbf{0}, \Gamma_{ij})$.

Clearly, both ρ and τ measures are nonlinear functions in γ_{ij}, the (i,j)-th entry of the association matrix Γ. To visualize their relationships, for each fixed γ_{ij}, the Monte Carlo method is employed to compute ρ_{ij} and τ_{ij} numerically. Figure 6.2 displays the relationships between the τ versus ν and ρ versus ν, respectively, where Monte Carlo simulation size is chosen to be $M = 5000$. It is found that in the Gaussian copula setting (6.5) ν and ρ are effectively very close to each other, and ν and τ are positively "correlated" in the sense that increasing the value of ν measure results in a similar increase for the value of τ measure, and *vice versa*.

Note that the normal scoring ν measure may also be used for the association of discrete random variables, although the above interpretation is not applicable in the discrete distribution case. Some supportive evidence for this extension can be drawn from Examples 6.5 and 6.6 where the bivariate binomial and Poisson are studied in detail. Hence, in the following development

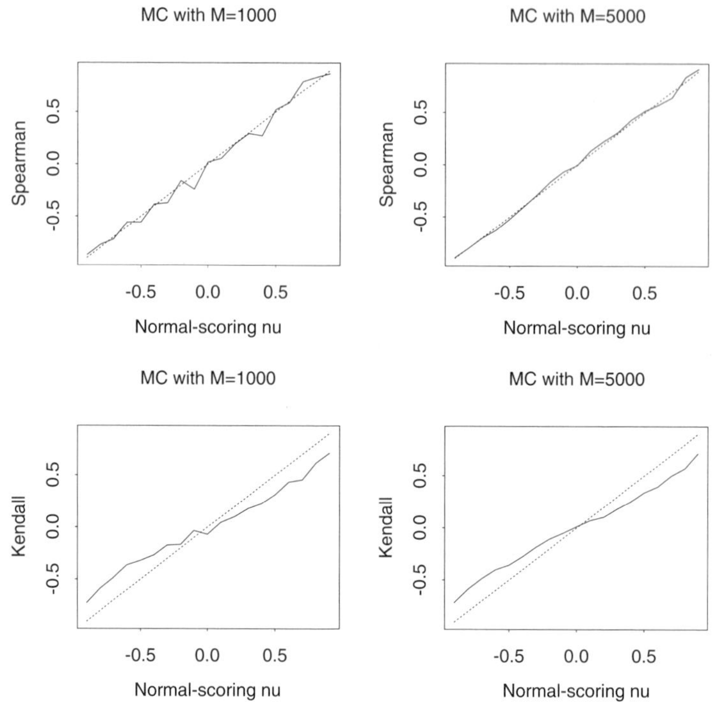

Fig. 6.2. Spearman's ρ versus normal scoring ν (top panel) and Kendall's τ versus normal scoring ν (bottom panel).

of VGLMs, the measure ν is assumed well-defined in both the continuous and discrete cases.

We now give three examples, two of which are discrete models.

Example 6.5 (Multivariate Binary Model).

Let Y_j, $j = 1, \ldots, n$ be n binary random variables with the probability of success π_j. The CDF of Y_j is

$$F_j(y_j) = \begin{cases} 0, & y_j < 0 \\ 1 - \pi_j, & 0 \leq y_j < 1 \\ 1, & y_j \geq 1. \end{cases}$$

The n-variate probability mass function is given by (6.8), which defines the cell probabilities uniquely as long as the Gaussian copula C_Φ and the association matrix Γ are given. In particular, when $n = 2$, the bivariate probability mass function is of the form

$$\mathrm{P}(Y_1 = y_1, Y_2 = y_2) = C_\gamma(u_1, u_2) - C_\gamma(u_1, v_2) - C_\gamma(v_1, u_2) + C_\gamma(v_1, v_2), \tag{6.10}$$

where $u_j = F_j(y_j)$ and $v_j = F_j(y_j - 1)$. Here C_γ is the bivariate Gaussian copula which is parametrized by a single association parameter $\gamma \in (-1, 1)$. It follows that the four cell probabilities are given by

$$P(Y_1 = y_1, Y_2 = y_2) = \begin{cases} C_\gamma(1-\pi_1, 1-\pi_2), & \text{if } y_1 = 0, y_2 = 0 \\ 1 - \pi_1 - C_\gamma(1-\pi_1, 1-\pi_2), & \text{if } y_1 = 0, y_2 = 1 \\ 1 - \pi_2 - C_\gamma(1-\pi_1, 1-\pi_2), & \text{if } y_1 = 1, y_2 = 0 \\ \pi_1 + \pi_2 + C_\gamma(1-\pi_1, 1-\pi_2) - 1, & \text{if } y_1 = 1, y_2 = 1. \end{cases} \tag{6.11}$$

To make the use of this model to the regression analysis of correlated binary data, marginal expectations are specified via the logit model, $\text{logit}(\pi_j) = \eta_j$ or the probit model, $\Phi^{-1}(\pi_j) = \eta_j$, where $\eta_j = \mathbf{x}_j^T \boldsymbol{\beta}$ is the linear predictor and $\mathbf{x}_j$ is a vector of covariates. This leads to a bivariate logistic model or a multivariate probit model, respectively. In particular, the probit link results in $1 - \pi_j = \Phi(-\eta_j)$, and therefore $C_\gamma(1-\pi_1, 1-\pi_2) = \Phi_2(-\eta_1, -\eta_2|\gamma)$, where Φ_2 denotes the CDF of bivariate normal with the standard normal marginals and correlation coefficient γ.

In effect, the bivariate probit model can be interpreted as a probit model with the latent variable representation. Let (Z_1, Z_2) be the latent normal vector satisfying $Z_j = \mathbf{x}_j^T \boldsymbol{\beta} + \epsilon_j$, $j = 1, 2$, where $(\epsilon_1, \epsilon_2) \sim \text{MVN}_2(0, 0, 1, 1, \gamma)$, and define $Y_i = 0$, if $Z_i \leq 0$; 1, otherwise. Then the point probability $P(Y_1 = 0, Y_2 = 0) = \Phi_2(-\mathbf{x}_1^T \boldsymbol{\beta}, -\mathbf{x}_2^T \boldsymbol{\beta}|\gamma)$, identical to the first expression of (6.11). It is easy to prove that the other three point probabilities are the same as the rest in (6.11). In this case, the correlation parameter γ in (6.10) is identical to that from the latent bivariate normal distribution via dichotomization. This implies that the association parameter γ_{ij} (6.8) can be interpreted as the *tetrachoric correlation* (Drasgow, 1988; Harris, 1988) or more generally as *polychoric correlation* given by Olsson (1979) and Anderson and Pemberton (1985). This argument can be extended to a general multivariate probit model, because each γ_{ij} in the matrix Γ is in fact a pairwise association measuring the dependence between two components only.

For the bivariate binary model, the lower and upper Fréchet bounds are given, respectively, in the first and second lines of the following two-way array,

	$0 \leq y_1 < 1$	$y_1 \geq 1$
$0 \leq y_2 < 1$	$\max\{0, 1 - \pi_1 - \pi_2\}$	$1 - \pi_2$
	$1 - \max\{\pi_1, \pi_2\}$	$1 - \pi_2$
$y_2 \geq 1$	$1 - \pi_1$	1
	$1 - \pi_1$	1

and the bounds otherwise equal zero. It is easy to show that the bivariate binary model attains these two bounds when γ equals -1 and 1, respectively.

Example 6.6 (Multivariate Poisson Model).

Let $Y_j, j = 1, \ldots, n$ be n Poisson random variables with marginal mean parameter μ_j. Similarly, the joint probability mass function of $\mathbf{Y} = (Y_1, \ldots, Y_n)^T$

is given by (6.8). To link this model to a set of covariates $\mathbf{x}_j$ in the context of VGLMs, a log-linear model is assumed for each of marginal expectations μ_j via $\log(\mu_j) = \mathbf{x}_j^T \boldsymbol{\beta}$.

It is known that the stochastic representation is another approach in the literature (e.g., Joe, 1997, Section 7.2) to constructing a multivariate Poisson distribution. As seen in Example 4.1, this method constructs a bivariate Poisson random vector as $(Y_1, Y_2) = (Z_1 + Z_{12}, Z_2 + Z_{12})$ where Z_1, Z_2, Z_{12} are independent Poisson with parameters $\lambda_1, \lambda_2, \lambda_{12}$. Although this construction seems much simpler, it can only allow positive dependence, whereas the copula-based distribution (6.8) can accommodate both positive and negative association.

A simple comparison of the two constructions is considered on the basis of conditional expectations as follows. It is easy to prove that

$$\begin{aligned} \mathrm{E}(Z_1 + Z_{12} | Z_2 + Z_{12} = y_2) &= \lambda_1 + \frac{\lambda_{12}}{\lambda_2 + \lambda_{12}} y_2 \\ &= \mu_1 + r\sqrt{\frac{\mu_1}{\mu_2}}(y_2 - \mu_2), \end{aligned} \tag{6.12}$$

as a linear function in y_2, where r is the Pearson correlation coefficient of (Y_1, Y_2) equal to $\lambda_{12}/\sqrt{\lambda_1 + \lambda_{12}}\sqrt{\lambda_2 + \lambda_{12}}$ and $\mu_j = \lambda_j + \lambda_{12}$ are given marginal means. For the copula-based construction, the conditional mean is

$$\mathrm{E}(Y_1 | Y_2 = y_2) = \sum_{y_1=0}^{\infty} y_1 P(Y_1 = y_1, Y_2 = y_2)/P(Y_2 = y_2), \tag{6.13}$$

where the joint probability mass function $P(Y_1 = y_1, Y_2 = y_2)$ is given by (6.10). It is relatively easy to compute this function numerically, although its closed form expression is unavailable. A comparison between the two conditional expectations are illustrated in Figure 6.3, where the two margins are set to be the same.

To better understand this comparison, in Figure 6.3, a linear approximation to the conditional expectation (6.13) is derived as follows. This approximation takes a form similar to (6.12), given by

$$\mathrm{E}(Y_1 | Y_2 = y_2) \approx \mu_1 + \gamma K(\mu_1)\psi(y_2, \mu_2), \tag{6.14}$$

where $K(\mu_1) = \sum_{y_1=0}^{\infty} \phi\{q_1(y_1)\}$ and

$$\psi(y_2, \mu_2) = \frac{\phi\{q_2(y_2 - 1)\} - \phi\{q_2(y_2)\}}{F_2(y_2) - F_2(y_2 - 1)},$$

where ϕ is the standard normal density. The approximation (6.14) is obtained simply by the Taylor expansion of (6.13) around $\gamma = 0$. This expansion is, for $u_j = F_j(y_j),\ j = 1, 2,$

$$C_\gamma(u_1, u_2) = F_1(y_1)F_2(y_2) + \phi(q_1)\phi(q_2)\gamma + O(\gamma^2).$$

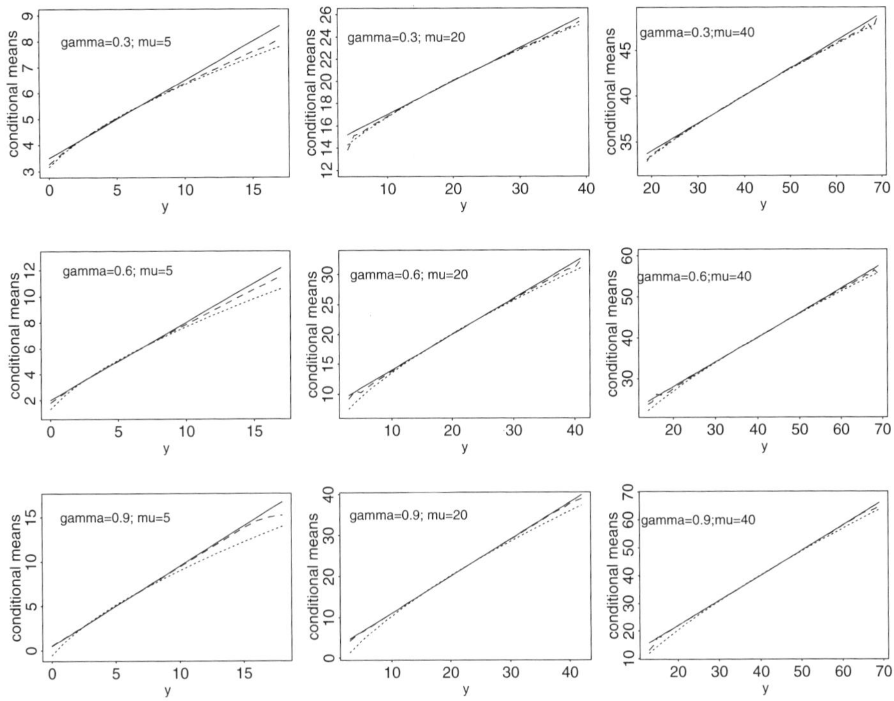

Fig. 6.3. Two exact conditional means and a linear approximation represented, respectively, by solid line, dashed line, and dotted line.

In the comparison between (6.14) and (6.12), their difference is governed by a leading term, $\sqrt{\mu} - K(\mu)$, which is positive at $\mu = 1, 2, \ldots$, and monotone decreasing to zero as μ goes to the infinity. For example, it equals 0.0225, 0.0127, 0.0099 when $\mu = 10, 30, 50$, respectively.

Figure 6.3 contains nine plots with all possible combinations of (r, γ, μ) for $r = \gamma = 0.3, 0.6, 0.9$ and $\mu = 5, 20, 40$, and each graph consists of three lines corresponding to the linear conditional mean (6.12), the conditional mean (6.13), and the approximation (6.14), respectively, represented by solid line (—), dashed line (- - -), and dotted line ($\cdots\cdot$). Clearly, when marginal means are not small, say 20 or bigger as seen in the figure, the two exact conditional means are almost identical within fairly reasonably large ranges of y_2 around the means, and the approximation is also shown fairly close to the two exact cases, although near the tails there are some small departures.

For small marginal means (equal to 5 in the figure), the two exact conditional means are still close enough to each other, and the approximation appears to almost overlap with the two exact cases at low y values but start to go away from them when y is far from the mean μ (in this figure, the going-away begins approximately at 2μ).

This comparison sheds light on the interpretation of the correlation parameter γ in the copula-based construction. At least numerically there exits the closeness between the association parameter γ and the Pearson correlation r in the stochastic representation.

Example 6.7 (Multivariate Gamma Model).

Let Y_j, $j = 1, \ldots, n$ be n gamma random variables, and $Y_j \sim Ga(\mu_j, \sigma_j^2)$ where μ_j and σ_j^2 are the mean and dispersion parameters, respectively. Clearly, the n-variate joint density of $\mathbf{Y} = (Y_1, \ldots, Y_n)^T$ is given by (6.7). With connection to VGLMs, we assume $g(\mu_j) = \mathbf{x}_j^T \boldsymbol{\beta}$ for each marginal expectation and a constant dispersion $\sigma_j^2 = \sigma^2$, $j = 1, \ldots, n$, where g is the link function which may be chosen to be either the reciprocal link or the log link in the context of gamma regression. Note that when $\sigma^2 = 1$, a family of multivariate exponential distributions is produced by the copula method. In Section 6.5.1, an example of 5-variate exponential distributions is shown.

6.4 Simultaneous Maximum Likelihood Inference

6.4.1 General Theory

Suppose data $(\mathbf{Y}_1, X_1), \ldots, (\mathbf{Y}_K, X_K)$ are independently sampled from an n-variate MED distribution,

$$\mathbf{Y}_i | X_i = (\mathbf{x}_{i1}, \ldots, \mathbf{x}_{in}) \sim \mathrm{MED}_n(\boldsymbol{\mu}_i, \boldsymbol{\sigma}_i^2, \Gamma), \; i = 1, \ldots, K$$

where response vector $\mathbf{Y}_i = (Y_{i1}, \ldots, Y_{in})^T$ has the mean vector $\boldsymbol{\mu}_i = (\mu_{i1}(\mathbf{x}_{i1}) \ldots, \mu_{in}(\mathbf{x}_{in}))^T$ and the dispersion vector $\boldsymbol{\sigma}_i^2 = (\sigma_{i1}^2, \ldots, \sigma_{in}^2)^T$, in which the j-th component $\sigma_{ij}^2 = \sigma_j^2 / w_{ij}$ with a known positive weight w_{ij} and dispersion σ_j^2, $j = 1, \ldots, n$. Here $\mathbf{x}_{ij}$ is a p-element vector of covariates associated with subject i for component j, and $X_i = (\mathbf{x}_{i1}, \ldots, \mathbf{x}_{in})$ is a $p \times n$ matrix of covariates. Moreover, the marginal mean μ_{ij} follows a marginal GLM, $g_j(\mu_{ij}) = \eta_j(\mathbf{x}_{ij})$ with linear predictor $\eta_{ij} = \mathbf{x}_{ij}^T \boldsymbol{\beta}_j$ and link function g_j, $j = 1, \ldots, n$. The primary task is to establish a simultaneous maximum likelihood inference for all model parameters $\boldsymbol{\theta} = (\boldsymbol{\beta}, \boldsymbol{\sigma}^2, \Gamma)$.

In many cases, the above general model formulation may become more specific. For example, one may assume a VGLM (6.1) takes a common regression parameter vector $\boldsymbol{\beta}$, namely $\boldsymbol{\beta}_j = \boldsymbol{\beta}$ for all j, in the situation of longitudinal or clustered data analysis, where the population-average effects of covariates are of interest. In addition, the association matrix Γ may be further parametrized by a parameter vector $\boldsymbol{\alpha}$, denoted by $\Gamma(\boldsymbol{\alpha})$, following the structure of interchangeable, AR-1, or 1-dependence. In this case, all model parameters are denoted by $\boldsymbol{\theta} = (\boldsymbol{\beta}, \boldsymbol{\sigma}^2, \boldsymbol{\alpha})$. Moreover, for convenience, all weights are set to be $w_{ij} = 1$ in the rest of the chapter.

Let the log-likelihood function of the given model be

$$\ell(\boldsymbol{\theta}; \mathbf{Y}) = \sum_{i=1}^{K} \ell_i(\boldsymbol{\theta}; \mathbf{y}_i). \tag{6.15}$$

Then, the MLE of $\boldsymbol{\theta}$ is

$$\widehat{\boldsymbol{\theta}} = \operatorname{argmax}_{\boldsymbol{\theta}} \ell(\boldsymbol{\theta}; \mathbf{Y}).$$

The biggest challenge for the application of the MLE in VGLMs arises from the numerical complexity and difficulty associated with the procedure of searching for the MLE $\widehat{\boldsymbol{\theta}}$. First of all, in most of situations, analytically deriving the second order derivatives of the log-likelihood (6.15) with regard to the model parameters is very tedious, so the related numerical solution of this optimization problem has to be made with the utility of the first order derivatives of $\ell(\boldsymbol{\theta}; \mathbf{Y})$, i.e., the scores $\dot{\ell}(\boldsymbol{\theta}; \mathbf{Y})$. Therefore, the popular Newton-Raphson or the Fisher scoring algorithm becomes unavailable, and alternative algorithms are called for help. Among many available algorithms, this book suggests two highly competent algorithms to overcome this numerical hurdle. The first algorithm is the so-called Maximization By Parts (MBP) (Song et al., 2005), which is recommended to deal with VGLMs with continuous vector outcomes. The second algorithm is a Gauss-Newton type algorithm (Ruppert, 2005) that works well for VGLMs with discrete outcomes or mixed outcomes. Both algorithms will be introduced in Section 6.5 in detail.

Under some mild regularity conditions, the standard MLE theory ensures that the MLE $\widehat{\boldsymbol{\theta}}$ is consistent and asymptotically normal. When the second order derivatives of the log-likelihood are not available, the observed Fisher information is estimated by using the following sandwich form:

$$\widehat{\mathbf{i}} = \left[\mathbf{H}(\widehat{\boldsymbol{\theta}})\mathbf{B}^{-1}(\widehat{\boldsymbol{\theta}})\mathbf{H}(\widehat{\boldsymbol{\theta}})\right]^{-1} = \mathbf{H}^{-1}(\widehat{\boldsymbol{\theta}})\mathbf{B}(\widehat{\boldsymbol{\theta}})\mathbf{H}^{-1}(\widehat{\boldsymbol{\theta}}), \tag{6.16}$$

where $\mathbf{H}(\theta)$ is the numerical Hessian derived from numerical differentiation via differencing, which approximates the observed Fisher information, and $\mathbf{B}(\theta) = \frac{1}{K}\sum_{i=1}^{K} \dot{\ell}_i(\boldsymbol{\theta}; \mathbf{y}_i)\dot{\ell}_i(\boldsymbol{\theta}; \mathbf{y}_i)^T$, which is a sample variance estimate of the variance of the score vector. This (6.16) consistently estimates the asymptotic covariance matrix.

In the application of VGLMs for correlated discrete outcomes or correlated mixed outcomes, numerically evaluating multivariate normal CDFs is required. Genz's (1992) algorithm available in the `R` software package `mvtnorm` is able to compute the normal CDF of 100 dimensions or lower.

6.4.2 VGLMs for Correlated Continuous Outcomes

When the margins are all continuous, the log-likelihood function for parameter $\boldsymbol{\theta} = (\boldsymbol{\beta}, \boldsymbol{\sigma}^2, \boldsymbol{\alpha})$ takes the form

$$\ell(\boldsymbol{\theta}; \mathbf{Y}) = -\frac{K}{2} \ln |\Gamma| + \sum_{i=1}^{K} \sum_{j=1}^{n} \log f(y_{ij}; \boldsymbol{\beta}, \sigma_j^2)$$
$$+\frac{1}{2} \sum_{i=1}^{K} \mathbf{q}_i^T(\mathbf{y}_i; \boldsymbol{\beta}, \boldsymbol{\sigma}^2) \left(I_n - \Gamma^{-1}\right) \mathbf{q}_i(\mathbf{y}_i; \boldsymbol{\beta}, \boldsymbol{\sigma}^2)$$

where $\mathbf{q}_i(\mathbf{y}_i; \boldsymbol{\beta}, \boldsymbol{\sigma}^2) = (q_{i1}, \ldots, q_{in})^T$ with components $q_{ij} = \Phi^{-1}(F_{ij}(y_{ij}))$ and F_{ij} is the marginal CDF of $\mathrm{ED}\left(\mu_{ij}, \sigma_j^2\right)$. The scores for $\boldsymbol{\beta}$ and σ_j^2 are given by, respectively,

$$\dot{\ell}_{\boldsymbol{\beta}} = \sum_{i=1}^{K} D_i^T \mathrm{diag}^{-1}[\sigma_1^2 V(\mu_{i1}), \ldots, \sigma_n^2 V(\mu_{in})](\mathbf{y}_i - \boldsymbol{\mu}_i)$$
$$+ \quad \sum_{i=1}^{n} Q_{i,\boldsymbol{\beta}}^T (I_n - \Gamma^{-1}) \mathbf{q}_i,$$
$$\dot{\ell}_{\boldsymbol{\sigma}^2} = \sum_{i=1}^{K} \sum_{j=1}^{n} \frac{\dot{f}_{\boldsymbol{\sigma}^2}(y_{ij}; \boldsymbol{\beta}, \sigma_j^2)}{f(y_{ij}; \boldsymbol{\beta}, \sigma_j^2)} + \sum_{i=1}^{K} Q_{i,\boldsymbol{\sigma}^2}^T (I_n - \Gamma^{-1}) \mathbf{q}_i, \; j = 1, \ldots, n,$$

where $D_i^T = \partial \boldsymbol{\mu}_i^T / \partial \boldsymbol{\beta}$, $Q_{i,\boldsymbol{\beta}}^T = \partial \mathbf{q}_i^T / \partial \boldsymbol{\beta}$, and $Q_{i,\boldsymbol{\sigma}^2}^T = \partial \mathbf{q}_i^T / \partial \boldsymbol{\sigma}^2$. Finally, the score for $\boldsymbol{\alpha}$ is $\dot{\ell}_{\boldsymbol{\alpha}} = K \dot{w}_{\boldsymbol{\alpha}}(\boldsymbol{\alpha}|A)$, where

$$w(\boldsymbol{\alpha}|A) = -\frac{1}{2} \ln |\Gamma(\boldsymbol{\alpha})| - \frac{1}{2} \mathrm{tr}\left\{\Gamma(\boldsymbol{\alpha})^{-1} A\right\} \tag{6.17}$$

with $A = \frac{1}{K} \sum_{i=1}^{K} \mathbf{q}_i \mathbf{q}_i^T$. Section 6.4.4 gives the details for derivative $\dot{w}_{\boldsymbol{\alpha}}(\boldsymbol{\alpha}|A)$.

It is worth noting that the expression of score vector $\dot{\ell}_{\boldsymbol{\beta}}$ differs from that of the ED GEE (5.12). On the basis of the above score equation, the GEE is proposed by absorbing the second term into the diagonal matrix of the first term. Song (2000a) showed that this absorption is valid only approximately for small dispersion parameters, i.e., $\max_j \sigma_j^2 \to 0$.

6.4.3 VGLMs for Correlated Discrete Outcomes

When the margins are all discrete, the log-likelihood is

$$\ell(\boldsymbol{\theta}; \mathbf{Y}) = \sum_{i=1}^{K} \ln f(\boldsymbol{\theta}; \mathbf{y}_i),$$

where $f(\cdot)$ is specified by (6.8) and parameter $\boldsymbol{\theta} = (\boldsymbol{\beta}, \boldsymbol{\sigma}^2, \boldsymbol{\alpha})$. Note that in some cases such as for the binomial and Poisson distributions, the dispersion parameter $\boldsymbol{\sigma}^2$ is known, so $\boldsymbol{\theta} = (\boldsymbol{\beta}, \boldsymbol{\alpha})$. Then, the score with respect to θ_l is given by

$$\dot{\ell}_{\theta_l}(\boldsymbol{\theta}) = \sum_{i=1}^{K} \frac{\dot{f}_{\theta_l}(\boldsymbol{\theta}; \mathbf{y}_i)}{f(\boldsymbol{\theta}; \mathbf{y}_i)}.$$

With $\theta_l = \boldsymbol{\beta}$, the numerator is

$$\begin{aligned}\dot{f}_{\boldsymbol{\beta}}(\boldsymbol{\beta}; \mathbf{y}_i) &= \sum_{j_1=1}^{2} \cdots \sum_{j_n=1}^{2} (-1)^{j_1+\ldots+j_n} \sum_{t=1}^{n} \frac{\partial C(u_{i,1,j_1}, \ldots, u_{i,n,j_n} | \Gamma(\boldsymbol{\alpha}))}{\partial u_{i,t,j_t}} \frac{\partial u_{i,t,j_t}}{\partial \boldsymbol{\beta}}, \\ &= \sum_{j_1=1}^{2} \cdots \sum_{j_n=1}^{2} (-1)^{j_1+\ldots+j_n} \sum_{t=1}^{n} \frac{\dot{\Phi}_{n,t,j_t} \left(\Phi^{-1}(u_{i,1,j_1}), \ldots, \Phi^{-1}(u_{i,n,j_n})\right)}{\phi\left(\Phi^{-1}(u_{i,t,j_t})\right)} \\ &\quad \times \frac{\partial u_{i,t,j_t}}{\partial \boldsymbol{\beta}},\end{aligned}$$

where $\dot{\Phi}_{n,t,j_t}(\mathbf{u})$ is the first order derivative of n-variate normal CDF $\Phi_n(\mathbf{y})$ with respect to y_t. Suppressing subscript i,

$$\frac{\partial u_{t,j_t}}{\partial \boldsymbol{\beta}} = \frac{\partial F_t(y; \mu_t)}{\partial \boldsymbol{\beta}} = \frac{\partial F_t(y; \mu_t)}{\partial \mu_t} \frac{\partial \mu_t}{\partial \boldsymbol{\beta}} = \frac{\partial F_t(y; \mu_t)}{\partial \mu_t} \{\dot{g}(\mu_t)\}^{-1} \mathbf{x}_t,$$

with $y = y_t - 1$ if $j_t = 1$ and $y = y_t$ if $j_t = 2$, $t = 1, \ldots, n$.

When $\theta_l = \boldsymbol{\alpha}$, the numerator is

$$\dot{f}_{\boldsymbol{\alpha}}(\boldsymbol{\beta}; \mathbf{y}_i) = \sum_{j_1=1}^{2} \cdots \sum_{j_n=1}^{2} (-1)^{j_1+\ldots+j_n} \frac{\partial C(u_{i,1,j_1}, \ldots, u_{i,n,j_n} | \Gamma(\boldsymbol{\alpha}))}{\partial \boldsymbol{\alpha}},$$

with

$$\frac{\partial C(u_{i,1,j_1}, \ldots, u_{i,n,j_n} | \Gamma(\boldsymbol{\alpha}))}{\partial \boldsymbol{\alpha}} = \int_{-\infty}^{\Phi^{-1}(u_{i,1,j_1})} \cdots \int_{-\infty}^{\Phi^{-1}(u_{i,n,j_n})} \frac{\partial \ln \phi_n(\mathbf{x}; \boldsymbol{\alpha})}{\partial \boldsymbol{\alpha}} \times \phi_n(\mathbf{x}; \boldsymbol{\alpha}) d\mathbf{x}$$

where $\phi_n(\mathbf{x}; \boldsymbol{\alpha}) = \phi_n(x_1, \ldots, x_n; \boldsymbol{\alpha})$ is the density of $\text{MVN}_n(0, \Gamma(\boldsymbol{\alpha}))$. Note that

$$\frac{\partial \ln \phi_n(x_1, \ldots, x_n; \boldsymbol{\alpha})}{\partial \boldsymbol{\alpha}} = \frac{\partial w(\boldsymbol{\alpha}|A)}{\partial \boldsymbol{\alpha}}$$

where $w(\boldsymbol{\alpha}|A)$ is given in (6.17) with matrix $A = (x_1, \ldots, x_n)^T (x_1, \ldots, x_n)$. The derivative of $w(\boldsymbol{\alpha}|A)$ with respect to $\boldsymbol{\alpha}$ with various association structures is given in the next Section 6.4.4.

6.4.4 Scores for Association Parameters

The scores for the association parameter $\boldsymbol{\alpha}$ are derived for either an unstructured or a structured Γ matrix. With a given matrix A, let $\mathbf{a}_i$ be vectors that satisfy $A = \sum_{i=1}^{K} \mathbf{a}_i \mathbf{a}_i^T$.

Example 6.8 (Unstructured Correlation).

When $\Gamma = (\gamma_{jl})$ is fully unspecified, $\boldsymbol{\alpha}$ contains $n(n-1)/2$ distinct unknown parameters. Then, the derivatives are given by

$$\begin{aligned}\frac{\partial}{\partial \gamma_{jl}} w(\boldsymbol{\alpha}|A) &= -\frac{1}{2}\text{tr}\Gamma^{-1}\left(\frac{\partial \Gamma}{\partial \gamma_{jl}}\right) + \frac{1}{2}\text{tr}\Gamma^{-1}\left(\frac{\partial \Gamma}{\partial \gamma_{jl}}\right)\Gamma^{-1}A \\ &= -\frac{1}{2}\text{tr}\left(\frac{\partial \Gamma}{\partial \gamma_{jl}}\right)(\Gamma^{-1} - \Gamma^{-1}A\Gamma^{-1}) \\ &= -d_{jl}\end{aligned}$$

where the matrix $D = (d_{jl}) = \Gamma^{-1} - \Gamma^{-1}A\Gamma^{-1}$, and $\partial\Gamma/\partial\gamma_{jl}$ is a matrix with all elements zero except the (j,l)-th and (l,j)-th, which are one.

Example 6.9 (Interchangeable Structure).

The interchangeable structure gives $\Gamma = \Gamma(\alpha) = \alpha \mathbf{1}\mathbf{1}^T + (1-\alpha)I_n$ for $\alpha \in (-\frac{1}{(n-1)}, 1)$, where $\mathbf{1}$ is an n-dimensional column vector of ones. It follows from Olkin and Pratt (1958) that

$$\Gamma^{-1}(\alpha) = \frac{1}{1-\alpha}I_n - \frac{\alpha}{(1-\alpha)\{1+(n-1)\alpha\}}\mathbf{1}\mathbf{1}^T,$$

and

$$\begin{aligned}w(\alpha|A) = -\frac{1}{2}\log|\Gamma(\alpha)| - \frac{1}{2(1-\alpha)}\sum_{i=1}^K\sum_{j=1}^n a_{ij}^2 + \\ \frac{\alpha}{2(1-\alpha)\{1+(n-1)\alpha\}}\sum_{i=1}^K\left(\sum_{j=1}^n a_j\right)^2.\end{aligned}$$

Note that

$$\frac{\partial}{\partial\alpha}\log|\Gamma(\alpha)| = \text{tr}\left\{\Gamma^{-1}(\alpha)\frac{\partial\Gamma(\alpha)}{\partial\alpha}\right\} = -\frac{n(n-1)\alpha}{(1-\alpha)\{1+(n-1)\alpha)\}}.$$

Thus, the derivative with respect to α is given by

$$\begin{aligned}\dot{w}_\alpha(\boldsymbol{\alpha}|A) = -\frac{1}{2(1-\alpha)}\Bigg[\frac{\sum_{i=1}^K\sum_{j=1}^n a_{ij}^2}{1-\alpha} - \\ \frac{1+(n-1)\alpha^2}{(1-\alpha)(1+(n-1)\alpha)^2}\sum_{i=1}^K\left(\sum_{j=1}^n a_j\right)^2 - \frac{n(n-1)\alpha}{1+(n-1)\alpha}\Bigg].\end{aligned}$$

Example 6.10 (AR-1 Structure).

For the AR-1 structure, the (i,j)-th element of $\Gamma(\alpha)$ is $\alpha^{|i-j|}$, $\alpha \in (-1,1)$. The inverse matrix takes the form (*e.g.* Chaganty, 1997)

$$\Gamma^{-1}(\alpha) = \frac{1}{1-\alpha^2}(I_n + \alpha^2 M_2 - \alpha M_1)$$

where $M_2 = \text{diag}(0,1,\ldots,1,0)$ and M_1 is a tridiagonal matrix with 0 on the main diagonal and 1 on the upper and lower diagonals. It is easy to show that

$$\text{tr}\left\{\frac{\partial \Gamma(\alpha)}{\partial \alpha}\right\} = 0, \ \text{tr}\left\{M_1 \frac{\partial \Gamma(\alpha)}{\partial \alpha}\right\} = 2(n-1), \ \text{tr}\left\{M_2 \frac{\partial \Gamma(\alpha)}{\partial \alpha}\right\} = 0.$$

Thus the derivative with respect to α is

$$\begin{aligned}\dot{w}_\alpha(\boldsymbol{\alpha}|A) = & -\frac{1}{2(1-\alpha^2)^2}\{-2(n-1)\alpha(1-\alpha^2) + \\ & 2\alpha \sum_{i=1}^K \mathbf{a}_i^T (I_n + M_2)\mathbf{a}_i - (1+\alpha^2)\sum_{i=1}^K \mathbf{a}_i^T M_1 \mathbf{a}_i\}.\end{aligned}$$

Example 6.11 (1-Dependence Structure).

1-dependence correlation structure corresponds to a matrix $\Gamma(\alpha)$ that is tridiagonal with 1 on the main diagonal and α on the upper and lower diagonals. It is well known that the eigenvalues and eigenvectors of the matrix are, respectively (*e.g.* Chaganty, 1997),

$$r_j(\alpha) = 1 + 2\alpha \cos\left\{\frac{j}{n+1}\pi\right\}, \ j = 1,\ldots,n$$

and

$$\mathbf{a}_j = [\sin\{j\pi/(n+1)\},\ldots,\sin\{jn\pi/(n+1)\}]^T, j = 1,\ldots,n.$$

Using Gram-Schmidt orthogonalization procedure for the $\mathbf{a}_j$'s, one can construct an orthonormal matrix P, and define $\mathbf{b}_j = P^T \mathbf{a}_j = (b_{i1},\ldots,b_{in})^T$. Note that matrix Γ may be decomposed as $\Gamma = P\text{diag}(r_1,\ldots,r_n)P^T$. So, $\log|\Gamma(\alpha)| = \sum_{j=1}^n \log r_j(\alpha)$. It follows that the score function for α is given by

$$\dot{w}_\alpha(\boldsymbol{\alpha}|A) = -\frac{1}{2}\sum_{i=1}^K \sum_{j=1}^n \left(\frac{1}{n} - b_{ij}^2\right)\frac{r_j(\alpha)-1}{\alpha r_j^2(\alpha)}.$$

6.5 Algorithms

This section introduces two useful algorithms to search for the MLE $\widehat{\boldsymbol{\theta}}$ in VGLMs. Both algorithms do not require the second order derivatives of the log-likelihood function, which are hard to derive analytically.

6.5.1 Algorithm I: Maximization by Parts

Maximization by parts (MBP) algorithm is an iterative numerical procedure that searches for the solution to a system of nonlinear score equations, without using the second order derivatives of the log-linear likelihood function. This algorithm starts with a selected additive partition of a complex log-likelihood function,

$$\ell(\boldsymbol{\theta}) = \ell_w(\boldsymbol{\theta}) + \ell_e(\boldsymbol{\theta}),$$

with the corresponding score function given by

$$\dot{\ell}(\boldsymbol{\theta}) = \dot{\ell}_w(\boldsymbol{\theta}) + \dot{\ell}_e(\boldsymbol{\theta}),$$

where $\dot{\ell}_w(\boldsymbol{\theta})$ is called the *working* log-likelihood (or the log-likelihood of a working model), and $\dot{\ell}_e(\boldsymbol{\theta})$ is termed as the remainder log-likelihood. Usually, the piece of $\ell_w(\boldsymbol{\theta})$ is so chosen that (a) the resulting $\dot{\ell}_w(\boldsymbol{\theta})$ is an unbiased inference function, and (b) the second order derivative $\ddot{\ell}_w(\boldsymbol{\theta})$ is easy to handle.

Take the example of the copula generated joint distribution (6.7), in which the likelihood function is formed as

$$L(\theta) = \prod_{i=1}^{K} \left\{ c(F_1(y_{i1};\mu_1), \ldots, F_n(y_{in};\mu_n)|\Gamma) \prod_{j=1}^{n} f_j(y_{ij};\mu_j) \right\}.$$

Naturally, its log-likelihood function can be written in the additive form with

$$\ell_w(\boldsymbol{\theta}) = \sum_{i=1}^{K}\sum_{j=1}^{n} \ln f_j(y_{ij};\mu_j)$$

$$\ell_e(\boldsymbol{\theta}) = -\frac{K}{2}\ln|\Gamma| + \frac{1}{2}\sum_{i=1}^{K} \mathbf{q}_i(\boldsymbol{\theta})^T (I_n - \Gamma^{-1})\mathbf{q}_i(\boldsymbol{\theta}). \tag{6.18}$$

Note that $\ell_w(\boldsymbol{\theta})$ is the likelihood function under the independence correlation structure ($\Gamma = I_n$) and only involves the marginal parameters μ_j or $\boldsymbol{\beta}$ in the regression setting, and $\ell_e(\boldsymbol{\theta})$ contains all parameters.

Direct maximization of $\ell(\boldsymbol{\theta})$ is tedious and numerically unstable, because marginal parameters $\mu_1, \ldots, \mu_n$ appear in $\ell_e(\boldsymbol{\theta})$ through complicated normal scores $\mathbf{q}_i(\boldsymbol{\theta})$. It is obviously straightforward to handle ℓ_w by computing its first and second order derivatives, but hard to derive the second order derivatives of ℓ_e. Despite being consistent, the estimator $\boldsymbol{\theta}^1$, as the solution to $\dot{\ell}_w(\boldsymbol{\theta}) = 0$, can have low efficiency since only part of the full log-likelihood function is used. To increase the efficiency, it is necessary to utilize the information in the second piece ℓ_e.

The MBP algorithm requires that the evaluation of the first order derivative $\dot{\ell}_e$ is available, which is certainly the case for the VGLMs, with all scores listed in Section 6.4.2. Then, the MBP algorithm proceeds as follows:

STEP 1: Solve $\dot{\ell}_w(\boldsymbol{\theta}) = 0$ for $\boldsymbol{\theta}^1$.
STEP k: Solve $\dot{\ell}_w(\boldsymbol{\theta}) = -\dot{\ell}_e(\boldsymbol{\theta}^{k-1})$ to produce estimate $\boldsymbol{\theta}^k$, $k = 2, 3, \ldots$. Liao and Qaqish (2005) suggested a one-step Newton-Raphson update at this iteration:

$$\boldsymbol{\theta}^k = \boldsymbol{\theta}^{k-1} - \{\ddot{\ell}_w(\boldsymbol{\theta}^{k-1})\}^{-1}\dot{\ell}(\boldsymbol{\theta}^{k-1}),$$

where $\ddot{\ell}_w(\boldsymbol{\theta}^{k-1})$ is the Hessian matrix of the working model evaluated at the previous update $\boldsymbol{\theta}^{k-1}$. When this Hessian matrix is replaced by the corresponding $\mathrm{E}\{\ddot{\ell}_w\}$, this one-step formula becomes a one-step Fisher-scoring update.

Song et al. (2005) proved that under the condition of information dominance this sequence of updates $\{\boldsymbol{\theta}^k\}$ will converge to the MLE, namely the solution to the score equation $\dot{\ell}(\boldsymbol{\theta}) = 0$. The information dominance condition is a requirement as to how the likelihood function $\ell(\boldsymbol{\theta})$ can be partitioned. In effect, this condition requires that the ℓ_w piece should contain more information about the parameter $\boldsymbol{\theta}$ than the other piece ℓ_e does. Technically speaking, that is $||\mathbf{i}_w^{-1}\mathbf{i}_e|| < 1$, where $\mathbf{i}_w = -K^{-1}\mathrm{E}\ddot{\ell}_w(\boldsymbol{\theta}_0)$ and $\mathbf{i}_e = -K^{-1}\mathrm{E}\ddot{\ell}_e(\boldsymbol{\theta}_0)$.

To better appreciate the MBP algorithm, let us study an example below.

Example 6.12 (Bivariate Exponential Model). Consider exponential margins with densities $f_j(y_j; \lambda_j) = \lambda_j \exp(-\lambda_j y_j)$, $\lambda_j > 0$, $j = 1, 2$, where λ_j are the rate parameters and $\mu_j = 1/\lambda_j$ are the means. Here $\boldsymbol{\theta} = (\boldsymbol{\theta}_1, \theta_2)$ with $\boldsymbol{\theta}_1 = (\lambda_1, \lambda_2)$ and $\theta_2 = \rho$, the association parameter in the bivariate Gaussian copula. The likelihood function for the independence model is

$$\ell_w(\boldsymbol{\theta}_1) = [K \ln \lambda_1 - \lambda_1 \sum_{i=1}^K y_{i1}] + [K \ln \lambda_2 - \lambda_2 \sum_{i=1}^K y_{i2}].$$

Let $\bar{y}_j = K^{-1}\sum_{i=1}^K y_{ij}, j = 1, 2$ and at iteration k let $\bar{\Delta}^k = (\bar{\Delta}_1^k, \bar{\Delta}_2^k)^T$, where $\bar{\Delta}^k = \Delta(\boldsymbol{\theta}^k)/K$, in which $\Delta(\boldsymbol{\theta})$ is given by

$$\begin{aligned} \Delta(\boldsymbol{\theta}) \equiv \frac{\partial \ell_e(\boldsymbol{\theta}_1, \rho)}{\partial \boldsymbol{\theta}_1} &= -\frac{\rho}{1-\rho^2}\left\{\rho\frac{\partial A(\boldsymbol{\theta}_1)}{\partial \boldsymbol{\theta}_1} - 2\frac{\partial B(\boldsymbol{\theta}_1)}{\partial \boldsymbol{\theta}_1}\right\} \\ \frac{\partial \ell_e(\boldsymbol{\theta}_1, \rho)}{\partial \rho} &= \frac{K\rho}{1-\rho^2} - \frac{1}{(1-\rho^2)^2}\{\rho A(\boldsymbol{\theta}_1) - (1+\rho^2)B(\boldsymbol{\theta}_1)\}. \end{aligned}$$

Here $A(\boldsymbol{\theta}_1) = \sum_{i=1}^K [q_{i1}(\lambda_1)^2 + q_{i2}(\lambda_2)^2]$, and $B(\boldsymbol{\theta}_1) = \sum_{i=1}^K q_{i1}(\lambda_1)q_{i2}(\lambda_2)$. Then, updates are given by

$$\begin{aligned} \boldsymbol{\theta}_1^1 &= \{\bar{y}_1^{-1}, \bar{y}_2^{-1}\} \\ \boldsymbol{\theta}_1^k &= \{(\bar{y}_1 + \bar{\Delta}_1^{k-1})^{-1}, (\bar{y}_2 + \bar{\Delta}_2^{k-1})^{-1}\}, \text{ for } k \geq 2. \end{aligned} \tag{6.19}$$

To update ρ, it only requires solving a third order polynomial equation.

However, when ρ is high, say equal to 0.9 or larger, the MBP algorithm based on the partition (6.18) using the independence model for the ℓ_w fails to produce convergent updates. This failure is due wholly to an inappropriate selection of ℓ_w that does not satisfy the information dominance condition. A better working model should incorporate a certain correlation among the components. Song et al. (2005) suggested a new partition given as follows:

$$\ell_w(\boldsymbol{\theta}) = -\frac{1}{2}\sum_{i=1}^{K}(\mathbf{y}_i - \boldsymbol{\mu})^T \Gamma_0^{-1}(\mathbf{y}_i - \boldsymbol{\mu}) + \sum_{i=1}^{K}\sum_{j=1}^{n} \ln f_j(y_{ij};\lambda_j)$$

$$\ell_e(\boldsymbol{\theta}) = -\frac{K}{2}\ln|\Gamma| + \frac{1}{2}\sum_{i=1}^{K}\mathbf{q}_i^T(I_n - \Gamma^{-1})\mathbf{q}_i + \frac{1}{2}\sum_{i=1}^{K}(\mathbf{y}_i - \boldsymbol{\mu})^T \Gamma_0^{-1}(\mathbf{y}_i - \boldsymbol{\mu}),$$

where Γ_0 is a known association matrix, and $\boldsymbol{\mu} = (1/\lambda_1, \ldots, 1/\lambda_n)^T$ is the vector of marginal means. With regard to the choice of Γ_0, one might obtain some useful clues from the one-step estimate Γ^1 using $\hat{\lambda}_j$ from the independence working model.

Given $(\boldsymbol{\mu}^1, \rho^1)$, the update for $\boldsymbol{\mu}$ is

$$\boldsymbol{\mu}^2 = \bar{\mathbf{y}} - K^{-1}\{I_n + G\Gamma_0^{-1}\}^{-1}\sum_{i=1}^{K} Q_i\{I_n - \Gamma(\rho^1)^{-1}\}\mathbf{q}_i - \{\Gamma_0 G^{-1} + I_n\}^{-1}(\bar{\mathbf{y}} - \boldsymbol{\mu}^1), \tag{6.20}$$

where $Q_i = Q_i(\boldsymbol{\mu}^1)$ is an n-dimensional diagonal matrix with the jth diagonal element $\partial q_{ij}/\partial\lambda_j = y_{ij}e^{-\lambda_j^1 y_{ij}}/\phi(q_{ij}(\lambda_j^1))$, and $G = G(\boldsymbol{\alpha}^1)$ is another n-dimensional diagonal matrix with the jth diagonal element $(\lambda_j^1)^{-2}$. As in the case of independence working model, the update for the association parameter ρ is the real root of the following third order polynomial,

$$\rho^3 + a_2\rho^2 + a_1\rho + a_0 = 0,$$

where

$$\begin{aligned} a_2 &= \frac{s_2}{n} - \frac{s_1}{n(n-1)} - \frac{n-2}{n-1} \\ a_1 &= \frac{2s_2}{n(n-1)} - \frac{1}{n-1} \\ a_0 &= \frac{s_2 - s_1}{n(n-1)^2}, \end{aligned}$$

with $s_1 = \frac{1}{K}\sum_{i=1}^{K}\{\mathbf{1}^T\mathbf{q}_i(\boldsymbol{\mu}^1)\}^2$ and $s_2 = \frac{1}{K}\sum_{i=1}^{K}\mathbf{q}_i(\boldsymbol{\mu}^1)^T\mathbf{q}_i(\boldsymbol{\mu}^1)$.

The result of one simulation study is presented here to demonstrate the MBP algorithm. Refer to Song et al. (2005) for more simulation studies. This simulation aims not only to assess the performance of the MBP algorithm itself, but also to compare with the second algorithm, a Gauss-Newton (G-N) type algorithm introduced in the next Section 6.5.2. For the marginal expectations, a simple log-linear model is used, $\lambda_j = e^{\theta_j}, j = 1, \ldots, n$. And for the association parameter, a logit model is specified, $\rho = 2H(\theta_{n+1}) - 1$, with

$H(x) = e^x/(1+e^x)$. The true parameters are set $K = 10, n = 5$, $\mu_j = \alpha_j = 1$, $j = 1, \ldots, n$, so that $\theta_j = 0, j = 1, \ldots, n$. In addition, the interchangeable correlation structure was assumed for the true Γ.

In the simulation study, Γ_0 is specified as an interchangeable structure with a pre-specified $\rho_0 = 0.95$, while the true association matrix Γ is interchangeable with the true $\rho = 0.9$. The structure of Γ_0 may be specified in other types, such as AR-1, and the ρ_0 can be set at a different value, say 0.5. However, the closer the pre-fixed Γ_0 to the true Γ, the faster the MBP algorithm converges.

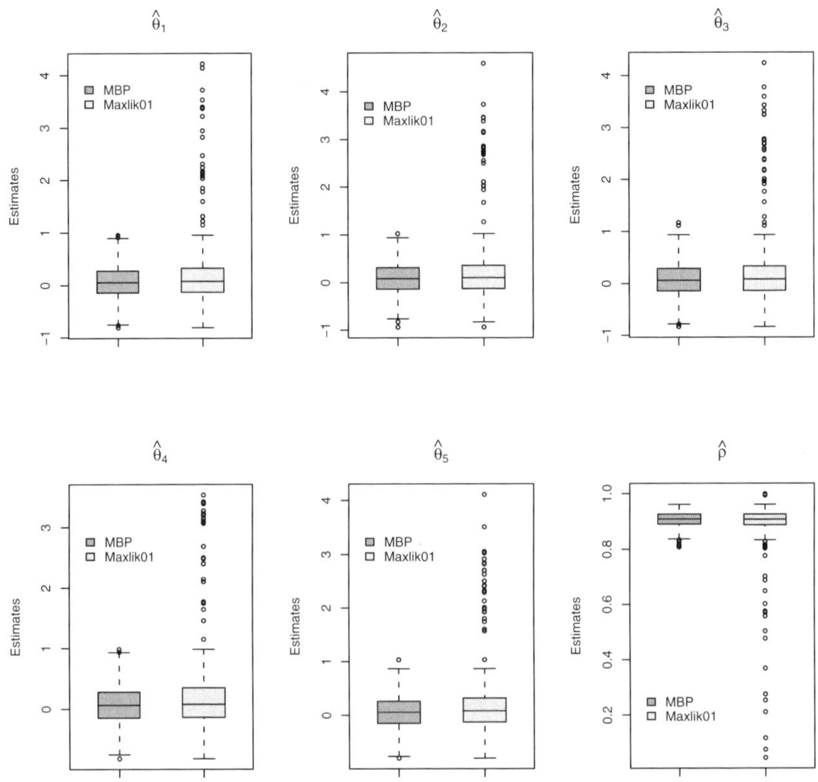

Fig. 6.4. Side-by-side boxplots comparing MBP and G-N over 500 replicates, with $\rho = 0.9$ and $K = 10$. The working correlation is interchangeable with $\rho_0 = 0.95$.

To compare the variation of the two methods, Figure 6.4 gives side-by-side boxplots for each model parameter based on 500 replications. Note that the plots are based on the transformed θ_j parameters. Both methods give the medians very close to the true values ($\theta_j = 0.0, j = 1, \ldots, 5$ and $\rho = 0.9$), but the G-N algorithm appears to be more variable with noticeably many outliers. This simulation actually considers a tough setup, in which 6 parameters are estimated by 10 data points. Clearly, the MBP handles this estimation remarkably well.

Song et al. (2005) concluded that (a) when the sample size is small and/or the dimension of the parameters is large, the MBP is an appealing algorithm to search for the MLE in VGLMs with continuous outcomes; (b) the G-N algorithm performs well when the sample size, K, is large; and (c) when good starting values are assigned to begin the G-N algorithm, its search is usually efficient and stable.

6.5.2 Algorithm II: Gauss-Newton Type

A Gauss-Newton type algorithm is suggested to search for the MLE $\widehat{\boldsymbol{\theta}}$ in the VGLMs for discrete outcomes or mixed outcomes. This optimization procedure uses only the first order derivatives of the log-likelihood functions. This algorithm works well when the sample size K is relatively large and reasonable starting values are used.

The key step of this algorithm is to take step-halving, which guarantees a steady increase in the likelihood from the previous iteration. Precisely, the $(k+1)^{th}$ iteration proceeds as

$$\boldsymbol{\theta}^{k+1} = \boldsymbol{\theta}^k + \delta\{\mathbf{B}(\boldsymbol{\theta}^k)\}^{-1}\dot{\ell}(\boldsymbol{\theta}^k),$$

where $\mathbf{B}(\boldsymbol{\theta}) = \frac{1}{K}\sum_{i=1}^K \dot{\ell}_i(\boldsymbol{\theta}; \mathbf{y}_i)\dot{\ell}_i(\boldsymbol{\theta}; \mathbf{y}_i)^T$, and δ is the step-halving term that is chosen as follows: starting at 1 (or 1/4, say, whichever is appropriate), it halves each time until $\ell(\boldsymbol{\theta}^{k+1}) > \ell(\boldsymbol{\theta}^k)$ holds in one iteration. Finally, the algorithm stops when the increase in the likelihood is no longer possible or the difference between two consecutive updates is smaller than a pre-specified precision level.

6.6 An Illustration: VGLMs for Trivariate Discrete Data

This section exemplifies two VGLMs for discrete outcomes, one for trivariate binary outcomes and the other for trivariate count outcomes. In both cases, the asymptotic relative efficiency is compared with the popular quasi-likelihood GEE method. Note that the GEE is quasi-likelihood approach, which claims to improve the estimation efficiency from the independent data analysis. However, the theory of GEE has not indicated if the gained efficiency improvement is satisfactory, which requires a comparison of the GEE's efficiency to a certain upper bound, if available. It is known that the upper bound is set by the MLE. With the availability of the copula-based VGLMs, it is possible to obtain such an upper bound and hence to assess the GEE's improvement in the efficiency gain. In the analysis of correlated binary data, there are two multivariate parametric models, the binary VGLM and the QEM representation. It is of interest to know which one gives better estimation efficiency, if the marginal logistic models are specified the same.

6.6.1 Trivariate VGLMs

For simplicity, consider the interchangeable correlation structure. The trivariate probability mass function f is obtained from (6.8) as

$$\begin{aligned} f(\mathbf{Y}_i;\boldsymbol{\theta}) &= P(Y_{i1}=y_{i1}, Y_{i2}=y_{i2}, Y_{i3}=y_{i3}) \\ &= \sum_{j_1=1}^{2}\sum_{j_2=1}^{2}\sum_{j_3=1}^{2}(-1)^{j_1+j_2+j_3}C(u_{i,1,j_1}, u_{i,2,j_2}, u_{i,3,j_3}|\alpha). \end{aligned}$$

The parameter vector is then $\boldsymbol{\theta}=(\boldsymbol{\beta},\alpha)$ as all dispersion parameters $\sigma_{ij}^2=1$.

Let $\mathbf{u}_{i,j_1,j_2,j_3} = (u_{i,1,j_1}, u_{i,2,j_2}, u_{i,3,j_3})$, and let $\dot{f}_{\theta_l}(\cdot)$ be the first order derivative of density f with respect to θ_l. Then, the scores are

$$\begin{aligned} \dot{\ell}_{\theta_l}(\boldsymbol{\theta}) &= \sum_{i=1}^{K}\dot{f}_{\theta_l}(\mathbf{y}_i;\boldsymbol{\theta})/f(\mathbf{y}_i;\boldsymbol{\theta}) \\ &= \sum_{i=1}^{K}\sum_{j_1=1}^{2}\sum_{j_2=1}^{2}\sum_{j_3=1}^{2}\left\{(-1)^{j_1+j_2+j_3}\dot{C}_{\theta_l}(\mathbf{u}_{i,j_1,j_2,j_3}|\alpha)\right\}/f(\mathbf{y}_i;\boldsymbol{\theta}). \end{aligned}$$

Moreover, by the chain rule, the scores with respect to $\theta_l=\beta_j$ are given by

$$\frac{\partial C(\mathbf{u}_{i,j_1,j_2,j_3}|\alpha)}{\partial \beta_j} = \sum_{t=1}^{3}\frac{\partial C(\mathbf{u}_{i,j_1,j_2,j_3}|\alpha)}{\partial u_{i,t,j_t}}\frac{\partial u_{i,t,j_t}}{\partial \beta_j},$$

where the first factor on the right-hand side takes the following forms:

$$\begin{aligned} \frac{\partial C(\mathbf{u}_{i,j_1,j_2,j_3}|\alpha)}{\partial u_{i,1,j_1}} &= \Phi_2\left\{\Delta_\alpha(u_{i,2,j_2}, u_{i,1,j_1}), \Delta_\alpha(u_{i,3,j_3}, u_{i,1,j_1}); \alpha\right\}, \\ \frac{\partial C(\mathbf{u}_{i,j_1,j_2,j_3}|\alpha)}{\partial u_{i,2,j_2}} &= \Phi_2\left\{\Delta_\alpha(u_{i,1,j_1}, u_{i,2,j_2}), \Delta_\alpha(u_{i,3,j_3}, u_{i,2,j_2}); \alpha\right\}, \\ \frac{\partial C(\mathbf{u}_{i,j_1,j_2,j_3}|\alpha)}{\partial u_{i,3,j_3}} &= \Phi_2\left\{\Delta_\alpha(u_{i,1,j_1}, u_{i,3,j_3}), \Delta_\alpha(u_{i,2,j_2}, u_{i,3,j_3}); \alpha\right\} \end{aligned}$$

with $\Delta_\alpha(u_{i,t,j_t}, u_{i,s,j_s}) = \frac{\Phi^{-1}(u_{i,t,j_t})-\alpha\Phi^{-1}(u_{i,s,j_s})}{\sqrt{1-\alpha^2}}$. On the other hand, the second factor is given by

$$\frac{\partial u_{i,t,j_t}}{\partial \beta_j} = \frac{\partial u_{i,t,j_t}}{\partial \mu_{it}} \frac{x_{itj}}{\dot{g}_t(\mu_{it})}.$$

Note that derivatives $\partial u_{i,j,j_t}/\partial \mu_{it}$ can have closed form expressions when certain marginal distributions are assumed. For example, the Bernoulli margin for binary data leads to

$$\frac{\partial u_{i,t,1}}{\partial \mu_{it}} = -1[y_{it} = 0], \quad \frac{\partial u_{i,t,2}}{\partial \mu_{it}} = -1[y_{it} = 1],$$

where $1[A]$ denotes the indicator function on set A. And the Poisson margin for count data gives

$$\frac{\partial u_{i,t,1}}{\partial \mu_{it}} = F_{it}(y_{it} - 1) - F_{it}(y_{it}), \quad \frac{\partial u_{i,t,2}}{\partial \mu_{it}} = F_{it}(y_{it} - 2) - F_{it}(y_{it} - 1),$$

where $F_{it}(\cdot)$ is the Poisson CDF with mean μ_{it}.

Similarly, for the association parameter α,

$$\frac{C(\mathbf{u}_{i,j_1,j_2,j_3}|\alpha)}{\partial \alpha} = \int_{-\infty}^{\Phi^{-1}(u_{i,1,j_1})} \int_{-\infty}^{\Phi^{-1}(u_{i,2,j_2})} \int_{-\infty}^{\Phi^{-1}(u_{i,3,j_3})} \frac{\partial}{\partial \alpha}\{\ln \phi_3(z_1, z_2, z_3|\alpha)\} \times \phi_3(z_1, z_2, z_3|\alpha) dz_1 dz_2 dz_3 \tag{6.21}$$

with

$$\frac{\partial}{\partial \alpha}\{\ln \phi_3(z_1, z_2, z_3|\alpha)\} = -\frac{1}{2(1-\alpha)} \left[\frac{z_1^2 + z_2^2 + z_3^2}{1-\alpha} - \frac{1+2\alpha^2}{(1-\alpha)(1+2\alpha)^2}(z_1 + z_2 + z_3)^2 - \frac{6\alpha}{1+2\alpha} \right].$$

The integral in (6.21) will be evaluated using the Gaussian-Hermite quadrature method discussed in Section 7.4.

Note that even for the 3-dimensional model, analytic derivation of the second order derivatives of the log-likelihood appears already very cumbersome, so the Gauss-Newton type algorithm introduced in Section 6.5.2 is adopted to search for the MLE in the discrete VGLMs.

6.6.2 Comparison of Asymptotic Efficiency

Now a comparison of the asymptotic relative efficiency between the VGLMs and the GEEs is conducted for trivariate binary data and trivariate count data, respectively. The focus is only on the regression parameters $\boldsymbol{\beta}$, since the association parameter α is usually treated as a nuisance parameter in the GEEs. The asymptotic relative efficiency (ARE) takes the form

$$\text{ARE}(\boldsymbol{\beta}) = \text{diag}\{\text{Var}_{vglm}\}[\text{diag}\{\text{Var}_{gee}\}]^{-1}, \tag{6.22}$$

where Var_{gee} is the Godambe information matrix of the GEE estimator $\widehat{\boldsymbol{\beta}}_{gee}$, and Var_{vglm} is the asymptotic covariance of the ML estimator $\boldsymbol{\beta}_{vglm}$ from the VGLM.

Example 6.13 (Trivariate Logit Model). The first comparison is based on correlated binary data generated by a hypothetical clinical trial in which a binary response is repeatedly measured over three time periods. Following Fitzmaurice et al. (1993), at each trial period, placebo ($x_t = 0$) or an active drug ($x_t = 1$) is assumed to be randomly assigned amongst the subjects, and all the eight possible covariate configurations have equal probability of occurrence. A logistic model for the marginal expectation is specified as

$$\text{logit}(\mu_{it}) = \beta_0 + \beta_1 x_{it} + \beta_2(t-2), \; t = 1, 2, 3,$$

where $\beta_0 = 0$, and $\beta_1 = \beta_2 = 0.5$. For such a simple model, fortunately the closed form expressions of Var_{gee} and Var_{vglm} can be analytically derived, respectively. Hence, the ARE comparison in this case does not rely on any simulated data or parameter estimates, but does depend only on the design of the experiment.

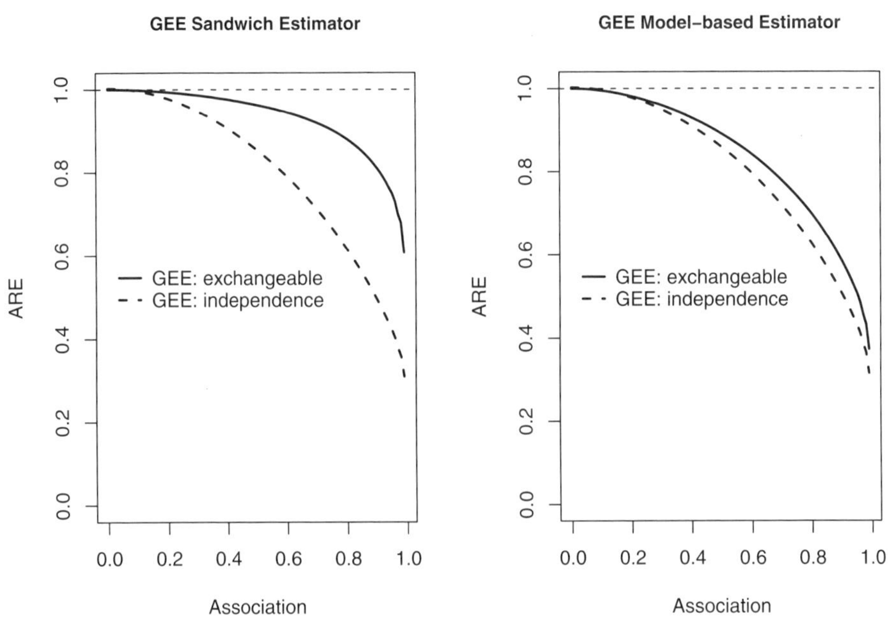

Fig. 6.5. Asymptotic efficiencies of the VGLM estimator of the slope parameter β_1 relative to the estimators, respectively, from the GEE and from the log-linear model representation under the trivariate logistic model.

The left panel of Figure 6.5 displays the ARE for the estimator of treatment effect β_1 as a function of the association parameter $\alpha \in [0, 1)$, with the interchangeable structure for the VGLM. For the GEEs, both the interchangeable and independence structures are considered in the comparison, respectively. Evidently, the estimator from the VGLM is more efficient than the GEE estimator, especially when the association is high. It is worth mentioning that under the logit link, the resulting GEEs are indeed coincident with the score equations derived from the QEM representation (6.3). The right panel of Figure 6.5 shows the ARE of the Var_{vglm} versus the model-based asymptotic covariance matrix (namely the Fisher information matrix given by the QEM), indicating similar efficiency gain when the VGLM is used to fit the data from this designed experiment. The amount of ARE between the VGLM and the QEM can vary from one experiment to another.

Example 6.14 (Trivariate Log-linear Poisson Model). The second comparison takes place between a trivariate Poisson VGLM and the corresponding GEEs, under the interchangeable structure for the VGLM and the interchangeable and independence structures for the GEEs, respectively. Unlike the first comparison, here the Var_{vglm} has no closed form expression, which therefore has to be estimated from simulated data. In order to minimize the effect of simulation variation and achieve high precision for the calculation of Var_{vglm}, a large sample size $K = 500$ is chosen. Data were generated from the 3-variate Poisson VGLM with the following marginal log-linear model:

$$\log(\mu_{ij}) = \beta_0 + \beta_1 x_{ij}, \ j = 1, 2, 3, i = 1, \ldots, K, \tag{6.23}$$

where the true values were $\beta_0 = \beta_1 = 0.5$, and the values of covariate x_{i1} were generated randomly according to uniform $U(0, 1)$.

The average ARE over 500 simulations for β_1 is plotted in Figure 6.6 at each of 20 grid points with 0.05 apart in $[0, 1)$. Figure 6.6 clearly indicates that high association leads to low ARE. This implies that when correlated count data are sampled from a 3-dimensional copula Poisson model, the estimator of β_1 from the GEEs does not make satisfactory efficiency gain, even if the interchangeable correlation has been incorporated into the inference procedure, especially when the association parameter α is bigger than 0.5.

6.7 Data Examples

This section presents three data examples that further demonstrate the usefulness of the VGLMs. The first is a bivariate logit regression model for a two-period cross-over trial data, the second is a 4-variate log-linear model for seasonal hospital visits, and the last one is a bivariate VGLM for mixed outcomes of binary and normal responses.

6.7.1 Analysis of Two-Period Cross-Over Trial Data

Data arising from cross-over clinical trials can be analyzed by the proposed VGLMs. A typical two-period cross-over clinical trial aims to compare two

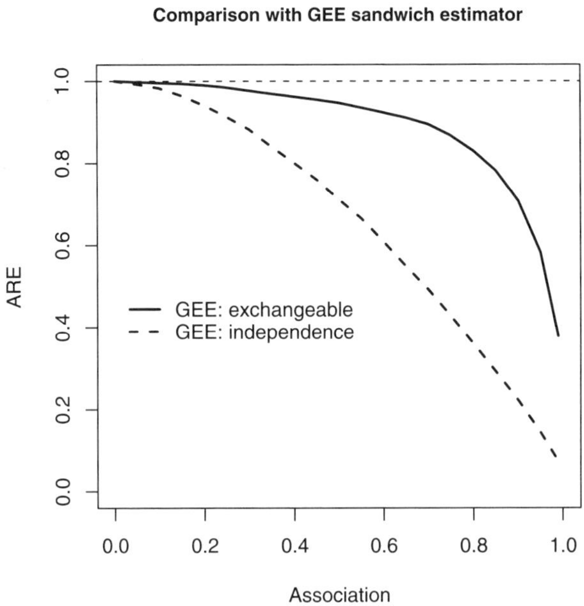

Fig. 6.6. Average asymptotic efficiency of the VGLM estimator relative to the estimator from the GEE under the trivariate Poisson model over 500 replications, in which each cluster contains 500 trios.

drugs, say A and B, in which each patient serves as his or her own control. Drugs are administrated over two periods with A/B or B/A sequence combinations. Thus, the two measurements from each patient collected at the two periods form a bivariate dataset.

This example re-analyzes the data of Example 8.1 from Diggle et al. (2002). The data, originally reported by Jones and Kenward (1989), contains results from a 2×2 cross-over trial on cerebrovascular deficiency in which an active drug (A) and a placebo (B) were compared. Sixty-seven individuals were involved in the trial. See more details of the data description in Diggle et al. (2002). A key empirical finding in the data was that both the drug group and placebo group show strong within-subject association. The data were analyzed previously by Diggle et al. (2002) using the QEM approach, in which a marginal insignificance of the treatment effect was found. Here the VGLM is applied to fit the data in the hope to improve the efficiency. For individual i at drug administration period j, let Y_{ij} be the response variable with 1 indicating a normal electrocardiogram reading and 0 otherwise. The marginal expectations were specified as follows:

$$\text{logit}(\mu_{ij}) = \beta_0 + \beta_1 x_{ij1} + \beta_2 x_{ij2} + \beta_3 x_{ij1} x_{ij2}, \ j = 1, 2, i = 1, \ldots, 67,$$

where $\mu_{ij} = P(Y_{ij} = 1)$, and two covariates, `treatment` (x_1) and `period` (x_2), are defined as $x_{ij1} = 1$ for active drug (A) and 0 otherwise, and $x_{ij2} = 1$ for period 2 and 0 otherwise.

Table 6.1. Estimated regression coefficients (standard errors) and Z-statistics from the VGLM and GEE.

	VGLM		GEE	
Variable	$\hat{\beta}$ (Std Err)	Z	$\hat{\beta}$(Std Err)	Z
Intercept	.43 (.36)	1.20	.43 (.36)	1.21
Treatment (x_1)	1.17 (.59)	1.98	1.11 (.57)	1.93
Period (x_2)	.17 (.51)	.32	.18 (.51)	.35
Interaction (x_1x_2)	−1.09 (.98)	−1.11	−1.20 (.98)	−1.04

The results of both VGLM and GEEs are reported in Table 6.1. The estimate of the association parameter by the VGLM was 0.89. Based on the ARE study in Section 6.6.2, such a high association will lead to some efficiency gain by the VGLM over the GEEs. As a result, the VGLM detected stronger evidence ($Z = 1.98$) for the effect of the active drug than the GEEs ($Z = 1.93$). In general, the efficiency gain will be elevated as the dimension of the outcomes increases.

6.7.2 Analysis of Hospital Visit Data

This example illustrates a 4-variate Poisson VGLM for a longitudinal data provided by Karim and Zeger (1988). Also see Table 9.22 in Davis (2002). The response vector $\mathbf{y}$ consists of quarterly numbers of hospital visits over a one year period for a child age at four or younger, and three baseline covariates are `age` (x_1 in months), `sex` ($x_2 = 1$ for girl and 0 for boy), and `maternal smoking status` ($x_3 = 1$ for yes and 0 for no). The data are reported from 73 children. The central task of this analysis is to investigate the relationship of the expected quarterly number of hospital visits as a function of these baseline covariates and hence identify which factors are statistically significant for the frequency of hospital visit.

The box-plots (not shown) of the response outcomes over the four seasons clearly suggest that the average number of visits in the first season (January to March) appears to be different from that in the other seasons. Therefore, an additional covariate x_4 is added to indicated the first season ($x_4 = 1$) and 0 for the rest of seasons. The four marginal means, $\mu_{ij} = \mathrm{E}(Y_{ij}), j = 1, \ldots, 4$, used in both VGLM and GEEs, are specified as follows:

$$\log(\mu_{ij}) = \beta_0 + \beta_1 x_{1ij} + \beta_2 x_{2ij} + \beta_3 x_{3ij} + \beta_4 x_{4ij}, j = 1, \ldots, 4.$$

A preliminary exploration on correlation structure indicates that interchangeable structure seems reasonable and hence is used in both VGLM and GEEs. The results are given in Table 6.2, where the 4-dimensional normal CDF was evaluated by using Joe's (1995) numerical recipe.

Table 6.2. Results from the analysis of hospital visit data using the VGLM and GEE

	VGLM		GEE	
Variable	$\hat{\beta}$ (Std Err)	Z	$\hat{\beta}$(Std Err)	Z
Intercept	$-.46$ (.10)	-4.39	-.43 (.42)	-1.03
Age(x_1)	.03 (.04)	1.13	.003 (.007)	.39
Smoking(x_3)	.15 (.10)	1.50	.15 (.27)	.56
Season(x_4)	.61 (.10)	5.90	.56 (.16)	3.48

The results given by the two methods are similar. The two methods found that the seasonal effect is significant, namely children less than 4 years old tend to visit hospital more frequently during the winter period (January to March) than during the rest of a year. Also, both methods are in agreement that the effects of both sex and maternal smoking covariates are not significant for the average frequency of child's hospital visits. This example indicates that when the statistical significance (or p-value) is not around the boundary, the GEE and the VGLM would be very likely to provide a similar conclusion about the effect of a covariate.

6.7.3 Analysis of Burn Injury Data

To demonstrate the flexibility of the VGLMs, a VGLM is employed to analyze the burn injury data that involve two response variables of mixed types. Reported in Fan and Gijbels (1996), the data contain 981 cases of burn injuries, and two response variables, the disposition of death, and the total burn area, and are modeled jointly as a function of patients' age. The severity of burn injury is measured by $Y_1 = \log(\text{burn area} + 1)$, which is a continuous response variable. The disposition Y_2 is a binary response with 1 for death from burn injury and 0 for survival. It is of interest to investigate how age (x) affects the severity of burn injury and the probability of death. To do this, two marginal mean models are specified as follows:

$$\begin{aligned} \mu_{i1} &= \beta_{01} + \beta_{11} x_i = \mathbf{x}_1^T \boldsymbol{\beta}_1 \\ \text{logit}(\mu_{i2}) &= \beta_{02} + \beta_{12} x_i = \mathbf{x}_2^T \boldsymbol{\beta}_2, \end{aligned}$$

where $\mu_{i1} = \mathrm{E}(Y_{i1}|x_i)$ is the expected log-burn area, and $\mu_{i2} = P(Y_{i2} = 1|x_i)$ is the probability of death from burn injury for patient i, given the age of the patient. Note that this is not a longitudinal data but a correlated data with different marginal response variables. In this case, it imposes the two regression models having different regression coefficients ($\boldsymbol{\beta}_1$ and $\boldsymbol{\beta}_2$), and different link functions (the identity and the logit).

Suppressing the subject index, it follows from (6.9) that the joint density of $\mathbf{Y} = (Y_1, Y_2)$ is given as follows:

$$f(y_1, y_2) = \begin{cases} \phi(y_1; \mu_1, \sigma_1^2)\{1 - \Delta_\alpha(\mu_2, z_1)\}, & \text{if } y_2 = 0, \\ \phi(y_1; \mu_1, \sigma_1^2)\Delta_\alpha(\mu_2, z_1), & \text{if } y_2 = 1, \end{cases} \tag{6.24}$$

where $\phi(\cdot; \mu_1, \sigma_1^2)$ is the density of $N(\mu_1, \sigma_1^2)$, $z_1 = (y_1 - \mu_1)/\sigma_1$, and $\Delta_\alpha(a, b) = \Phi\left(\frac{\Phi^{-1}(a) - \alpha b}{\sqrt{1-\alpha^2}}\right)$. An advantage of this joint copula modeling is that it avoids the artificial bimodal mixture of two normal margins, which is usually incurred by a conditional modeling approach, such as Fitzmaurice and Laird's (1995). That is, suppose that conditionally on a binary response Y_2,

$$Y_1 \mid Y_2 = l \sim N(\mu_l, \sigma_l^2), \; l = 0, 1.$$

Then it is easy to show that the marginal density of Y_1 takes a mixture of two normals,

$$f(y_1) = P(Y_2 = 0)\Phi(y_1; \mu_0, \sigma_0^2) + P(Y_2 = 1)\Phi(y_1; \mu_1, \sigma_1^2).$$

For the burn injury data $\{\mathbf{y}_i, (\mathbf{x}_{i1}, \mathbf{x}_{i2})\}, i = 1, \ldots, K$, the log-likelihood for $\boldsymbol{\theta} = (\boldsymbol{\beta}_1, \boldsymbol{\beta}_2, \sigma_1^2, \alpha)$ is given by

$$\begin{aligned} \ell(\boldsymbol{\theta}) &= \sum_{i \in S_0} \ln[\phi(y_{i1}; \mu_{i1}, \sigma_1^2)\{1 - \Delta_\alpha(\mu_{i2}, z_{i1})\}] + \\ &\quad \sum_{i \in \bar{S}_0} \ln[\phi(y_{i1}; \mu_{i1}, \sigma_1^2)\Delta_\alpha(\mu_{i2}, z_{i1})] \\ &= \sum_{i=1}^{K} \ln \phi(y_{i1}; \mu_{i1}, \sigma_1^2) + \sum_{i \in S_0} \ln\{1 - \Delta_\alpha(\mu_{i2}, z_{i1})\} + \sum_{i \in \bar{S}_0} \ln \Delta_\alpha(\mu_{i2}, z_{i1}), \end{aligned}$$

where $S_0 = \{i : y_{i2} = 0\}$ and $\bar{S}_0 = \{i : y_{i2} = 1\}$ are subsets of indices for survived and dead subjects, respectively.

Both joint model and individual univariate models are applied to fit the data, in which the Gauss-Newton type algorithm is implemented for the VGLM. The results are summarized in Table 6.3.

The estimated association parameter α by the VGLM was 0.80, which indicates strong association between these two responses. Overall, the point estimates obtained by the VGLM and the separate univariate models are very similar to each other. However, the VGLM appears to gain much efficiency in comparison to the separate univariate analysis. More specifically, the effect

Table 6.3. The estimates and standard errors obtained from the analysis of the burn injury data, where both joint model and separate univariate models are applied.

		VGLM			Univariate Models		
Model	β	$\hat{\beta}$	Std Err	Z	$\hat{\beta}$	Std Err	Z
Linear	Intercept	6.6980	.0479	139.73	6.7118	.0690	97.24
(log(burn area+1))	Age	.0039	.0012	3.16	.0035	.0018	1.97
Logit	Intercept	−4.0521	.1658	−24.44	−3.6891	.2342	−17.78
(death)	Age	.0527	.0028	19.13	.0509	.0046	11.07

of age on the burn severity is found to be statistically significant (p-value = 0.0016) by the VGLM but only marginally significant (p-value = 0.0488) by the univariate linear regression model. So, ignoring a strong association between the response variables will greatly reduce the power of the statistical inference. In conclusion, the joint modeling approach is clearly preferred, and the copula method provides a reasonable means of such joint modeling.

7

Mixed-Effects Models: Likelihood-Based Inference

7.1 Introduction

Mixed effects models (MEMs) provide another class of models for the analysis of correlated data. It is a conditional modeling approach that essentially specifies a fully parametric probability model, in which maximum likelihood estimation and inference can be established. Comparing to the marginal modeling approach, the availability of the MLE in the MEMs arguably makes such a modeling approach favorable. In the meantime, because it is a fully parametric approach, checking model assumptions is essential. Weiss and Lazaro (1992), among others, has investigated model diagnostics in linear mixed-effects models, but there has been not much progress so far in the literature for the development of model diagnostics in generalized linear mixed-effects models.

As a conditional modeling approach, the MEMs use latent variables to characterize subject-specific heterogeneity and to introduce correlation among correlated outcomes, such as within-cluster correlation for clustered data or serial correlation for longitudinal data.

Figure 7.1 gives a graphic representation of the conditional model for cluster i; that is, given a latent variable (or vector) $\mathbf{b}_i$, the outcomes $Y_{i1}, \ldots, Y_{in_i}$ of cluster i are conditionally independent and distributed according to a parametric model, such as a dispersion model. It is worth pointing out that such a specification of the conditional modeling is just one way to introduce correlation. For example, the copula-based modeling approach in Chapter 6 provides another way to incorporate correlation in a joint parametric model. Moreover, with the involvement of latent variables in MEMs, model interpretation is not as straightforward as that in marginal GLMs in Chapter 5 or vector GLMs in Chapter 6. The following terms involved in MEMs need to be properly interpreted:

(a) the meaning of latent variables $\mathbf{b}_1, \ldots, \mathbf{b}_K$;
(b) the form of induced correlation structure by the included latent variables;

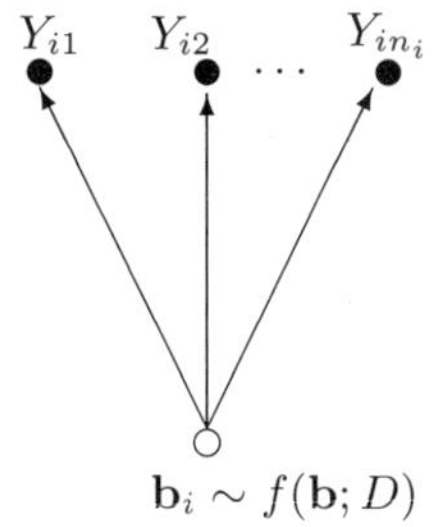

Fig. 7.1. Graphic representation of conditional modeling for cluster/subject i.

(c) the meaning of the regression coefficient $\boldsymbol{\beta}$ present in the conditional distribution of $Y_{ij}|\mathbf{b}_i$.

Another important issue pertains to the relation between MEMs and marginal GLMs or vector GLMs, which requires us to clarify the differences or similarities among these three types of modeling approached on modeling and interpretation. That is, with a given correlated data, which modeling approach should one take to analyze the data?

What are the latent variables $\mathbf{b}_1, \ldots, \mathbf{b}_K$? To answer, let us consider a simple example that concerns the estimation of mortality rate based on a sample of 12 hospitals that performed cardiac surgeries on babies. The data are reported in Table 7.1.

Table 7.1. Cardiac surgeries on babies performed in 12 hospitals.

	Hospital											
	A	B	C	D	E	F	G	H	I	J	K	L
n_i	47	148	119	810	211	196	148	215	207	97	256	360
y_i	0	18	8	46	8	13	9	31	14	8	29	24
$\frac{y_i}{n_i}$	0	0.12	0.07	0.06	0.04	0.07	0.06	0.14	0.07	0.08	0.11	0.07

Here, n_i denotes the number of operations performed at hospital i, and y_i is a realization regarding the number of deaths out of n_i surgeries. Let $Y_i = \sum_{j=1}^{n_i} Y_{ij}$, where Y_{ij} is the binary indicator of death (1) or survival (0), with π_i being the probability of death at hospital i. Because surgeries were very likely operated by the same group of surgeons in a hospital, outcomes of patients' survival status were inevitably clustered with hospitals, which can lead to overdispersion, i.e., $\text{Var}(y_i) > \pi_i(1-\pi_i)/m_i$, and other complications violating the standard assumption of *i.i.d.* observations. A simple population-average model (a marginal GLM) for the data takes the form of

$$\text{logit}(\pi_{ij}) = \beta_0, \ i = 1, \ldots, 12$$

where β_0 represents the population-average log-odds of death due to operation failure in the study population of all hospitals. Obviously, the estimated probability is

$$\hat{\pi}_0 = \frac{\sum_i y_i}{\sum_i n_i} = 0.074,$$

which means that the population average probability of operation failure for cardiac surgery is 7.4%. This result may be of interest for general citizens, but it is certainly not of interest to parents who are in a position to decide to which hospital to send their child for surgical operation. Parents are apparently more interested in the hospital-specific failure rate, so they are able to compare and choose the one with the lowest risk. Decisions may be made based directly on the observed percentages in Table 7.1. But, such empirical evidence is masked by sampling errors; that is, that hospitals C, F, I, and L all had a 7% failure rate could due to the sampling chance, or that hospital D (having a 6% failure rate) was better than hospital C (having a 7% failure rate) due to chance, and perhaps the true failure rates are no difference between the two hospitals. Proper statistical models can help to deal with this difficulty. In addition, it is easy to observe that the empirical failure rates in Table 7.1 vary substantially over the 12 hospitals, and such heterogeneity pertains to the hospital-specific performance of cardiac surgeons and other factors of individual hospitals.

In effect, the analysis using the marginal model is unable to provide the assessment of surgical performance differentiation specific for each of the hospitals, apart from the average performance of the all hospitals. As a result, it is useful to develop a model addressing the heterogeneity across the hospitals. A model that fulfills such a need may be written as

$$\text{logit}(\pi_{ij}) = \beta_0 + b_i, \ i = 1, \ldots, 12,$$

where β_0 represents the population-average failure rate, and $b_1, \ldots, b_{12}$ are regarded as of a random sample drawn from the study population of all hospitals, representing hospital-specific effects. In general, they are referred to *the subject-specific effects* or *random effects*.

An implicit assumption in such a model specification is that given random effect b_i, $Y_{i1}, \ldots, Y_{in_i}$ are conditionally independent, as illustrated by Figure 7.1. More precisely, the model may be expressed in a hierarchical form:

$$Y_{ij}|b_i \overset{ind.}{\sim} \text{Bi}(1, \pi_{ij}^b),\ j = 1, \ldots, n_i,$$
$$b_i \overset{iid}{\sim} f(b; D),\ i = 1, \ldots, 12,$$

with

$$\text{logit}(\pi_{ij}^b) = \beta_0 + b_i,\ i = 1, \ldots, 12,$$

where $\pi_i^b = P(\text{failure of surgery } j \text{ at hospital } i)$, and $f(\cdot; D)$ is a density function for the study population of hospitals, with mean 0 and a certain parameter D.

Treating parameters as random variates is not new to most of statisticians, as this is the essence of Bayesian inference. So, in this setting of mixed effects models, the random effects b_i can be treated as parameters with a prior $f(\cdot; D)$, should the Bayesian inference be adopted. Consequently, the intercept β_0 will also have to be treated as a random variable with a certain (preferably a non-informative) prior. In contrast, the frequentist inference would view the b_i as non-stochastic parameters representing realizations of a random sample from a study population. This treatment is also familiar to most of statisticians, as it resembles the practice of prediction. The difference in interpreting the random effects needs to be made clear, because it determines which type of inferential method is used in data analysis. This chapter focuses on frequentist methods and the next chapter discusses Bayesian inference based on Markov chain Monte Carlo (MCMC).

To generalize the above discussion to the correlated data setting where multiple outcomes are measured from one subject, first note that in the marginal GLM, the regression model is specified as $g(\mu_{ij}) = \mathbf{x}_{ij}^T \boldsymbol{\beta}$, where the regression parameter $\boldsymbol{\beta}$ holds fixed over all subjects, leading to a population average interpretation. This modeling structure has to be relaxed in order to incorporate heterogeneity across subjects, and then naturally the regression parameter demands to vary from one cluster/subject to another. Therefore, the resulting cluster/subject-dependent regression coefficients would give rise to a subject-specific interpretation. As a result, a regression model with subject-specific coefficients takes the form

$$g\left\{\mu_{ij}|\mathbf{b}_i)\right\} = \mathbf{x}_{ij}^T \boldsymbol{\beta}_i, \tag{7.1}$$

where $\boldsymbol{\beta}_i$ depends on subject i.

For an example of longitudinal data, to study infant growth, the weight of baby i at age t_j is recorded as Y_{ij}, associated with some covariates $x_{ij1}, \ldots, x_{ijp}$. A simple linear mixed model with only random intercepts may be expressed as

$$\text{E}(Y_{ij}) = \beta_{i0} + x_{ij1}\beta_1 + \ldots + x_{ijp}\beta_p$$

where only intercepts are subject-dependent in the form of $\beta_{i0} = \beta_0 + b_i$. The random effects, $b_1, \ldots, b_K$, may be essentially thought of as a random sample from a distribution of infant weight at birth.

7.2 Model Specification

Let us first consider linear mixed effects models (or linear random effects models) for normal correlated responses Y_{ij}. For the ease of exposition, imagine a longitudinal dataset collected from a clinical trial that aims to make a comparison between a new treatment to a standard treatment. A linear mixed-effects model (LMM) may be specified in a hierarchical form as follows:

Stage I : the response variable Y_{ij} is expressed as a linear model with subject-specific regression coefficients:

$$Y_{ij} = \beta_{i0} + \beta_{i1} t_j + \varepsilon_{ij},\ j = 0, 1, \ldots, n_i; i = 1, \ldots, K,$$

where t_0 is the baseline visit time (prior to randomization), and $\varepsilon_{ij} \stackrel{iid}{\sim} N(0, \sigma^2)$. Note that here β_{i0} represents the average effect for cluster i, and β_{i1} represents the average rate of change in time for cluster i.

Stage II : the subject-specific regression coefficients are specified as follows:

$$\begin{aligned}\beta_{i0} &= \alpha_0 + \boldsymbol{\alpha}_1^T \mathbf{x}_i + b_{i0},\\ \beta_{i1} &= \gamma_0 + \gamma_1 z_i + b_{i1}\end{aligned}$$

where $z_i = 1$ denotes the new treatment and 0 the standard treatment, and $\mathbf{x}_i$ denotes a set of baseline covariates such as age and baseline disease severity as well as possibly z_i. In addition,

$$\begin{pmatrix} b_{i0} \\ b_{i1} \end{pmatrix} \stackrel{iid}{\sim} \mathrm{MVN}_2\left(\begin{pmatrix} 0 \\ 0 \end{pmatrix}, D = \begin{pmatrix} D_{11} & D_{12} \\ D_{21} & D_{22} \end{pmatrix}\right),\ i = 1, \ldots, K,$$

where $D_{12} = D_{21}$. In this model, γ_0 represents the average rate of change in the standard treatment group ($z_i = 0$), while $(\gamma_0 + \gamma_1)$ indicates the average rate of change in the new treatment group ($z_i = 1$). Moreover, parameter $\boldsymbol{\alpha}$ characterizes the relation of the response on the baseline covariates.

Combining the two hierarchies together leads to one equation given by

$$Y_{ij} = \alpha_0 + \boldsymbol{\alpha}_1^T \mathbf{x}_i + \gamma_0 t_j + \gamma_1 (z_i t_j) + b_{i0} + b_{i1} t_j + \varepsilon_{ij},$$

which consists of three primary components:

Modeling component	Expression
Component of fixed effects	$\alpha_0 + \boldsymbol{\alpha}_1^T \mathbf{x}_i + \gamma_0 t_j + \gamma_1 (z_i t_j)$
Component of random effects	$b_{i0} + b_{i1} t_j$
Component of measurement noise	ε_{ij}

Here, (b_{i0}, b_{i1}) varies across subjects, characterizing heterogeneity due to unmeasured factors in the trial. Such heterogeneity across subjects gives rise to a certain within-cluster dependence.

It is easy to show that the first two moments of the model are

$$\begin{aligned} \mathrm{E}(Y_{ij}) &= \alpha_0 + \boldsymbol{\alpha}_1^T \mathbf{x}_i + \gamma_0 t_j + \gamma_1 (z_i t_j) \\ \mathrm{Var}(Y_{ij}) &= D_{11} + 2t_j D_{12} + t_j^2 D_{22} + \sigma^2 \\ \mathrm{cov}(Y_{ij}, Y_{ij'}) &= D_{11} + (t_j + t_{j'}) D_{12} + t_j t_{j'} D_{22}. \end{aligned}$$

If the model contains only the random intercept b_{i0}, then $D_{12} = D_{21} = D_{22} = 0$, and hence the covariance is $\mathrm{cov}(Y_{ij}, Y_{ij'}) = D_{11}$, which is the same for all pairs of (j, j'). This implies the interchangeable (or compound symmetry) correlation structure. If the random effect b_{1i} is present, then the covariance is time-dependent. However, the form of the time dependence is different from the standard AR-1 structure. Therefore, the induced correlation structure from the LMM seems somewhat awkward.

Also, the marginal mean (or the first moment) is

$$\mathrm{E}(Y_{ij}) = \alpha_0 + \boldsymbol{\alpha}_1^T \mathbf{x}_i + \gamma_0 t_j + \gamma_1 (z_i t_j),$$

which suggests that the regression coefficients in the component of the fixed effects have a marginal (population-average) interpretation. Unfortunately, this property does not retain in the generalized linear mixed models for non-normal responses. To see this, let us consider a simple Poisson random effects model with only the random intercept and one covariate z_i of treatment: $Y_{ij}|b_{0i} \stackrel{ind.}{\sim} \mathrm{Po}(\mu_{ij}^b)$, where the mean μ_{ij}^b follows a GLM of the form

$$\log(\mu_{ij}^b) = \beta_0^* + \beta_1^* z_i + b_{0i},$$

and $b_{0i} \stackrel{iid}{\sim} N(0, \sigma_b^2), i = 1, \ldots, K$. Then, the marginal mean is

$$\mu_{ij} = \mathrm{E}\{\mathrm{E}(Y_{ij}|b_{0i})\} = e^{\beta_0^* + \beta_1^* z_i} e^{\sigma_b^2/2},$$

which leads to the following marginal log-linear model:

$$\log(\mu_{ij}) = \left(\frac{\sigma_b^2}{2} + \beta_0^* \right) + \beta_1^* z_i.$$

Clearly, the intercept in the marginal model is different from the intercept in the conditional model. This means that β_0^* has no marginal (or population-average) interpretation. Although β_1^* appears to be the same in both models, when an additional random effect $b_{1i} z_i$ is added in the model, β_1^* will lose the marginal interpretation due to the similar argument.

In the logistic mixed effects model with only the random effect b_{0i}, the marginal mean has no closed form expression. That is,

$$\mu_{ij} = \int \frac{\exp(\beta_0^* + \beta_1^* z_i + b_{i0})}{1 + \exp(\beta_0^* + \beta_1^* z_i + b_{i0})} \phi(b_{i0}; \sigma_b^2) db_{i0}$$
$$\neq \frac{\exp(\beta_0^* + \beta_1^* z_i)}{1 + \exp(\beta_0^* + \beta_1^* z_i)}.$$

The situation of shifting by a constant in the Poisson mixed-effects model does not repeat in this logistic model case. Neuhaus et al. (1991) showed that if the variance of the random effects $D = \text{Var}(\mathbf{b}_i)$ is positive definite, then the fixed effects β_l^* in an MEM and regression coefficients β_l in the marginal model counterpart satisfy the following relation:

(a) $|\beta_l| \leq |\beta_l^*|, l = 1, \ldots, p$;
(b) the equality holds if and only if $\beta_l^* = 0$; and
(c) the discrepancy between β_l and β_l^* increases as $D = \text{Var}(\mathbf{b}_i)$ (or the main diagonal elements) increases.

In the logistic mixed effects model with single random intercepts, Zeger et al. (1988) obtained a more accurate assessment of the relationship. That is, the marginal log-odds can be approximated by

$$\text{logit}(\boldsymbol{\mu}_i) \approx (1 + c^2 D_{11})^{-1/2} \boldsymbol{\tau}_i$$

where $c = \frac{16\sqrt{3}}{15\pi}$ and $\boldsymbol{\tau}_i$ the vector of the marginal linear predictors. Thus, an approximate relation is given by

$$\beta_l \approx \frac{\beta_l^*}{\sqrt{1 + 0.346 D_{11}}}, \quad l = 1, \ldots, p.$$

Because of this discrepancy, it is necessary to distinguish the regression coefficients β_l^* in the mixed effects model from those β_l in the marginal GLM. Equivalently, the ratio of estimates obtained from the mixed effects models and from the marginal model is approximately

$$\frac{\beta_l^*}{\beta_l} \approx \sqrt{1 + 0.346 D_{11}}, \tag{7.2}$$

which is bigger than 1, unless $D_{11} = 0$ or little heterogeneity across subjects is present in the population.

As a matter of fact, in most of generalized linear mixed models, the regression coefficients $\beta_l^*, l = 1, \ldots, p$ have only the conditional interpretation.

Now a formal introduction to the generalized linear mixed effects model (GLMM) is given as follows:

(i) Given random effects $\mathbf{b}_i$, the responses $Y_{i1}, \ldots, Y_{in_i}$ are mutually independent, and
$$Y_{ij} | \mathbf{b}_i \sim \text{DM}(\mu_{ij}^b, \sigma^2)$$
where the conditional location parameter μ_{ij} is assumed to follow a generalized linear model given by

$$\eta_{ij}^b = g(\mu_{ij}^b) = \mathbf{x}_{ij}^T \boldsymbol{\beta} + \mathbf{z}_{ij}^T \mathbf{b}_i, \tag{7.3}$$

where g is the link function, where $\boldsymbol{\beta}$ is a p-dimensional vector of fixed effects, and $\mathbf{z}_{ij}$ is a q-dimensional subvector of $\mathbf{x}_{ij}$.

(ii) The random effects, $\mathbf{b}_1, \ldots, \mathbf{b}_K$, are *i.i.d.* according to a multivariate density $f(\mathbf{b}; D)$.

Following the literature, this chapter assumes density f is $\mathrm{MVN}_q(0, D)$, where $D = D(\boldsymbol{\tau})$ is a positive-definite covariance matrix, which may be further parametrized by a parameter vector $\boldsymbol{\tau}$ of variance components. The set of parameters to be estimated is $\boldsymbol{\theta} = (\boldsymbol{\beta}, \boldsymbol{\tau}, \sigma^2)$.

The linear mixed-effects model is a special case of the GLMM with the identity link function, $g(\mu) = \mu$, for the normal response. The model takes the form

$$Y_{ij} = \mathbf{x}_{ij}^T \boldsymbol{\beta} + \mathbf{z}_{ij}^T \mathbf{b}_i + \varepsilon_{ij}$$

where ε_{ij} are *i.i.d.* $N(0, \sigma^2)$. The model can also be expressed in matrix notation as follows:

$$\mathbf{Y}_i = X_i \boldsymbol{\beta} + Z_i \mathbf{b}_i + \boldsymbol{\varepsilon}_i,$$

where $\mathbf{Y}_i = (Y_{i1}, \ldots, Y_{in_i})^T, X_i^T = (\mathbf{x}_{i1}, \ldots, \mathbf{x}_{in_i}), Z_i^T = (\mathbf{z}_{i1}, \ldots, \mathbf{z}_{in_i})$, and $\boldsymbol{\varepsilon}_i = (\varepsilon_{i1}, \ldots, \varepsilon_{in_i})^T$. It follows that

$$\begin{aligned} \mathrm{E}(\mathbf{y}_i) &= X_i \boldsymbol{\beta} \\ \mathrm{Var}(\mathbf{y}_i) &= Z_i D Z_i^T + \sigma^2 I_{n_i}. \end{aligned}$$

In the current literature, the distribution of random effects is routinely assumed to be normal, mainly for mathematical convenience (Neuhaus et al., 1992; Neuhaus and Kalbfleisch, 1998). In recent years, violation of such normality has been reported in many data analyses. For examples, Pinherio et al. (2001) pointed out that the distribution of random effects appeared to have heavier tails than the normal in their orthodontic data analysis, Zhang and Davidian (2001) found that the random intercepts followed a positively skewed distribution in their model for Framingham cholesterol data, and Ishwaran and Takahara (2002) indicated that the distribution of random effects deviated from normality due to negative skewness and positive kurtosis in their analysis of chronic renal disease data. The widespread use of the mixed effects models with normal random effects is, in part, motivated by the fact that under general regularity conditions, estimates of the fixed effects and variance components obtained under the normally distributed random effects remain consistent and asymptotically normal, even if the assumption of normality is violated (Beal and Sheiner, 1988; Verbeke and Lesaffre, 1997). However, deviations from normality can adversely affect the efficiency of estimates of the fixed effects, and may also adversely affect the estimation of subject-specific random effects. Although some works have tried to fix some problems due to the violation of the normality assumption, such as Song et al. (2007), considering robust mixed effects models with t-distributed random effects, some further investigations are worth being explored in this area.

7.3 Estimation

One of statistical tasks in the GLMMs is to estimate the parameters $\boldsymbol{\theta}$, as well as the random effects $\mathbf{b}_i$ when the subject-specific effects are of interest. Because the GLMM is a fully parametric model, the MLE would naturally be the choice of estimation.

Roughly speaking, the development of MLE is driven by whether the random effects are of interest and need to be explicitly estimated. When $\mathbf{b}_i$ are not of central focus, one may take an approach that treats these random effects as nuisance parameters and integrates them out. As a result, the resulting estimation procedure is effectively based on the marginal likelihood, in which the estimation of fixed effects $\boldsymbol{\beta}$ can be carried out with no need of estimating $\mathbf{b}_i$. For instance, the MLE based on the quadrature numerical method and the MLE based on Monte Carlo EM (MCEM) are of this kind.

In a similar spirit, the conditional MLE proceeds by conditioning the random effects out of the problem based on relevant sufficient statistics. For example, consider the random intercept model

$$g(\mu_{ij}^b) = b_{0i} + \beta_0 + \beta_1(\text{age at entry})_i + \beta_2(\text{follow-up time})_j,$$

where only β_2 is of interest and interpretable since the unknown b_{0i} make the comparison of across-subject effects (i.e., β_1) impossible, unless individual b_{0i} is known *a priori*.

Another type of approach assumes that the subject-specific coefficients are themselves of interest, in which both $\mathbf{b}_i$ and $\boldsymbol{\beta}$ (fixed effects) are estimated simultaneously. Given the nature of random effects as being realizations of random variables, it is sensible to invoke "posterior" density $f(\mathbf{b}_i|\text{data})$ for the estimation of $\mathbf{b}_i$, and consequently either the mean or the mode of the posterior can be used as the point estimation for the $\mathbf{b}_i$. Methods of this kind include Breslow and Clayton's (1993) approximate inference based on penalized quasi-likelihood (PQL) and Bayesian approach based on Markov chain Monte Carlo (MCMC) algorithm.

Besides those mentioned above, there are other types of methods such as the simulated MLE (McCulloch and Searle, 2001) and the hierarchical likelihood estimation (Lee and Nelder, 1996 and 2004) developed in the literature. Also see Vonesh and Carter (1987), Schall (1999), and McGilchrist (1994) for some *ad hoc* estimation procedures. MLE in the GLMMs is still currently an active research area in the field of longitudinal or clustered data analysis.

To begin the discussion of MLE, let us first write down the likelihood function of the GLMM. Clearly, the i-th piece likelihood for subject/cluster i is

$$\begin{aligned} L_i(\boldsymbol{\theta}) &= \int_{\mathcal{R}^q} f(\mathbf{y}_i|\mathbf{b}_i)\phi(\mathbf{b}_i; D)d\mathbf{b}_i \\ &= \int_{\mathcal{R}^q} \prod_{j=1}^{n_i} f(y_{ij}|\mathbf{b}_i)\phi(\mathbf{b}_i; D)d\mathbf{b}_i, \end{aligned} \tag{7.4}$$

and the log-likelihood for the full data is

$$\ell(\boldsymbol{\theta}) = \log L(\boldsymbol{\theta}) = \sum_{i=1}^{K} \log L_i(\boldsymbol{\theta}). \tag{7.5}$$

Then, without estimating $\mathbf{b}_i$, the MLE of $\boldsymbol{\theta}$ can be obtained by directly maximizing the log-likelihood $\ell(\boldsymbol{\theta})$ with respect to $\boldsymbol{\theta}$, namely

$$\widehat{\boldsymbol{\theta}} = \arg\max_{\boldsymbol{\theta}} \ell(\boldsymbol{\theta}).$$

In the LMM that assumes $Y_{ij}|\mathbf{b}_i \sim N(\mu_{ij}^b, \sigma^2)$, with

$$\mu_{ij}^b = \mathbf{x}_{ij}^T \boldsymbol{\beta} + \mathbf{z}_{ij}^T \mathbf{b}_i,$$

the integral in (7.4) has a closed form expression. This is because it takes effectively a normal-normal convolution, so the marginal distribution must remain normal. That is, marginally $\mathbf{Y}_i \sim \text{MVN}_{n_i}(X_i\boldsymbol{\beta}, Z_i D Z_i^T + \sigma^2 I_{n_i})$. Furthermore, the MLE of the fixed effects $\boldsymbol{\beta}$ can be obtained by the iterative weighted least squares, and the variance component parameters can be estimated by either MLE or the restricted MLE (Harville, 1977). SAS PROC MIXED provides a comprehensive numerical package to fit correlated normal data by the linear mixed-effects model. Interested readers may refer to McCulloch and Searle (2001) for more technical details regarding inference in the linear mixed-effects models.

However, when data are not normal, the ML inference becomes more complicated. For the instance of binary regression model,

$$\text{logit}(\pi_{ij}^b) = \mathbf{x}_{ij}^T \boldsymbol{\beta} + \mathbf{z}_{ij}^T \mathbf{b}_i,$$

the likelihood $L(\boldsymbol{\theta})$ is proportional to

$$\prod_{i=1}^{K} \int_{\mathcal{R}^q} \prod_{j=1}^{n_i} \left\{\pi_{ij}^b(\boldsymbol{\beta})\right\}^{y_{ij}} \left\{1 - \pi_{ij}^b(\boldsymbol{\beta})\right\}^{1-y_{ij}} |D|^{-q/2} \exp\left(-\frac{1}{2}\mathbf{b}_i^T D^{-1} \mathbf{b}_i\right) d\mathbf{b}_i,$$

where the integral does not have a closed form expression.

Because of the involvement of unobserved random effects, the resulting likelihood of the GLMM usually appears intricate, so the related numerical implementation in the search of MLE $\widehat{\boldsymbol{\theta}}$, as well as in the estimation of random effects if relevant, could be challenging. The challenge arises mainly from three aspects:

- Optimization or maximization is not trivial, when the score $\dot{\ell}(\boldsymbol{\theta})$ is high-dimensional. In the GLMMs, the second order derivatives $\ddot{\ell}(\boldsymbol{\theta})$ may be hard to derive analytically or even if available appear very complicated, so the application of Newton-Raphson or Fisher-scoring algorithm is generally not feasible. Instead, searching for MLE may have to be done by algorithms relying only on its scores or numerical second derivatives. For example, the current SAS PROC NLMIXED adopts a dual quasi-Newton algorithm to search for the solution to the score equations.

- Evaluation of the q-dimensional integral is troublesome, especially when the dimension q is high. The available numerical evaluation methods like the quadrature method can only handle low dimensional ($q \leq 3$, say) integral effectively. When the q is high, the number of quadratures required in the evaluation can increase astronomically, which hampers the use of any quadrature method literally. In this situation, Markov chain Monte Carlo (MCMC) algorithm is suggested to deal with the integration evaluation. Note that MCMC implementation requires specifying prior distributions for all model parameters, which results in a Bayesian inference in nature. In the aspect of model specification, some additional assumptions on priors, besides those already made for the GLMM, are imposed, which will have to be checked thoroughly by sensitivity analysis. Unfortunately, sensitivity analysis can appear to be somewhat subjective and computationally intensive, and it can easily add extra complexity upon model fitting. This extra difficulty might be even greater than the original one.
- For any algorithm proposed to search for MLE, its computational efficiency and numerical stability needs to be seriously examined, especially when some approximations are invoked in the development of that algorithm.

This chapter is devoted to an introduction to a few popular inference methods in the GLMMs, where the integration in the evaluation of likelihood functions as well as their derivatives has no closed form expressions. Some methods have been implemented in various software packages and hence are ready to be used for data analysis.

Finally, it is worth noting that as a fully parametric modeling approach, assumptions made on the MEMs have to be validated via, *say*, residual analysis. In the current literature, systematic model diagnostics in the MEMs are largely absent, although some efforts have been made.

7.4 MLE Based on Numerical Integration

When the dimension of the random effects $\mathbf{b}_i$ is low, say $q \leq 5$, one may directly apply a quadrature numerical integration method to evaluate the integration in the likelihood function (7.4). Among several quadrature methods, the Gauss-Hermite quadrature method is illustrated in this section. In essence, quadrature methods differ by how and how many quadrature points are chosen in the evaluation of integration. SAS PROC NLMIXED uses the adaptive Gaussian quadrature method (Lange, 1999) to select quadrature points.

Let us begin by considering one-dimensional random effects, namely $q = 1$ corresponding to the GLMM with only random intercepts. The GLMM can be expressed in a hierarchical form as follows:

$$\begin{aligned} Y_{ij}|b_i &\overset{ind.}{\sim} \mathrm{MD}(\mu_{ij}^b, \sigma^2) \\ b_i &\overset{i.i.d.}{\sim} N(0, \tau) \end{aligned}$$

with

$$g(\mu_{ij}^b) = \mathbf{x}_{ij}^T \boldsymbol{\beta} + b_i.$$

Let $\boldsymbol{\theta} = (\boldsymbol{\beta}, \tau, \sigma^2)$. The resulting likelihood function is then

$$\begin{aligned} L(\boldsymbol{\theta}) &= \prod_{i=1}^{K} \int_{\mathcal{R}} \prod_{j=1}^{n_i} f(y_{ij}|b_i) f(b_i) db_i \\ &= \prod_{i=1}^{K} \left\{ \prod_{j=1}^{n_i} c(y_{ij}; \sigma^2) \right\} \int_{\mathcal{R}} \exp\left\{ -\frac{1}{2\sigma^2} \sum_{j=1}^{n_i} d(y_{ij}; \mu_{ij}^b) \right\} \frac{e^{-b_i^2/(2\tau)}}{\sqrt{2\pi\tau}} db_i. \end{aligned}$$

Clearly, suppressing index i, the following form of the integral needs to be evaluated

$$\begin{aligned} \int_{\mathcal{R}} h(b) \frac{e^{-b^2/(2\tau)}}{\sqrt{2\pi\tau}} db &= \int_{\mathcal{R}} h(\sqrt{2\tau} v) \frac{e^{-v^2}}{\sqrt{\pi}} dv \\ &= \int_{\mathcal{R}} h^*(v) e^{-v^2} dv, \end{aligned}$$

where

$$\begin{aligned} h^*(v) &= h(\sqrt{2\tau} v)/\sqrt{\pi}, \quad \text{with} \\ h(v) &= \exp\left\{ -\frac{1}{2\sigma^2} \sum_{j=1}^{n_i} d(y_{ij}; \mu_{ij}^b) \right\}. \end{aligned}$$

This standardization is necessary in order to apply standardized quadrature points and weights that are generated from Hermite orthogonal polynomials. The Gauss-Hermite quadrature evaluation of the integral takes the form

$$\int_{\mathcal{R}} h^*(v) e^{-v^2} dv \approx \sum_{k=1}^{M} h^*(v_k) w_k, \tag{7.6}$$

where v_k are the quadrature points and w_k are weights, as illustrated in Figure 7.2. Basically, the sum of areas of many vertical retangles approximates the area under the curve $h^*(\cdot)$.

Relevant questions raised here are which quadrature points v_k and weights w_k should be chosen and how many. The number of quadrature points, M, is pre-specified, which determines the precision (or order) of this approximation. Practical experience suggests that in the context of the GLMM, such an approximation is usually unsatisfactory when $M \leq 10$ and is good enough when $M \geq 20$. The adaptive Gaussian quadrature method builds in an iterative procedure that automatically selects a number of quadrature points to achieve a pre-specified precision. In contrast, the Gauss-Hermite quadrature method has to fix the M beforehand. See more details in Lange (1999).

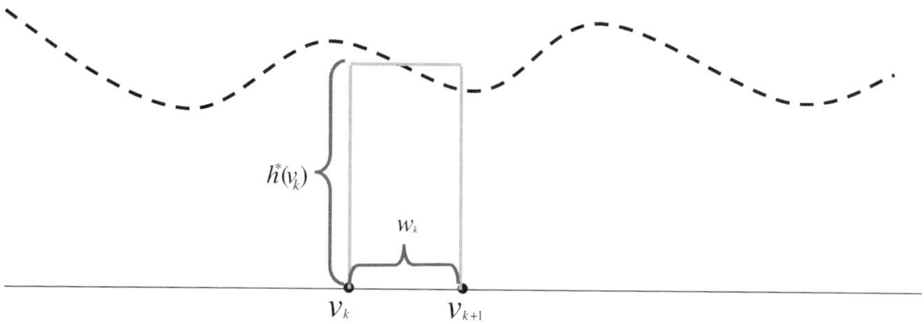

Fig. 7.2. Diagram for Gauss-Hermite quadrature evaluation of integration.

The values of v_k and w_k are determined such that the approximation $\sum_{k=1}^{M} h^*(v_k)w_k$ will give the exact answer to the integral when the integrand $h^*(\cdot)$ are all orthogonal polynomials up to degrees $(2M-1)$. This implies that

$$v_k = k\text{th zero (root) of Hermite polynomial } H_m(v)$$
$$w_k = \frac{2^{-1}m!\sqrt{\pi}}{m^2\{H_{m-1}(v_k)\}^2},$$

where $H_m(\cdot)$ is the Hermite orthogonal polynomial of order m.

Abramowitz and Stegun's (1970) Handbook of Mathematical Functions lists the values of v_k and w_k for $M \leq 25$, and a large M, `MATLAB` software package can generate desirable quadrature points and weights easily. For example, Table 7.2 lists these values for the case of $M = 3, 4, 5$, respectively.

To see how this method works, take an example of an integral with the integrand being a polynomial of order 2,

$$\int_{-\infty}^{\infty} (1+v^2)e^{-v^2}dv = \frac{3}{2}\sqrt{\pi} = 2.65868.$$

The Gauss-Hermite quadrature method with $M = 3$ gives

$$\begin{aligned}\int_{-\infty}^{\infty} (1+v^2)e^{-v^2}dv &= \{1+(-1.22474487)^2\}(0.29540898)\\ &\quad +(1+0^2)(1.18163590)\\ &\quad +(1+1.22474487^2)(0.29540898)\\ &= 2.65868,\end{aligned}$$

as expected. Consider an intergrand of a sixth order polynomial,

$$\int_{-\infty}^{\infty} (1+v^6)e^{-v^2}dv = \frac{23}{8}\sqrt{\pi} = 5.0958.$$

The same 3-point quadrature formula gives a result of 3.76645, a very poor approximation. However, the 4-point quadrature formula gives an almost exact answer. Interested readers can try this calculation themselves.

Table 7.2. Gauss-Hermite quadrature points and weights.

M	v_k	w_k
3	−1.22474487	0.29540898
	0.0	1.18163590
	1.22474487	0.29540898
4	−1.65068012	0.08131284
	−0.52464762	0.80491409
	0.52464762	0.80491409
	1.65068012	0.08131284
5	−2.02018287	0.01995324
	−0.95857246	0.39361932
	0.0	0.94530872
	0.95857246	0.39361932
	2.02018287	0.01995324

When the quadrature formula is used to evaluate the likelihood (7.4), the approximation takes the form

$$\prod_{i=1}^{K}\left\{\prod_{j=1}^{n_i} c(y_{ij};\sigma^2)\right\}\sum_{k=1}^{M}\frac{1}{\sqrt{\pi}}\exp\left\{-\frac{1}{2\sigma^2}\sum_{j=1}^{n_i} d(y_{ij};\mu_{ij}^{\sqrt{2}\tau v_k})\right\} w_k.$$

The logarithm of the approximate likelihood becomes

$$\begin{aligned}\ell_{gq}(\boldsymbol{\theta}) &= \sum_{i=1}^{K}\sum_{j=1}^{n_i}\log c(y_{ij};\sigma^2) \\ &\quad + \sum_{i=1}^{K}\log\sum_{k=1}^{M}\frac{1}{\sqrt{\pi}} w_k \exp\left\{-\frac{1}{2\sigma^2}\sum_{j=1}^{n_i} d(y_{ij};\mu_{ij}^{\sqrt{2}\tau v_k})\right\}.\end{aligned}$$

Therefore, the estimator of $\boldsymbol{\theta}$ is given by

$$\widehat{\boldsymbol{\theta}}_{gq} = \arg\max_{\boldsymbol{\theta}} \ell_{gq}(\boldsymbol{\theta}).$$

Various optimization procedures are possibly applied to search for the maximizer of the $\ell_{gq}(\boldsymbol{\theta})$. For example, Newton-Raphson algorithm may be implemented in some simple cases. `SAS` PROC NLMIXED uses a dual quasi-Newton algorithm to carry out the maximization. Alternatively, one may use a Gauss-Newton type algorithm introduced in Section 6.5.2. As pointed out before, the advantage of this Gauss-Newton type algorithm is that it does not require the second order derivatives of ℓ_{gq}.

The quadrature numerical method is in principle applicable to an integration of any dimension. However, the related computational burden increases dramatically as the dimension increases. Essentially, the multi-dimensional quadrature formula is based on Cartesian product Gaussian-Hermite quadrature rules (Davis and Rabinowitz, 1984), which carries out 1-dimensional Gauss-Hermite allocation. With the loss of generality, we assume matrix D is diagonal, or the components of $\mathbf{b}_i$ are independent; otherwise, the multivariate normal may be reparametrized as a product of conditional normals through Cholesky decomposition of the variance-covariance matrix. Let $m_j = w_j e^{b_j^2}\sqrt{2D_{jj}}, j = 1, \ldots, q$. Then the q-dimensional integral can be approximated by

$$\int_{\mathcal{R}^q} h(\mathbf{b})db_1 \cdots db_q \approx \sum_{i_q} m_{i_q}^{(q)} \cdots \sum_{i_2} m_{i_2}^{(2)} \sum_{i_1} m_{i_1}^{(1)} h(\sqrt{2D_{11}}b_1, \cdots, \sqrt{2D_{qq}}b_q).$$

There are some caveats for the application of Gauss-Hermite quadrature formula:

(a) the integrand $h^*(\cdot)$ has to be "centered". For example,

$$\int e^{2va-a^2} e^{-v^2} dv = \sqrt{\pi}, \text{ for all } a.$$

If one uses the 5-point quadrature to evaluate this un-centered integrand, the errors are reported in Table 7.3. It is noted that this uncentral parameter a affects the precision of the evaluation substantially.

Table 7.3. Errors in the 5-point quadrature evaluation.

a	Error
$a = 0$	0.0
$a = 1$	0.001
$a = 2$	0.240
$a \to \pm\infty$	$\sqrt{\pi}$

(b) Integrand $h^*(\cdot)$ should be kind of smooth function. The approximation can be very poor if the integrand has jumps. This is because the underlying theory is based on a polynomial-based approximation to an analytic function.

Example 7.1 (Analysis of Multiple Sclerosis Data).

This example illustrates the use of `SAS` PROC NLMIXED to analyze the multiple sclerosis data, which was introduced in Section 1.3.6. In Section 5.6.2, a marginal model was applied to assess the population-average treatment effect, adjusted by baseline disease duration and a second-order polynomial of time. The conclusion was that the test drug does not help to reduce the risk of exacerbation of the disease.

Now the GLMM is applied to fit the data, in order to examine patient-specific treatment effects on the risk of exacerbation. Besides `treatment`, `duration` of disease, `time`, and `squared time`, this GLMM includes an additional baseline covariate initial `EDSS` (Expanded Disability Status Scale) scores, which is regarded as another potential covariate of interest. In this analysis, instead of treating the three dosage levels (placebo, low, and high) as one ordinal covariate, two dummy variables for the treatment are set as follows, with the reference to high dose:

$$\text{Ptrt} = \begin{cases} 1, \text{Placebo} \\ 0, \text{Otherwise}, \end{cases} \quad \text{Ltrt} = \begin{cases} 1, \text{Low Dose} \\ 0, \text{Otherwise}. \end{cases}$$

In order to specify the mixed effects model, in particular the structure of random effects, a preliminary analysis was conducted, in which individual logistic regression was run for each subject and produced 52 sets of estimates of the regression coefficients. After an exhaustive screening, covariate `EDSS` was the only one significant covariate in all individual analyses, suggesting that a good attention should be paid to this variable. Figure 7.3 shows the histograms of estimated intercepts (in the top panel) and of estimated coefficients of `EDSS` (in the bottom panel).

Both histograms appear to be unimodal, and the distribution of intercepts is slightly left skewed while the distribution of EDSS' slopes is a little right skewed. Thus, the distributions of random effects related to these two terms may be assumed to be approximately normal. Moreover, the GLMM may take the form

$$\begin{aligned} \text{logit}\pi_{ij} &= \beta_0 + \beta_1 \text{t}_j + \beta_2 \text{t}_j^2 + \beta_3 \text{EDSS}_i + \beta_4 \text{Ptrt}_i + \beta_5 \text{Ltrt}_i + \beta_6 \text{dur}_i \\ &\quad + b_{i0} + \text{EDSS}_i b_{i1}, \; j = 1, \ldots, 17, i = 1, \ldots, 52, \end{aligned}$$

where $\pi_{ij} = \text{prob}(Y_{ij} = 1|\mathbf{x}_{ij})$ is the probability of exacerbation at the j-th visit for patient i. Three scenarios concerning the structure of the random effects were considered:

(a) only the random intercepts $b_{i0} \sim N(0, D_{11})$ is included;
(b) both random effects b_{i0} and b_{i1} are included, but assumed to be independent, namely

$$(b_{i0}, b_{i1}) \sim \text{MVN}_2\left(\mathbf{0}, \begin{pmatrix} D_{11} & 0 \\ 0 & D_{22} \end{pmatrix}\right);$$

(c) the general case,

$$(b_{i0}, b_{i1}) \sim \text{MVN}_2\left(\mathbf{0}, \begin{pmatrix} D_{11} & D_{12} \\ D_{21} & D_{22} \end{pmatrix}\right).$$

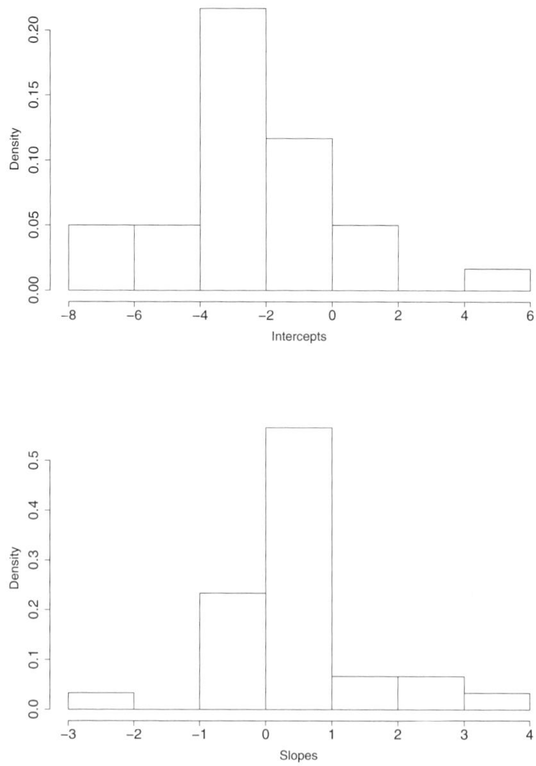

Fig. 7.3. Histograms of estimated intercepts (the top panel) and slopes for covariates EDSS (the bottom panel) obtained in individual logistic regression, each for one subject.

`SAS` PROC NLMIXED was employed to fit the three models and the results were summarized in Table 7.4. For the purpose of comparison, a marginal GLM was also fit using `SAS` PROC GENMOD with interchangeable working correlation structure.

As indicated in Table 7.4, three covariates, time^2, EDSS, and duration of disease, are statistical significant; covariate time is marginally significant. However, treatment is not significant, although the positive sign of the estimates suggests a potential suppression of high dose against the risk of exacerbation. Based on Model (c), one unit increase in EDSS will result in an increase in the odds of exacerbation by $\exp(0.3069) = 1.36$. Also, given the other covariates withheld, the odds ratio of exacerbation between a patient who had a disease history of $(T + 1)$ years and a patient who had a disease

Table 7.4. Estimates and standard errors from the GLMM and the marginal GLM for the multiple sclerosis trial data, using SAS PROC NLMIXED and SAS PROC GENMOD with interchangeable structure.

Parameter	Model(a)	Model(b)	Model(c)	MGLM
intercept	−1.6300(.5177)	−1.6348(.5241)	−1.6499(.5337)	−1.5691(.4834)
time	−0.0307(.0151)	−0.0306(.0151)	−0.0296(.0152)	−0.0302(.0133)
time^2	0.0002(.0001)	0.0002(.0001)	0.0002(.0001)	0.0002(.0001)
EDSS	0.2933(.0850)	0.2935(.0854)	0.3069(.0983)	0.2885(.0891)
Ptrt	0.2488(.3159)	0.2536(.3295)	0.2397(.3210)	0.2442(.2982)
Ltrt	0.4564(.3040)	0.4620(.3248)	0.4503(.3103)	0.4332(.3071)
dur	−0.0435(.0213)	−0.0433(.0215)	−0.0459(.0208)	−0.0446(.0219)
D_{11}	0.1381(.1531)	0.1306(.2148)	0.7943(1.042)	–
D_{12}	–	–	−0.2230(.3347)	–
D_{22}	–	0.0007(.0143)	0.0653(.1007)	–

history of T years is $\exp(-0.0459) = 0.955$. In other words, patients who have longer disease history are more likely to suffer from exacerbation.

For all of the three mixed effects models, the variance components are not statistically significant. This means that the patients recruited in this clinical trial did not express any significant heterogeneity, or the intercepts $\beta_1 + b_{i0}$ and the slope of EDSS $\beta_3 + b_{i1}$ were sampled from a highly homogeneous population. In this case, the marginal analysis based on the MGLM would give almost the same results as those from the mixed effects models, and moreover, the estimates given in the mixed effects models have marginal interpretation.

Precisely, ratios of the estimates from Model (a) and the estimates from the MGLM are ranged from 0.98 to 1.04, compared to the ratio 1.02 given by equation (7.2). One may test the null hypothesis of the ratio equal to 1. The application of the δ-method for the function given in equation (7.2) leads to the asymptotic standard error

$$\mathrm{se}(D_{11}) \times \frac{0.173}{\sqrt{0.346\hat{\sigma}^2 + 1}} = 0.0259.$$

The resulting Z-statistic is 0.91, which does not show any evidence that this ratio would be different from 1.

7.5 Simulated MLE

Recall the likelihood function of a GLMM is $L(\boldsymbol{\theta}) = \prod_{i=1}^{K} L_i(\boldsymbol{\theta})$ where the i-th piece of likelihood function is

$$L_i(\boldsymbol{\theta}) = \int_{\mathcal{R}^q} \prod_{j=1}^{n_i} f(y_{ij}|\mathbf{b}_i) f(\mathbf{b}_i; D) d\mathbf{b}_i.$$

Geyer and Thompson (1992) suggested a direct Monte Carlo approximation of the integral by the method of importance sampling. The importance sampling scheme requires us to choose a proposal distribution, say $h(\cdot)$, of the random effects from which one is able to draw random samples. Denote a random sample of M realizations by

$$\mathbf{b}_i^{(1)}, \ldots, \mathbf{b}_i^{(M)} \overset{i.i.d.}{\sim} h(\mathbf{b}_i).$$

Then the integral can be approximated as follows,

$$\begin{aligned} L_i(\boldsymbol{\theta}) &= \int_{\mathcal{R}^q} \prod_{j=1}^{n_i} f(y_{ij}|\mathbf{b}_i) \frac{f(\mathbf{b}_i; D)}{h(\mathbf{b}_i)} h(\mathbf{b}_i) d\mathbf{b}_i \\ &= \mathrm{E}_h \left\{ \prod_{j=1}^{n_i} f(y_{ij}|\mathbf{b}_i) \frac{f(\mathbf{b}_i; D)}{h(\mathbf{b}_i)} \right\} \\ &\approx \sum_{k=1}^{M} \prod_{j=1}^{n_i} f(y_{ij}|\mathbf{b}_i^{(k)}) \frac{f(\mathbf{b}_i^{(k)}; D)}{h(\mathbf{b}_i^{(k)})}. \end{aligned}$$

Then, the Monte Carlo version of the approximate likelihood is given by

$$L_{mc}(\boldsymbol{\theta}) = \prod_{i=1}^{K} \sum_{l=1}^{M} \prod_{j=1}^{n_i} f(y_{ij}|\mathbf{b}_i^{(l)}) \frac{f(\mathbf{b}_i^{(l)}; D)}{h(\mathbf{b}_i^{(l)})}.$$

Moreover, an estimator of $\boldsymbol{\theta}$ is obtained as

$$\widehat{\boldsymbol{\theta}}_{mc} = \arg\max_{\boldsymbol{\theta}} \log L_{mc}(\boldsymbol{\theta}).$$

An important property of this method is that the estimator $\widehat{\boldsymbol{\theta}}_{mc}$ is always consistent regardless of the choice of the importance sampling distribution $h(\mathbf{b}_i)$. However, the Monte Carlo approximation of the integral is sensitive to the choice of the $h(\mathbf{b}_i)$, which affects the precision of the asymptotic standard errors for the resulting estimates. Results can be highly variable for choice far from the optimal choice $h(\mathbf{b}_i) = f_{\mathbf{b}_i|\mathbf{Y}}(\mathbf{b}_i|\mathbf{Y})$, the conditional density of the random effects given the data (or the posterior). This optimal proposal density effectively gives a zero variance estimator of the Monte Carlo approximation of the form:

$$\sum_{k=1}^{M} \prod_{j=1}^{n_i} f(y_{ij}|\mathbf{b}_i^{(k)}) \frac{f(\mathbf{b}_i^{(k)}; D)}{h(\mathbf{b}_i^{(k)})}.$$

In addition, the Monte Carlo simulation brings extra variation/noise into the procedure of MLE, so the resulting standard errors (or asymptotic covariance matrix) of the estimates can be inflated. Booth and Hobert (1999) considered a logistic random effects model, in which they compared the simulated MLE and Monte Carlo EM algorithm. It is found that the Monte Carlo EM algorithm that will be presented in Section 7.7 performs better than the simulated MLE method.

7.6 Conditional Likelihood Estimation

Now consider the conditional likelihood approach (McCullagh and Nelder, 1989). The basic idea is to treat $\mathbf{b}_i$ as nuisance parameters, and to construct the so-called conditional likelihood, conditioning on the sufficient statistics for the $\mathbf{b}_i$. For the ease of exposition, this section concerns only a special subclass of discrete ED models with the dispersion parameter $\sigma^2 = 1$. It includes Poisson and Bernoulli distributions.

Treat $\mathbf{b} = (\mathbf{b}_1^T, \ldots, \mathbf{b}_K^T)^T$ as fixed parameters. The likelihood function for parameters $\boldsymbol{\theta} = (\boldsymbol{\beta}, \mathbf{b})$ is

$$L(\boldsymbol{\theta}) = \prod_{i=1}^{K}\prod_{j=1}^{n_i} f(y_{ij}|\boldsymbol{\beta}, \mathbf{b}_i) \propto \prod_{i=1}^{K}\prod_{j=1}^{n_i} \exp\left\{\theta_{ij} y_{ij} - \kappa(\theta_{ij})\right\}$$

where under the canonical link function g, the canonical parameter is given by

$$\begin{aligned}\theta_{ij} &= \tau^{-1}(\mu_{ij}) = \tau^{-1} \circ g^{-1}(\mathbf{x}_{ij}^T\boldsymbol{\beta} + \mathbf{z}_{ij}^T\mathbf{b}_i) \\ &= \mathbf{x}_{ij}^T\boldsymbol{\beta} + \mathbf{z}_{ij}^T\mathbf{b}_i.\end{aligned}$$

So, the likelihood is proportional to

$$L(\boldsymbol{\theta}) \propto \exp\left\{\boldsymbol{\beta}^T \sum_{ij} \mathbf{x}_{ij} y_{ij} + \sum_i \mathbf{b}_i^T \sum_j \mathbf{z}_{ij} y_{ij} - \sum_{ij} \kappa(\theta_{ij})\right\}.$$

It follows immediately that the sufficient statistics for $\boldsymbol{\beta}$ and $\mathbf{b}_i$ are $\sum_{ij} \mathbf{x}_{ij} y_{ij}$ and $\sum_j \mathbf{z}_{ij} y_{ij}$, respectively.

In the context of discrete ED models, given data $\mathbf{y}_i$, denote $\mathbf{a}_i = \sum_j \mathbf{x}_{ij} y_{ij}$ and $\mathbf{u}_i = \sum_j \mathbf{z}_{ij} y_{ij}$. Then, for subject i the conditional probability mass function is

$$
\begin{aligned}
f(\mathbf{y}_i | \sum_j \mathbf{z}_{ij} Y_{ij} = \mathbf{u}_i, \boldsymbol{\beta}) &= \frac{\mathrm{P}(\mathbf{Y}_i = \mathbf{y}_i, \sum_j \mathbf{z}_{ij} Y_{ij} = \mathbf{u}_i; \boldsymbol{\beta}, \mathbf{b}_i)}{\mathrm{P}(\sum_j \mathbf{z}_{ij} Y_{ij} = \mathbf{u}_i; \boldsymbol{\beta}, \mathbf{b}_i)} \\
&= \frac{\mathrm{P}(\sum_j \mathbf{x}_{ij} Y_{ij} = \mathbf{a}_i, \sum_j \mathbf{z}_{ij} Y_{ij} = \mathbf{u}_i; \boldsymbol{\beta}, \mathbf{b}_i)}{\mathrm{P}(\sum_j \mathbf{z}_{ij} Y_{ij} = \mathbf{u}_i; \boldsymbol{\beta}, \mathbf{b}_i)} \\
&= \frac{\sum_{R_{i1}} \exp(\boldsymbol{\beta}^T \mathbf{a}_i + \mathbf{b}_i^T \mathbf{u}_i)}{\sum_{R_{i2}} \exp(\boldsymbol{\beta}^T \sum_j \mathbf{x}_{ij} y_{ij} + \mathbf{b}_i^T \mathbf{u}_i)} \\
&= \frac{\sum_{R_{i1}} \exp(\boldsymbol{\beta}^T \mathbf{a}_i)}{\sum_{R_{i2}} \exp(\boldsymbol{\beta}^T \sum_j \mathbf{x}_{ij} y_{ij})}
\end{aligned}
$$

where R_{i1} and R_{i2} are two sub-sample spaces defined by

$$
\begin{aligned}
R_{i1} &= \left\{ \mathbf{y}_i | \ \sum_j \mathbf{x}_{ij} y_{ij} = \mathbf{a}_i, \sum_j \mathbf{z}_{ij} y_{ij} = \mathbf{b}_i \right\}, \\
R_{i2} &= \left\{ \mathbf{y}_i | \ \sum_j \mathbf{z}_{ij} y_{ij} = \mathbf{u}_i \right\}.
\end{aligned}
$$

This implies that the conditional likelihood for $\boldsymbol{\beta}$

$$
L(\boldsymbol{\beta} | \sum_j \mathbf{z}_{ij} Y_{ij} = \mathbf{u}_i) = \prod_{i=1}^{K} \frac{\sum_{R_{i1}} \exp(\boldsymbol{\beta}^T \mathbf{a}_i)}{\sum_{R_{i2}} \exp(\boldsymbol{\beta}^T \sum_j \mathbf{x}_{ij} y_{ij})}.
$$

The MLE of $\boldsymbol{\beta}$ is given by

$$
\widehat{\boldsymbol{\beta}} = \arg\max_{\boldsymbol{\beta}} \log L(\boldsymbol{\beta} | \sum_j \mathbf{z}_{ij} Y_{ij} = \mathbf{u}_i).
$$

Example 7.2 (Logistic Mixed Effects Model).

Consider a logistic GLMM with only the random intercepts given as follows:

$$
\mathrm{logit}(\pi_{ij}^b) = \beta_0 + b_i + \mathbf{x}_{ij}^T \boldsymbol{\beta},
$$

where $Y_{ij} | b_i \sim \mathcal{B}(1, \pi_{ij}^b)$ for a binary correlated data. Let $\gamma_i = \beta_0 + b_i$. Then the likelihood function can be expressed

$$
L(\boldsymbol{\theta}) = \prod_i \exp \left[\gamma_i \sum_j y_{ij} + \left(\sum_j y_{ij} \mathbf{x}_{ij}^T \right) \boldsymbol{\beta} - \sum_j \log \left\{ 1 + \exp(\gamma_i + \mathbf{x}_{ij}^T \boldsymbol{\beta}) \right\} \right].
$$

Let $\mathbf{y}_i^o$ be the vector of observed outcomes of $\mathbf{Y}_i$ for subject i. Accordingly, let $\mathbf{a}_i^o = \sum_j \mathbf{x}_{ij} y_{ij}^o$ and $y_{i+}^o = \sum_j y_{ij}^o$. Note that in this simple model, $\mathbf{z}_{ij} = 1$ for all (i, j), which implies that

$$R_{i2} = \left\{ \mathbf{y}_i \mid \sum_j y_{ij} = y_{i+}^o \right\}.$$

Clearly, th subspace R_{i2} contains $\binom{n_i}{y_{i+}^o}$ many elements, each satisfying $\sum_j y_{ij} = y_{i+}^o$, because y_{ij} is binary of either 0 or 1. Likewise, the other subspace is

$$R_{i1} = \left\{ \mathbf{y}_i \mid \sum_j \mathbf{x}_{ij} y_{ij} = \mathbf{a}_i^o, \sum_j y_{ij} = y_{i+}^o \right\}.$$

It is easy to see that in R_{i1} there exists a unique configuration that satisfies the two conditions, provided that $\mathbf{x}_{ij}$ are time-dependent covariates and $\left(X_i^T X_i\right)^{-1}$ exists. Obviously, this unique solution must equal to the observed $\mathbf{y}_i^o$. Therefore, in such a setting the conditional likelihood function reduces to

$$L(\boldsymbol{\beta}|\mathbf{a}_i^o, y_{i+}^o, i = 1, \ldots, K) = \prod_i \frac{\exp(\boldsymbol{\beta}^T \mathbf{a}_i^o)}{\sum_{R_{i2}} \exp(\boldsymbol{\beta}^T \sum_{k=1}^{y_{i+}^o} \mathbf{x}_{ik} y_{ik})}. \tag{7.7}$$

Maximizing this conditional likelihood (7.7) will lead to the MLE of the $\boldsymbol{\beta}$. In this approach, neither the intercept parameter β_0 nor the random intercept b_i is estimated. Breslow and Day (1980) derived a conditional likelihood of the same form as (7.7) in the context of stratified case-control studies. The numerical methods suggested in their paper can be applied to obtain the estimate of $\boldsymbol{\beta}$.

Another common scenario is that $\mathbf{x}_{ij}$ are all time-independent, namely $\mathbf{x}_{ij} = \mathbf{x}_i$. For this case, R_{i1} is identical to the configuration of R_{i2}, as long as $\mathbf{x}_i^T \mathbf{x}_i \neq 0$. This is because $\sum_j \mathbf{x}_i y_{ij} = \mathbf{a}_i^o = \mathbf{x}_i \sum_j y_{ij}^o$, which leads to $\sum_j y_{ij} = \sum_j y_{ij}^o = y_{i+}^o$. As a result, the first condition in the set R_{i1} is coincident with its second condition, which is the same as that given in set R_{i2}.

The focus of the conditional likelihood approach is to infer the fixed effects $\boldsymbol{\beta}$. It conditions out the random effects through the conditional likelihood on the sufficient statistics, so it does not provide estimation for the random effects. This method does not estimate any variance components parameters, either. An advantage of this conditional likelihood estimation is that it does not need any distributional assumptions for the random effects.

7.7 MLE Based on EM Algorithm

Treat the random effects $\mathbf{b}_i$ as missing data from $\text{MVN}_q(0, D)$. Then, EM-algorithm can be applied to carry out the likelihood inference. Readers who are not familiar with EM algorithm may first read Section 13.4.3 for an introduction in the context of maximum likelihood estimation in the presence of missing data. In effect, this method is relatively straightforward for the class of ED models, in which the score functions are in a linear form of residuals $y_i - \mu_i$. Hence this section focuses only on the ED models. Let $\boldsymbol{\theta} = (\boldsymbol{\beta}, D)$. Given the augmented data $(\mathbf{y}_i, \mathbf{b}_i), i = 1, \ldots, K$, the augmented likelihood function is

$$L(\boldsymbol{\theta}|\mathbf{y}, \mathbf{b}) = \prod_{i=1}^{K}\prod_{j=1}^{n_i} f(y_{ij}|\mathbf{b}_i, \boldsymbol{\beta}) f(\mathbf{b}_i; D)$$
$$\propto \prod_{i=1}^{K}\prod_{j=1}^{n_i} \exp\left[\lambda \left\{\theta_{ij} y_{ij} - \kappa(\theta_{ij})\right\}\right] f(\mathbf{b}_i; D),$$

where the additive ED model density is used.

With the choice of canonical link function g, the log-likelihood function then takes the form

$$\ell(\boldsymbol{\theta}) = \lambda \sum_{i=1}^{K} \left\{ \sum_{j=1}^{n_i} \theta_{ij}^b y_{ij} - \sum_{j=1}^{n_i} \kappa(\theta_{ij}^b) \right\} + \sum_{i=1}^{K} \log f(\mathbf{b}_i; D)$$
$$= \lambda \sum_{i=1}^{K} \left\{ \boldsymbol{\beta}^T \sum_{j=1}^{n_i} \mathbf{x}_{ij} y_{ij} + \mathbf{b}_i^T \sum_{j=1}^{n_i} \mathbf{z}_{ij} y_{ij} - \sum_{j=1}^{n_i} \kappa(\mathbf{x}_{ij}^T \boldsymbol{\beta} + \mathbf{z}_{ij}^T \mathbf{b}_i) \right\}$$
$$+ \sum_{i=1}^{K} \log f(\mathbf{b}_i; D).$$

It follows that the score equation for $\boldsymbol{\beta}$ is

$$\frac{\partial \ell(\boldsymbol{\theta})}{\partial \boldsymbol{\beta}} = \lambda \sum_{i=1}^{K}\sum_{j=1}^{n_i} \mathbf{x}_{ij} \left\{ y_{ij} - \mu_{ij}(\boldsymbol{\beta}, \mathbf{b}_i) \right\},$$

where

$$\mu_{ij}(\boldsymbol{\beta}, \mathbf{b}_i) = \mathrm{E}(Y_{ij}|\mathbf{b}_i) = g^{-1}(\mathbf{x}_{ij}^T \boldsymbol{\beta} + \mathbf{z}_{ij}^T \mathbf{b}_i).$$

Also, the augmented data score function for $D = D(\boldsymbol{\tau})$ is

$$\frac{\partial \ell(\boldsymbol{\theta})}{\partial \boldsymbol{\tau}} = \frac{\partial}{\partial \boldsymbol{\tau}} \left(-\frac{K}{2} \log |D| - \frac{1}{2} \sum_{i=1}^{K} \mathbf{b}_i^T D^{-1} \mathbf{b}_i \right)$$
$$= -\frac{K}{2} \mathrm{tr}\left(D^{-1} \frac{\partial D}{\partial \boldsymbol{\tau}} \right) + \frac{1}{2} \sum_{i=1}^{K} \mathbf{b}_i^T D^{-1} \frac{\partial D}{\partial \boldsymbol{\tau}} D^{-1} \mathbf{b}_i.$$

The E-step pertains to the calculation of an objective function, under "posterior" $f\left(\mathbf{b}_i|\mathbf{y}; \boldsymbol{\theta}^{(k)}\right)$, where $\boldsymbol{\theta}^{(k)}$ is an update value of the parameter from the previous iteration k. That is,

$$
\begin{aligned}
Q(\boldsymbol{\theta};\boldsymbol{\theta}^{(k)}) &= \int \log f(\boldsymbol{\theta}|\mathbf{y},\mathbf{b}) f(\mathbf{b}|\mathbf{y},\boldsymbol{\theta}^{(k)}) d\mathbf{b} \\
&\propto \lambda \sum_{i=1}^{K} \boldsymbol{\beta}^T \sum_{j=1}^{n_i} \mathbf{x}_{ij} y_{ij} - \lambda \sum_{j=1}^{n_i} \mathrm{E}_{\mathbf{b}_i|\mathbf{y}_i,\boldsymbol{\theta}^{(k)}} \kappa(\mathbf{x}_{ij}^T\boldsymbol{\beta} + \mathbf{z}_{ij}^T \mathbf{b}_i) \\
&\quad - \frac{K}{2}\log|D| - \frac{K}{2}\sum_{i=1}^{K} \mathrm{E}_{\mathbf{b}_i|\mathbf{y}_i,\boldsymbol{\theta}^{(k)}} \mathbf{b}_i^T D^{-1} \mathbf{b}_i .
\end{aligned}
$$

In order to evaluate the Q function, it is necessary to compute two expectations:

$$
\begin{aligned}
\mathrm{E}_{\mathbf{b}_i|\mathbf{y}_i,\boldsymbol{\theta}^{(k)}} \left\{\kappa(\mathbf{x}_{ij}^T\boldsymbol{\beta} + \mathbf{z}_{ij}^T\mathbf{b}_i)\right\} &= \int \kappa(\mathbf{x}_{ij}^T\boldsymbol{\beta} + \mathbf{z}_{ij}^T\mathbf{b}_i) f(\mathbf{b}_i|\mathbf{y}_i,\boldsymbol{\theta}^{(k)}) d\mathbf{b}_i \\
\mathrm{E}_{\mathbf{b}_i|\mathbf{y}_i,\boldsymbol{\theta}^{(k)}} \left(\mathbf{b}_i^T D^{-1}\mathbf{b}_i\right) &= \int \mathbf{b}_i^T D^{-1}\mathbf{b}_i f(\mathbf{b}_i|\mathbf{y}_i,\boldsymbol{\theta}^{(k)}) d\mathbf{b}_i .
\end{aligned}
$$

The M-step involves the maximization of the Q function with respect to the parameter $\boldsymbol{\theta}$ to obtain a new update value $\boldsymbol{\theta}^{(k+1)}$. That is,

$$
\boldsymbol{\theta}^{(k+1)} = \arg\max_{\boldsymbol{\theta}} Q(\boldsymbol{\theta};\boldsymbol{\theta}^{(k)}).
$$

This update can be done by solving the following two equations:

$$
\begin{aligned}
\frac{\partial Q(\boldsymbol{\theta};\boldsymbol{\theta}^{(k)})}{\partial \boldsymbol{\beta}} &= \lambda \sum_{i=1}^{K}\sum_{j=1}^{n_i} \mathbf{x}_{ij} \left\{ y_{ij} - \mathrm{E}_{\mathbf{b}_i|\mathbf{y}_i,\boldsymbol{\theta}^{(k)}} (\mu_{ij}^b) \right\} = 0 \\
\frac{\partial Q(\boldsymbol{\theta};\boldsymbol{\theta}^{(k)})}{\partial \boldsymbol{\tau}} &= -\frac{K}{2}\mathrm{tr}\left(D^{-1}\frac{\partial D}{\partial \boldsymbol{\tau}}\right) \\
&\quad + \frac{1}{2}\mathrm{tr}\left\{ D^{-1}\frac{\partial D}{\partial \boldsymbol{\tau}} D^{-1} \left(\sum_{i=1}^{K} \mathrm{E}_{\mathbf{b}_i|\mathbf{y}_i,\boldsymbol{\theta}^{(k)}} (\mathbf{b}_i\mathbf{b}_i^T) \right) \right\} \\
&= 0.
\end{aligned}
$$

Here the following two additional conditional expectations need to be computed in the E-step:

$$
\begin{aligned}
\mathrm{E}_{\mathbf{b}_i|\mathbf{y}_i,\boldsymbol{\theta}^{(k)}} \left(\mu_{ij}^b\right) &= \int g^{-1}(\mathbf{x}_{ij}^T\boldsymbol{\beta} + \mathbf{z}_{ij}^T\mathbf{b}_i) f(\mathbf{b}_i|\mathbf{y}_i,\boldsymbol{\theta}^{(k)}) d\mathbf{b}_i \\
\mathrm{E}_{\mathbf{b}_i|\mathbf{y}_i,\boldsymbol{\theta}^{(k)}} \left(\mathbf{b}_i\mathbf{b}_i^T\right) &= \mathrm{Var}_{\mathbf{b}_i|\mathbf{y}_i,\boldsymbol{\theta}^{(k)}}(\mathbf{b}_i) + \mathrm{E}_{\mathbf{b}_i|\mathbf{y}_i,\boldsymbol{\theta}^{(k)}}(\mathbf{b}_i)\mathrm{E}_{\mathbf{b}_i|\mathbf{y}_i,\boldsymbol{\theta}^{(k)}}(\mathbf{b}_i)^T .
\end{aligned}
$$

Iterate the E- and M- steps until a certain convergence criterion is satisfied.

In many cases, these conditional expectations in the E-step have no closed form expressions, and Monte Carlo simulation is suggested by Wei and Tanner (1990) to approximate them. The resulting algorithm is referred to as the Monte Carlo EM(MCEM) algorithm. That is, at each E-step, draw K random samples of size M each, $\mathbf{b}_i^{(1)}, \ldots, \mathbf{b}_i^{(M)} \overset{i.i.d.}{\sim} f(\mathbf{b}_i|\mathbf{y}_i, \boldsymbol{\theta}^{(k)})$, and then calculate the following sample means:

$$\mathrm{E}_{\mathbf{b}_i|\mathbf{y}_i,\boldsymbol{\theta}^{(k)}}\left(\mu_{ij}^b\right) \approx \frac{1}{M}\sum_{l=1}^{M} g^{-1}\left(\mathbf{x}_{ij}^T\boldsymbol{\beta}^{(k)} + \mathbf{z}_{ij}^T\mathbf{b}_i^{(l)}\right), \; i = 1, \ldots, K,$$

$$\mathrm{E}_{\mathbf{b}_i|\mathbf{y}_i,\boldsymbol{\theta}^{(k)}}(\mathbf{b}_i) \approx \overline{\mathbf{b}}_i = \frac{1}{M}\sum_{l=1}^{M} \mathbf{b}_i^{(l)}, \; i = 1, \ldots, K,$$

$$\mathrm{Var}_{\mathbf{b}_i|\mathbf{y}_i,\boldsymbol{\theta}^{(k)}}(\mathbf{b}_i) \approx \frac{1}{M}\sum_{l=1}^{M} (\mathbf{b}_i^{(l)} - \overline{\mathbf{b}}_i)(\mathbf{b}_i^{(l)} - \overline{\mathbf{b}}_i)^T, \; i = 1, \ldots, K.$$

Drawing random variates from the posteriors of the random effects may not always be trivial. When the dimension q is high, sampling can be difficult. McCulloch (1997) suggested a Gibbs sampling scheme to carry out Monte Carlo simulation from a Markov chain, with stationary distributions equal to the posteriors. This Markov chain Monte Carlo method basically turns a high-dimensional sampling problem into several low-dimensional sampling problems via Markov chains.

Any alternative to the Monte Carlo approach, the EM-algorithm may also be implemented through the iterative weighted least squares algorithm proposed by Green (1987). This is an approximate method, in which the theory of linear predictor (Chapter 9) is employed to estimate the expectations required in the E-step based only on the first two moments. This approximation works reasonably well for the estimation of the fixed effects $\boldsymbol{\beta}$, but may incur some bias in the estimation of variance component parameters.

Suppose that iteration k is completed, and that now one likes to update the estimates for $\boldsymbol{\beta}, D$ as well as $\mathbf{b}_i$ at iteration $k+1$. Following Harville's (1977) theory for the linear mixed-effects models, define a surrogate response

$$Y_{ij}^* = g(\mu_{ij}^b) + (Y_{ij} - \mu_{ij}^b)\dot{g}(\mu_{ij}^b),$$

and the covariance matrix of $\mathbf{Y}_i^* = (Y_{i1}^*, \ldots, Y_{in_i}^*)^T$ is

$$C_i = Q_i + Z_i D Z_i^T,$$

where $Z_i^T = (\mathbf{z}_{i1}, \ldots, \mathbf{z}_{in_i}), Q_i = \mathrm{diag}\left\{v_{ij}^b[\dot{g}(\mu_{ij}^b)]^2\right\}$, with $v_{ij}^b = \mathrm{Var}(Y_{ij}|\mathbf{b}_i) = \sigma^2 V(\mu_{ij}^b)$. Here $\dot{g}$ denotes the first order derivatives of link function g. Note that Q_i is the covariance matrix of $\boldsymbol{\varepsilon}_i^*$ whose j element is $(Y_{ij} - \mu_{ij}^b)\dot{g}(\mu_{ij}^b)$, and $V(\cdot)$ is the unit variance function of the data distribution $f(y_{ij}|\mathbf{b}_i)$. Then a linear mixed-effects model of the surrogate data is given by

$$\mathbf{Y}_i^* = X_i^T\boldsymbol{\beta} + Z_i\mathbf{b}_i + \boldsymbol{\varepsilon}_i^*$$

where $\boldsymbol{\varepsilon}_i^*$ has mean $\mathbf{0}$ and covariance matrix Q_i. According to Harville (1977), for fixed D, the update of $\boldsymbol{\beta}$ at iteration $k+1$ is obtained as the solution to equation

$$\left(\sum_i X_i^T C_i^{-1} X_i\right) \boldsymbol{\beta}^{(k+1)} = \sum_i X_i^T C_i^{-1} \mathbf{Y}_i^*,$$

where $\mathbf{Y}_i^*$ and C_i are updated by using the results from the previous iteration k.

In the mean time, updating $\mathbf{b}_i$ is given by the following best linear unbiased predictor (BLUP),

$$\widehat{\mathbf{b}}_i^{(k+1)} = D Z_i C_i^{-1} \left(\mathbf{Y}_i^* - X_i \widehat{\boldsymbol{\beta}}^{(k+1)}\right). \tag{7.8}$$

See details concerning BLUP in Chapter 9.

It follows immediately that an approximate covariance for $\widehat{\boldsymbol{\beta}}$ might be $\left(\sum_i X_i^T C_i^{-1} X_i\right)^{-1}$.

Updating the D matrix can be carried out by

$$\begin{aligned}
\widehat{D}^{(k+1)} &= K^{-1} \sum_i \mathrm{E}_{\mathbf{b}} \left(\mathbf{b}_i \mathbf{b}_i^T | \mathbf{y}_i, \boldsymbol{\theta}^{(k)}\right) \\
&= K^{-1} \sum_i \mathrm{E}_{\mathbf{b}} \left(\mathbf{b}_i | \mathbf{y}_i, \boldsymbol{\theta}^{(k)}\right) \mathrm{E}_{\mathbf{b}} \left(\mathbf{b}_i | \mathbf{y}_i, \boldsymbol{\theta}^{(k)}\right)^T \\
&\quad + K^{-1} \sum_i \mathrm{Var}_{\mathbf{b}} \left(\mathbf{b}_i | \mathbf{y}_i, \boldsymbol{\theta}^{(k)}\right),
\end{aligned}$$

where $\mathrm{E}_{\mathbf{b}}\left(\mathbf{b}_i | \mathbf{y}_i, \boldsymbol{\theta}^{(k)}\right)$ is estimated by the BLUP given in (7.8), and $\mathrm{Var}_{\mathbf{b}}(\mathbf{b}_i | \mathbf{y}_i)$ is estimated by the Mean Squared Error (MSE) of the BLUP, i.e., $(Z_i Q_i^{-1} Z_i + D^{-1})^{-1}$.

This version of approximate EM-algorithm may start with $\mathbf{b}_i = \mathbf{0}$ and $\boldsymbol{\beta}^0$ that is the estimate obtained from a cross-sectional GLM fit. Then the algorithm proceeds iteratively until convergence. This EM algorithm can be applied to obtain parameter estimation in nonlinear mixed effects models, see, for example, Lindstrom and Bates (1990).

7.8 Approximate Inference: PQL and REML

In comparison to the methods discussed above, this approximate inference can be developed in the GLMMs for the dispersion models. Both penalized quasi-likelihood (PQL) and restricted maximum likelihood (REML) can bypass the analytical difficulty arising from the nonlinearity of the score functions. More importantly, this PQL/REML method works for any dimensions of the random effects. Therefore, it is a rather general inference approach. It is worth pointing out that this approximate inference is similar to the MLE based only on 1-point quadrature evaluation of the integral in the likelihood.

Breslow and Clayton (1993) suggested using PQL to jointly estimating the fixed effects parameter $\boldsymbol{\beta}$ and random effects $\mathbf{b}_i$, and then applying the

REML to estimate the variance parameter $\boldsymbol{\tau}$. Their development was initially established within the family of ED models, but this inference tool is in fact available for the broader family of DM models. SAS PROC GLIMMIX furnishes PQL/REML inference in the GLMMs with ED models.

Denote $\mathbf{b} = (\mathbf{b}_1^T, \mathbf{b}_2^T, \cdots, \mathbf{b}_K^T)^T$ and $\mathbf{y} = (\mathbf{y}_1^T, \mathbf{y}_2^T, \cdots, \mathbf{y}_K^T)^T$ with $\mathbf{y}_i = (y_{i1}, y_{i2}, \cdots, y_{in_i})^T$ for each subject i. It is easy to see that the augmented likelihood based on $(\mathbf{y}, \mathbf{b})$ for parameters $\boldsymbol{\theta} = (\boldsymbol{\beta}, \boldsymbol{\tau})$ is proportional to

$$|D(\boldsymbol{\tau})|^{-K/2} \exp\left\{ -\frac{1}{2\sigma^2} \sum_{i=1}^{K} \sum_{j=1}^{n_i} d(y_{ij}, \mu_{ij}^b) - \frac{1}{2} \sum_{i=1}^{K} \mathbf{b}_i^T D^{-1}(\boldsymbol{\tau}) \mathbf{b}_i \right\},$$

where for convenience the dispersion parameter σ^2 is assumed to be known, such as in Binomial or Poisson mixed effects models, and otherwise it will be estimated separately with no use of the likelihood as done in Section 2.6.1.

Thus, subject to a constant the log-likelihood of $\boldsymbol{\theta}$ is given by

$$\ell(\boldsymbol{\theta}) \propto -\frac{K}{2} \log |D(\boldsymbol{\tau})| + \log \int_{\mathcal{R}^{Kq}} \mathrm{e}^{-\kappa(\mathbf{b})} d\mathbf{b}, \tag{7.9}$$

where $|\cdot|$ denotes the determinant of a matrix, and the exponent takes the form

$$\kappa(\mathbf{b}) = \frac{1}{2\sigma^2} \sum_{i=1}^{K} \sum_{j=1}^{n_i} d(y_{ij}, \mu_{ij}^b) + \frac{1}{2} \sum_{i=1}^{K} \mathbf{b}_i^T D^{-1}(\boldsymbol{\tau}) \mathbf{b}_i.$$

To deal with the Kq-dimensional integration in (7.9), Breslow and Clayton suggested to invoke Laplacian method for integral approximation at the saddlepoint of the $\kappa(\cdot)$ function. The basic idea of the Laplacian method is to approximate the $\kappa(\cdot)$ function by its second order Taylor expansion evaluated at the saddlepoint $\tilde{\mathbf{b}} = (\tilde{\mathbf{b}}_1^T, \ldots, \tilde{\mathbf{b}}_K^T)^T$ with the respective $\tilde{\mathbf{b}}_i$ being the solutions to the equations

$$\dot{\kappa}_{b_i}(\mathbf{b}) = 0, \ i = 1, 2, \cdots, K.$$

Note that in such a second order Taylor expansion, the quadratic term resembles a multivariate normal density and the linear term is zero at solution $\tilde{\mathbf{b}}$. This observation immediately leads to

$$\ell(\boldsymbol{\theta}) \approx -\frac{K}{2} \log |D(\boldsymbol{\tau})| - \frac{1}{2} \log |\ddot{\kappa}(\tilde{\mathbf{b}})| - \kappa(\tilde{\mathbf{b}}),$$

where the approximation pertains to the ignorance of the third or higher order terms in the Taylor expansion of $\kappa(\mathbf{b})$ around the $\tilde{\mathbf{b}}$. Breslow and Lin (1995) showed that PQL may lead to estimators that are asymptotically biased. Among several authors, Lin and Breslow (1996) and Raudenbush et al. (2000) derived PQL/REML using the high-order multivariate Laplacian approximation. This section only focuses on the second order approximation.

Some simple algebra gives the following derivatives,

$$
\begin{aligned}
\dot{\kappa}_{b_i}(\mathbf{b}) &= \frac{1}{2\sigma^2}\sum_{j=1}^{n_i} \dot{d}_{b_i}(y_{ij}, \mu_{ij}^b) + D^{-1}(\boldsymbol{\tau})\mathbf{b}_i, \\
\ddot{\kappa}_{b_i b_i^T}(\mathbf{b}) &= \frac{1}{2\sigma^2}\sum_{j=1}^{n_i} \ddot{d}_{b_i b_i^T}(y_{ij}, \mu_{ij}^b) + D^{-1}(\boldsymbol{\tau}), \\
\ddot{\kappa}_{b_i b_{i'}^T}(\mathbf{b}) &= 0, \ i \neq i'.
\end{aligned} \tag{7.10}
$$

Also it is easy to derive

$$
\ddot{d}_{b_i b_i^T}(y_{ij}, \mu_{ij}^b) = \frac{\ddot{d}_{\mu_{ij}\mu_{ij}}(y_{ij}, \mu_{ij}^b)}{\left\{\dot{g}\left(\mu_{ij}^b\right)\right\}^2}\mathbf{z}_{ij}^T\mathbf{z}_{ij} + \dot{d}_{b_i}(y_{ij}, \mu_{ij}^b)\frac{\partial}{\partial b_i^T}\left\{\mathbf{z}_{ij}\dot{g}\left(\mu_{ij}^b\right)\right\}, \tag{7.11}
$$

where the conditional expectation of the second term given $\mathbf{b}_i$ is zero because $\mathrm{E}\{\dot{d}_{b_i}(Y_{ij}, \mu_{ij}^b)|\mathbf{b}_i\} = 0$. It follows that

$$
\mathrm{E}\{\ddot{d}_{b_i b_i^T}(Y_{ij}, \mu_{ij}^b)|\mathbf{b}_i\} = w_{ij}\mathbf{z}_{ij}^T\mathbf{z}_{ij}
$$

with the weights w_{ij} equal to

$$
\begin{aligned}
w_{ij} &= \mathrm{E}\left[\frac{\ddot{d}_{\mu_{ij}\mu_{ij}}(Y_{ij}, \mu_{ij}^b)}{\left\{\dot{g}\left(\mu_{ij}^b\right)\right\}^2} \mid \mathbf{b}_i\right] \\
&= \frac{\mathrm{E}\left\{\ddot{d}_{\mu_{ij}\mu_{ij}}(Y_{ij}, \mu_{ij}^b)|\mathbf{b}_i\right\}}{2\sigma^2\left\{\dot{g}\left(\mu_{ij}^b\right)\right\}^2} \\
&= \frac{\mathrm{E}\left\{\dot{d}_{\mu_{ij}}^2(Y_{ij}, \mu_{ij}^b)|\mathbf{b}_i\right\}}{\left\{2\sigma^2\dot{g}\left(\mu_{ij}^b\right)\right\}^2}.
\end{aligned} \tag{7.12}
$$

In expression (7.11), the second term may be ignored due to the fact that it has expectation 0 and thus, in probability, is of lower order than the first term with respect to K. This gives rise to an approximation to (7.10) as follows,

$$
\begin{aligned}
\ddot{\kappa}_{b_i b_i^T}(\mathbf{b}) &\approx \sum_{j=1}^{n_i} \mathbf{z}_{ij}^T w_{ij}\mathbf{z}_{ij} + D^{-1}(\boldsymbol{\tau}) \\
&= Z_i W_i Z_i^T + D^{-1}(\boldsymbol{\tau})
\end{aligned} \tag{7.13}
$$

where $Z_i^T = (\mathbf{z}_{i1}, \cdots, \mathbf{z}_{in_i})$, and $W_i = \mathrm{diag}(w_{i1}, \ldots, w_{in_i})$, an $n_i \times n_i$ diagonal matrix with the j-th diagonal element given by (7.12).

Note that $\ddot{\kappa}(\mathbf{b})$ is a block-diagonal matrix with the i-th block equal to $\ddot{\kappa}_{b_i b_i^T}(\mathbf{b}), i = 1, 2, \cdots, K$, which is approximated according to (7.13). Therefore, the approximate log-likelihood of (7.9) can take the following form:

$$\ell(\boldsymbol{\theta}) \approx -\frac{1}{2}\sum_{i=1}^{K}\log|I + Z_i W_i Z_i^T D(\boldsymbol{\tau})|$$
$$-\frac{1}{2\sigma^2}\sum_{i=1}^{K}\sum_{j=1}^{n_i} d(y_{ij}, \mu_{ij}^b) - \frac{1}{2}\sum_{i=1}^{K}\tilde{\mathbf{b}}_i^T D^{-1}(\boldsymbol{\tau})\tilde{\mathbf{b}}_i. \qquad (7.14)$$

Estimates $(\widehat{\boldsymbol{\beta}}, \widehat{\mathbf{b}}) = (\widehat{\boldsymbol{\beta}}(\boldsymbol{\tau}), \widehat{\mathbf{b}}(\boldsymbol{\tau}))$, with $\widehat{\mathbf{b}}(\boldsymbol{\tau}) = \tilde{\mathbf{b}}(\widehat{\boldsymbol{\beta}}(\boldsymbol{\tau}))$, can be obtained by jointly maximizing Green's (1987) penalized quasi-likelihood of the form (provided that the first derivative of the first term or W_i in (7.14) *w.r.t.* μ_{ij}^b is ignorable)

$$\ell_{pq}(\boldsymbol{\beta}, \mathbf{b}) = -\frac{1}{2\sigma^2}\sum_{i=1}^{K}\sum_{j=1}^{n_i} d(y_{ij}, \mu_{ij}^b) - \frac{1}{2}\sum_{i=1}^{K}\mathbf{b}_i^T D^{-1}(\boldsymbol{\tau})\mathbf{b}_i. \qquad (7.15)$$

Differentiating the ℓ_{pq} with respect to $\boldsymbol{\beta}$ and $\mathbf{b}_i$, respectively, leads to the following quasi-score equations,

$$\sum_{i=1}^{K}\sum_{j=1}^{n_i}\frac{\mathbf{x}_{ij}}{\sigma^2\dot{g}(\mu_{ij}^b)}\delta(y_{ij};\mu_{ij}^b) = 0 \qquad (7.16)$$
$$\sum_{j=1}^{n_i}\frac{\mathbf{z}_{ij}}{\sigma^2\dot{g}(\mu_{ij}^b)}\delta(y_{ij};\mu_{ij}^b) = D^{-1}(\boldsymbol{\tau})\mathbf{b}_i,\ i = 1,\ldots,K, \qquad (7.17)$$

where the deviance score vector $\boldsymbol{\delta}_i = (\delta_{i1}, \cdots, \delta_{in_i})^T$ consists of elements $\delta_{ij} = \delta(y_{ij};\mu_{ij}^b) = -\frac{1}{2}\dot{d}(y_{ij};\mu_{ij}^b)$, $j = 1,\ldots,n_i$, $i = 1,\ldots,K$. It is known that the deviance scores are equal to $(y_{ij} - \mu_{ij}^b)/V(\mu_{ij}^b)$ for the ED models, but they in general are nonlinear functions in both y_{ij} and μ_{ij}^b.

An iterative procedure is used to solve the above equations for estimates $\hat{\boldsymbol{\beta}}$ and $\hat{\mathbf{b}}_i, i = 1,\ldots,K$. Given the values of these parameters from the previous iteration, define a surrogate vector $\mathbf{y}_i^* = (y_{i1}^*, \ldots, y_{in_i}^*)^T$ with the j-th component equal to

$$y_{ij}^* = \eta_{ij}^b + \frac{2\dot{g}(\mu_{ij}^b)}{\mathrm{E}\left\{\ddot{d}_{\mu_{ij}\mu_{ij}}(Y_{ij},\mu_{ij}^b)|\mathbf{b}_i\right\}}\delta(y_{ij};\mu_{ij}^b)$$
$$= \eta_{ij}^b + \frac{4\sigma^2\dot{g}(\mu_{ij}^b)}{\mathrm{E}\left\{\dot{d}_{\mu_{ij}}^2(Y_{ij},\mu_{ij}^b)|\mathbf{b}_i\right\}}\delta(y_{ij};\mu_{ij}^b).$$

Numerically, updating the parameters is equivalent to fitting a normal linear mixed-effects model,

$$\mathbf{Y}_i^* = X_i^T\boldsymbol{\beta} + Z_i\mathbf{b}_i + \boldsymbol{\varepsilon}_i^*,\ i = 1,\ldots,K,$$

where $\boldsymbol{\varepsilon}_i^* \sim \mathrm{MVN}_{n_i}(\mathbf{0}, W_i^{-1})$, and $\mathbf{b}_1,\ldots,\mathbf{b}_K$ are *i.i.d.* $\mathrm{MVN}_q(\mathbf{0}, D(\boldsymbol{\tau}))$. This update in the framework of LMMs can be done via SAS PROC MIXED.

As a matter of fact, Harville (1977) found that the solution to (7.16) and (7.17) can be obtained by first solving

$$\left(\sum_{i=1}^{K} X_i^T C_i^{-1} X_i\right) \widehat{\boldsymbol{\beta}} = \sum_{i=1}^{K} X_i^T C_i^{-1} \mathbf{Y}_i^*, \tag{7.18}$$

where $C_i = W_i^{-1} + Z_i D(\boldsymbol{\tau}) Z_i^T$ and then setting the BLUP

$$\widehat{\mathbf{b}}_i = D(\boldsymbol{\tau}) Z_i^T C_i^{-1} (\mathbf{Y}_i^* - X_i \widehat{\boldsymbol{\beta}}). \tag{7.19}$$

At the convergence, the approximate covariance for $\widehat{\boldsymbol{\beta}}$ is given by the matrix $\left(\sum_{i=1}^{K} X_i^T C_i^{-1} X_i\right)^{-1}$. It is worth pointing out that the inference functions defined by (7.16) and (7.17) are both insensitive to the variance components parameter $\boldsymbol{\tau}$ and the dispersion parameter σ^2. So, the efficiency of the estimators of $\boldsymbol{\tau}$ and σ^2 would have little effect on the asymptotic covariance of $\widehat{\boldsymbol{\beta}}$.

Now let us turn to the discussion of REML that provides an inference on the variance parameter $\boldsymbol{\tau}$ using an approximate profile quasi-likelihood function. Invoking the small-dispersion asymptotics of the Pearson residual type (Proposition 2.6), the unit deviance may be approximated by a quadratic polynomial function in the neighborhood of the PQL estimates $(\widehat{\boldsymbol{\beta}}, \widehat{\mathbf{b}}_i)$ as follows,

$$d(y_{ij}, \mu_{ij}^b) \simeq \frac{(y_{ij} - \mu_{ij}^b)^2}{V(\mu_{ij}^b)}, \quad \text{for small dispersion } \sigma^2.$$

Similar arguments in Breslow and Clayton (1993) lead to the REML of the form

$$\begin{aligned} \ell_1(\hat{\boldsymbol{\beta}}(\boldsymbol{\tau}), \boldsymbol{\tau}) \approx & -\frac{1}{2}\sum_{i=1}^{K} \log|C_i| - \frac{1}{2}\log\left|\sum_{i=1}^{K} X_i^T C_i^{-1} X_i\right| \\ & -\frac{1}{2}\sum_{i=1}^{K} (\mathbf{Y}_i^* - X_i\widehat{\boldsymbol{\beta}})^T C_i^{-1} (\mathbf{Y}_i^* - X_i\widehat{\boldsymbol{\beta}}). \end{aligned} \tag{7.20}$$

The REML estimate of $\boldsymbol{\tau}$ can then be obtained by solving the following estimating equations,

$$\mathbf{S}(\tau_k) = \frac{1}{2}\sum_{i=1}^{K}\left[(\mathbf{y}_i^* - X_i\widehat{\boldsymbol{\beta}})^T C_i^{-1} \frac{\partial C_i}{\partial \tau_k} C_i^{-1} (\mathbf{Y}_i^* - X_i\widehat{\boldsymbol{\beta}}) - \operatorname{tr}\left(P_i \frac{\partial C_i}{\partial \tau_k}\right)\right] = 0, \tag{7.21}$$

where $P_i = C_i^{-1} - C_i^{-1} X_i \left(\sum_{i=1}^{K} X_i^T C_i^{-1} X_i\right)^{-1} X_i^T C_i^{-1}$. The standard errors of the REML estimates $\hat{\boldsymbol{\tau}}$ are computed from the Fisher information matrix $\mathbf{j}$ whose (k, l)-th element is given by

$$[\mathbf{j}]_{kl} = \frac{1}{2}\sum_{i=1}^{K} \text{tr}\left(P_i \frac{\partial C_i}{\partial \tau_k} P_i \frac{\partial C_i}{\partial \tau_l}\right). \tag{7.22}$$

Note that if τ_k corresponds to the entry (a, b) in D, then $\partial C_i/\partial \tau_k$ equals to $[Z_i]_a [Z_i]_b^T$, where $[Z_i]_a$ denotes the a-th column of matrix Z_i.

The updating procedure of the Fisher-scoring algorithm takes form

$$\boldsymbol{\tau}^{(k+1)} = \boldsymbol{\tau}^{(k)} + \left\{\mathbf{j}(\boldsymbol{\beta}^{(k+1)}, \boldsymbol{\tau}^{(k)})\right\}^{-1} \mathbf{S}(\boldsymbol{\beta}^{(k+1)}, \boldsymbol{\tau}^{(k)}),$$

which produces the solution to equation (7.21) when convergence is achieved.

An important issue in the implementation of Fisher-scoring algorithm is the efficiency of searching for the solution. In order to ensure that covariance matrix D remains positive-definite along iterations, one may invoke Cholesky decomposition, $D = L^T L$, where the Cholesky factor L is an upper triangular matrix. Let $\boldsymbol{\alpha}$ denote the set of distinct elements of the L. The REML is now applied to $\boldsymbol{\alpha}$ and the related derivative *w.r.t.* $\boldsymbol{\alpha}$ can be derived easily. This reparametrization approach converts a constrained optimization problem to an unconstrained problem, by the insurance of a positive-definite estimate for D. As pointed out in the literature, this decomposition method facilitates numerical stability of the Fisher scoring algorithm (Lindstrom and Bates, 1988).

In a model where the dispersion parameter σ^2 is unknown to be estimated, estimating σ^2 may be separately carried out using the method of moments. To proceed, a certain moment property is needed. For instance, Proposition 2.17 suggests

$$\text{E}\left\{\ddot{d}_{\mu_{ij}\mu_{ij}}(Y_{ij}, \mu_{ij}^b)|\mathbf{b}_i\right\} = \frac{\text{E}\left\{\dot{d}_{\mu_{ij}}(Y_{ij}, \mu_{ij}^b)|\mathbf{b}_i\right\}}{2\sigma^2}.$$

It follows that

$$\sigma^2 = \frac{2\text{E}\left\{r_s^2(Y_{ij}, \mu_{ij}^b)\right\}}{\text{E}\left\{\ddot{d}_{\mu_{ij}\mu_{ij}}(Y_{ij}, \mu_{ij}^b)V(\mu_{ij}^b)\right\}}$$

where $r_s(\cdot)$ is the score residual given in Table 2.3. Thus, a consistent estimator of σ^2 is given by

$$\widehat{\sigma}^2 = \frac{2\sum_{i=1}^K \sum_{j=1}^{n_i} r_s^2(y_{ij}, \widehat{\mu}_{ij})}{\sum_{i=1}^K \sum_{j=1}^{n_i} \ddot{d}(y_{ij}, \widehat{\mu}_{ij})V(\hat{\mu}_{ij})}, \tag{7.23}$$

where the fitted values $\widehat{\mu}_{ij}$ could be either the marginal means or conditional means. The marginal means are usually preferred because the performance of the resulting estimator is numerically more stable.

Note that formula (7.23) can be simplified in the case of the ED models where $r_s = (y - \mu)$ and $\ddot{d} = 2/V(\mu)$. Therefore,

$$\widehat{\sigma}^2 = \frac{\sum_{i=1}^K \sum_{j=1}^{n_i} (y_{ij} - \widehat{\mu}_{ij})^2}{\sum_{i=1}^K n_i},$$

which is identical to the estimate derived directly from the variance-mean relation $\text{Var}(Y_{ij}) = \sigma^2 V(\mu_{ij})$ for the ED models. The denominator may be adjusted to $\sum_{i=1}^K n_i - p$ to account for the number of estimated fixed effects (including the intercept term), p, in the model when the marginal means are used.

For the GLMMs with non-ED models, some *ad hoc* moment properties may be used to formulate the dispersion parameter estimation, which may be different from (7.23). For example, Song and Tan (2000) suggested a moment property for the simplex distribution, $\text{E}d(y; \mu) = \sigma^2$. In the von Mises model, Artes et al. (2000) suggested a moment property $\text{E}\{\cos(Y - \mu)\} = A_1(\lambda)$, with $\lambda = 1/\sigma^2$, to estimate the dispersion parameter, where $A_1(\cdot)$ is the mean resultant length given in Section 2.6.4.

`SAS` PROC GLIMMIX has implemented the PQL/REML inference procedures discussed above. It can be applied to fit various GLMMs with ED models, including correlated binary data, correlated count data and correlated continuous data. The interpretation of results from this `SAS` Proc has no difference from that of the utility of `SAS` PROC NLMIXED as illustrated in Example 7.1. Which package to use in a given data analysis depends on the objective of the analysis; for example, `GLIMMIX` would be used if one wants to acquire estimates of random effects.

The following two examples consider two GLMMs that cannot be fitted by any of the currently available `SAS` procedures. One is the simplex GLMM for correlated continuous proportional outcomes, and the other is the von Mises GLMM for correlated circular outcomes. Because the generalization of the PQL/REML has been developed in the GLMMs for the DM family distributions, analyzing a broader range of data types becomes possible.

Example 7.3 (Simplex Mixed-Effects Models).

A simplex mixed effects model refers to a generalized linear mixed model for repeated continuous proportional data defined as follows:

$$Y_{ij}|\mathbf{b}_i \overset{ind.}{\sim} S^-(\mu_{ij}^b, \sigma^2), \; j = 1, \ldots, n_i, i = 1, \ldots, K,$$

with $g(\mu_{ij}^b) = \mathbf{x}_{ij}^T \boldsymbol{\beta} + \mathbf{z}_{ij}^T \mathbf{b}_i$, and $\mathbf{b}_i \overset{i.i.d.}{\sim} \text{MVN}_q(\mathbf{0}, D)$. Here $S^-(\mu, \sigma^2)$ denotes the simplex distribution with mean μ and dispersion parameter σ^2 (refer to Section 2.6.3 for the details), and as in the marginal GLM considered in Section 5.7.3, the g is chosen to be a logit link.

This simplex mixed-effects model is now applied to analyze the retinal surgery data introduced in Section 1.3.3. Let Y_{ij} be the measurement of the j-th gas volume for the i-th individual at day t_{ij} after the surgery. Recall these outcomes are percentages confined in the unitary interval $(0, 1)$. Refer to Section 5.7.3 for an analysis of the data using the marginal model, as well as the descriptions of the covariates, `gas` as standardized gas concentration level and `time` as days after eye surgery. See Section 5.7.3 for more discussion about the observations and covariates. Now, it is of interest to infer subject-specific effects of the covariates.

The conditional expectation of gas volume takes the form

$$\text{logit}(\mu^b_{ij}) = \beta_0 + b_{0i} + \beta_1 \log(\text{time}_{ij}) + \beta_2 \log^2(\text{time}_{ij}) + \beta_3 \text{gas}_i, \quad (7.24)$$

where the random intercept $b_{0i} \sim N(0, \tau_0)$. For comparison, another model is considered with one extra random effect term $b_{1i} \sim N(0, \tau_1)$ (independent of b_{0i}) as follows:

$$\text{logit}(\mu_{ij}) = \beta_0 + b_{0i} + \beta_1 \log(\text{time}_{ij}) + \beta_2 \log^2(\text{time}_{ij}) + (\beta_3 + b_{1i})\text{gas}_i. \quad (7.25)$$

The PQL/REML inference approach was first applied to fit the second model (7.25) and produced $\hat{\tau}_0 = 0.26(0.19)$ and $\hat{\tau}_1 = 0.09(0.25)$. These results indicate that none of the variance parameters is statistically significant, implying that the two random effects (b_{i0}, b_{i1}) together may over-parametrize the population heterogeneity. This is not surprising, as the estimated correlation from the marginal simplex model in Table 5.7 is about 0.3 for the exchangeable correlation or approximately 0.5 for the AR-1 correlation. Given such a mild within-subject correlation, the random intercepts alone seem to be sufficient to capture the heterogeneity and within subject correlation of the data. For the purpose of comparison, in this simplex model with 1-dimensional random effects, the method of MLE based on numerical integration given in Section 7.4 and the simple-minded fit of the normal LMM to the logit-transformed responses were also implemented to obtain parameter estimates.

Table 7.5 summarizes the results of model (7.24) using the approximate PQL/REML inference, the MLE method based on quadrature numerical evaluation and the naive method based on the normal LMM. For the MLE method, the number of quadrature points was set as 20 and 50, respectively, in the Gauss-Hermite quadrature evaluation of integrals. The naive analysis first takes a logit transformation directly on the response of gas volume percentage and then fits the logit-transformed response by the linear mixed-effects model.

The results of the MLE(20) and MLE(50) are very similar, meaning that 20 quadrature points may have already given satisfactory accuracy in the evaluation of integration. Overall, the results given by MLE and PQLs are very comparable. The estimate of the dispersion parameter σ^2 is 159.0, confirming again that the data are far from a normal distribution. Also, the variance component τ_0 is found to differ significantly from zero. For the fixed effects, the gas concentration level and the log.time^2 covariate are found clearly significant by both PQL and MLE methods. It indicates that the higher the concentration was, the longer it takes for the gas volume decreasing to a given level. Note that this significance of gas concentration was not found in the marginal GLM analysis either in Table 5.7 or in Table 5.8 (actually close to being significant).

Table 7.5. The analysis of the retinal surgery data by the simplex mixed-effects model using both PQL/REML approximate inference and quadrature numerical MLE. The results of the MLE are obtained with 20 and 50 quadrature points, respectively, and the results of the LMM are obtained by fitting the logit-transformed response to the normal LMM.

Method	Intercept	log(time)	$\log^2$(time)	Gas	τ_0	σ^2
PQL/REML	2.91(0.33)	0.06(0.34)	−0.35(0.09)	0.44(0.20)	0.283(0.03)	162.9
p-value	$< .0001$	.6904	$< .0001$	.0351	$< .0001$	
MLE(20)	3.12(0.32)	−0.13(0.49)	−0.33(0.14)	0.52(0.25)	0.991(0.12)	133.7
MLE(50)	3.13(0.33)	−0.15(0.51)	−0.32(0.15)	0.55(0.27)	0.996(0.12)	133.9
p-value	$< .0001$	.7641	.0340	.0393	$< .0001$	$< .0001$
LMM	3.46(0.38)	−0.01(0.28)	−0.39(0.07)	0.65(0.43)	1.650(0.35)	1.302
p-value	$< .0001$	.9624	$< .0001$	.1398		

Similar to the marginal GLM analysis, the log time covariate (log.time) was found not significant.

In the comparison to the results from the normal LMM, clearly this simple method failed to identify the gas concentration level as an important factor. This might be caused by the violation of normality assumption on the transformed responses. A Q-Q plot (not shown here) of the residuals from the naive LMM showed a curved pattern. A careful parametric modeling would help to better summarize the data and lead to a more powerful analysis.

Example 7.4 (von Mises Mixed Effects Models).

A GLMM is now applied to analyze the sandhopper orientation data introduced in Section 1.3.4. Sandhoppers who live in the beaches of the Mediterranean are driven by two main forces. First, they endeavor to stay away from the shore to avoid the risk of being swept away by waves, but they also stay close to the water to avoid the risk of dehydration under the sun. During the day, sandhoppers tend to stay burrowed in the sand. If they are displaced, the risk of dehydration is high. It is believed that sandhoppers will escape toward the sea, taking a course perpendicular to the shoreline known as the *theoretical escape direction* (TED). An experiment was performed at Castiglione della Pescaia beach in Italy, in which 65 sandhoppers were released sequentially five times. Each of their escape directions was recorded, along with other covariates such as wind, sun, and eye measurements. The sea was to the south-southwest, and the TED was 201°. Escape directions were measured as the compass direction, so 0° is north, 90° is east and so on. See Figure 7.4.

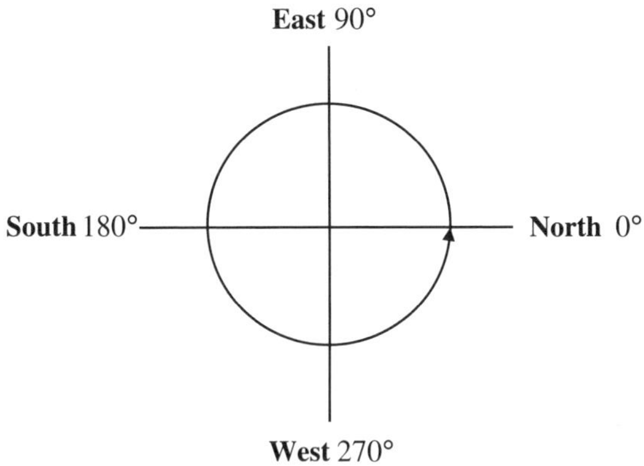

Fig. 7.4. Orientation for sandhoppers' escape direction.

Let Y_{ij} be the escape direction (in radians) of sandhopper i at the j-th release, $j = 1, \ldots, 5$ and $i = 1, \ldots, 65$. The fixed effects covariates in the GLMM include `Azimuth` for sun azimuth (in degrees), wind speeds measured in four categories (`OS` for Offshore, `LSE` for Longshore East, `LSW` for Longshore West, and Onshore), `Eye` for eye asymmetry, and `Release` for release number (equivalently time covariate). The wind speed categories are coded such that Onshore wind speed is the reference category. The GLMM contains only random intercepts. That is,

$$Y_{ij}|b_i \sim \text{vM}(\mu_{ij}^b, \lambda), \; i = 1, \ldots, 65; j = 1, \ldots, 5,$$

where

$$\begin{aligned} \mu_{ij}^b &= \mu_0 + 2\arctan(\eta_{ij}^b) \\ \eta_{ij}^b &= \beta_1 \text{Azimuth} + \beta_2 \text{Eye} + \beta_3 \text{OS} + \beta_4 \text{LSW} + \beta_5 \text{LSE} + \beta_6 \text{Release} + b_{0i} \\ & \quad b_{0,1}, \ldots, b_{0,65} \sim N(0, \tau_0). \end{aligned}$$

The PQL/REML approximation inference is implemented to obtain the parameter estimates and their standard errors. Table 7.6 lists the results.

This analysis suggests that all covariates, except Longshore East wind speed, have an effect on the escape direction of the sandhoppers. The covariate of the sun azimuth appears to influence a sandhopper's orientation the most. This means that as the sun moves more and more to the west, a sandhopper's escape direction will tend to increase along the circle of east, south, west, and north, as indicated in Figure 7.4. This finding is consistent with previous analyses of this data by Borgioli et al. (1999) and D'Elia et al. (2001).

Table 7.6. Results of sandhopper orientation data analysis by the von Mises GLMM based on the PQL/REML approximate inference approach.

Parameter	Estimate	Std Err	Z	p-value
Azimuth	0.006	0.001	4.734	< 0.001
Eye	−1.043	0.592	−1.762	0.039
OS	−0.665	0.232	−2.867	0.002
LSW	−1.134	0.388	−2.918	0.002
LSE	−0.187	0.344	−0.542	0.294
Release	0.058	0.022	2.620	0.004
τ_0	0.194	0.054	3.599	< 0.001
μ_0	2.093	–	–	–
λ	2.080	–	–	–

The release number also had an effect, which suggests that on later releases, escape direction tends to increase, which is clearly related to the movement of the sun during the day time; that is, because a later release was done at a later time, when the sun moved more to the west. The same pattern is found with the covariate of the sun azimuth.

The inclusion of the random effects is clearly necessary, as the variance parameter τ_0 is significantly different from zero, suggesting the presence of both heterogeneity across sandhoppers and within subject correlation among multiple releases. This implies that as expected, the repeated measurements of sandhoppers' escape directions are correlated.

7.9 SAS Software

7.9.1 PROC MIXED

SAS PROC MIXED implements the MLE in the LMMs, in which the inference is based on the multivariate normal likelihood, which is equivalent to the weighted least squares estimation with particular covariance structures. The random effects are estimated by the best linear unbiased predictor (BLUP) or the empirical Bayesian estimation. Harville (1977) presents a joint estimation procedure for both fixed effects $\boldsymbol{\beta}$ and random effects $\mathbf{b}_i$ and the REML for the variance component parameters. This is basically the same as the PQL/REML discussed in Section 7.8.

The SAS code for the analysis of the data from the hypothetical clinical trial example is listed below:

```
proc mixed data = clinic;
     class subj trt time;
```

```
      model y = age trt time trt*time;
      random subj (time);
run;
```

7.9.2 PROC NLMIXED

The SAS NLMIXED procedure fits both nonlinear mixed models and generalized linear mixed models. It implements the adaptive Gaussian quadrature numerical evaluation of integral over the random effects and a dual quasi-Newton algorithm as default. PROC NLMIXED is best suited for models with a single random effect, although it can reasonably compute two and three dimensions of the random effects as well. Currently, this procedure does not handle nested or crossed random effects.

Here is an illustration of PROC NLMIXED used in the analysis of multiple sclerosis trial data in Example 7.1 for model (c). In the use of this procedure, giving initial values is tricky. In this analysis, a sequence of steps was taken by starting with the simplest model with only one fixed-effects covariate EDSS and random intercepts. Using those as initial values one additional term was included at one time until all the terms were included. SAS failed to converge if the estimates from the marginal GLM were chosen as the initial values.

```
data msclero;
infile "P:\sclerdat.txt" delimiter=" ";
input number id time EDSS exacer dur Ptrt Ltrt n;
proc print;
run;

proc sort data=msclero;
by id;
run;

proc nlmixed data=msclero;
   parms b0= -2.59 b1=-0.0302 b2=0.000245 b3=0.3155
         b4=0.4945 b5=0.4503  b6=-0.0365
         s2b1=0.668 cb12=-0.1101 s2b2=0.024;
   eta = b0 + b1*time + b2*time*time + b3*EDSS + b4*Ptrt
         + b5*Ltrt + b6*dur
         + u1 + u2*EDSS;
   expeta = exp(eta);
   p = expeta/(1+expeta);
   model exacer ~ binomial(n,p);
   random u1 u2 ~ normal([0,0],[s2b1,cb12,s2b2])
         subject = id out=eb;
   predict eta out=eta;
run;
```

7.9.3 PROC GLIMMIX

SAS PROC GLIMMIX implements the PQL/REML inference based on Laplacian approximation in the GLMMs. This PROC is available in SAS Version 9.1.2 or above. Based on the same data and model setup in Section 7.9.2, the statement is given as follows.

```
proc glimmix data=one;
     class id trt;
     model y/n = trt visit / solution;
     random intercept visit / subject = id;
run;
```

Here n is the size of the binomial distribution. In order to use binomial distribution, syntax y/n is necessary; otherwise GLIMMIX will use the default of normal distribution.

8

Mixed-Effects Models: Bayesian Inference

8.1 Bayesian Inference Using MCMC Algorithm

A powerful method for handling the numerical integration is the Markov chain Monte Carlo (MCMC) algorithm, first investigated by Zeger and Karim (1991) in the context of generalized linear mixed effects models (GLMM). Nowadays, the availability of software such as BUGS (Bayesian Analysis Using Gibbs Sampling) or its Windows version `WinBUGS` has made related computation really easy to do and hence, greatly facilitated the popularity of Bayesian inference. BUGS is a free software developed by MRC Biostatistics Unit, the Institute of Public Health, Cambridge, which can be downloaded from the Internet at the URL address:

`http://www.mrc-bsu.cam.ac.uk/bugs/`

The primary usefulness of this software is to draw random samples from a Markov chain with a given stationary distribution f.

8.1.1 Gibbs Sampling: A Practical View

MCMC plays a leading role in dealing with high-dimensional objects in various statistical inference problems, including the evaluation of the high-dimensional integral required in the GLMMs or more generally, the generation of random samples in hierarchical models. The use of MCMC algorithm to enhance numerical evaluation of integration in GLMMs essentially results in a Bayesian inference, where all model parameters have to be assumed to follow certain pre-specified prior distributions. It is known that a Bayesian inference is based fully on posterior densities of parameters, or the conditional density $f(\theta_j|\text{data})$ for the j-th parameter θ_j, given data and other known quantities in a model.

Let us start with the Bayesian formulation of a GLMM. The posterior distribution $f(\boldsymbol{\theta}|\mathbf{y})$ can be expressed as follows:

$$f(\boldsymbol{\theta}|\mathbf{y}) \propto f(\boldsymbol{\theta}, \mathbf{y}) = \prod_{i=1}^{K} \int f(y_{ij}|\mathbf{b}_i, \boldsymbol{\beta}) f(\mathbf{b}_i|D) f(\boldsymbol{\theta}) d\mathbf{b}_i, \qquad (8.1)$$

where the normalizing constant is independent of the parameter $\boldsymbol{\theta}$. An estimate of $\boldsymbol{\theta}$ can be derived from $f(\boldsymbol{\theta}, \mathbf{y})$ alone, via a certain location measure of the posterior, such as posterior mode, mean, or median.

If the prior for $\boldsymbol{\theta}$, $f(\boldsymbol{\theta})$, is a constant, then the posterior in (8.1) is effectively proportional to the likelihood function (7.4), and hence the posterior mode is numerically identical to the maximum likelihood estimate. For this sake, flat priors or non-informative priors for $\boldsymbol{\beta}$ and D are usually preferred.

Throughout this section, the distribution of random effects assumed to be normal, namely $\mathbf{b}_1, \ldots, \mathbf{b}_K$ are *i.i.d.* q-dimensional $\text{MVN}_q(0, D)$. In principle, MCMC method can handle other types of distributions of random effects.

Typically, conjugate priors for $\boldsymbol{\beta}$ and D are specified as follows:

(a) For the fixed effects $\boldsymbol{\beta}$, assume each component $\beta_i \sim N(0, \varsigma)$, where the precision hyper-parameter ς, defined as the reciprocal of the variance parameter σ_β^2, is taken to be as small as 10^{-6}. This will effectively specify a normal prior with large variance, so the resulting prior density is approximately flat within a reasonable range of the parameter space.
(b) For the matrix D, when D is a diagonal matrix, $\text{diag}(D_{11}, \ldots, D_{qq})$, the prior for each of the diagonals is set as the inverse gamma or $1/D_{ii} \sim \Gamma(a, b)$, where both hyper-parameters a and b are set small, say 10^{-3} or 10^{-4}.
(c) When D takes a general form with non-zero off-diagonal elements, the prior is specified on its inverse matrix, namely $D^{-1} \sim \text{Wishart}(R, \kappa)$, where $\kappa \geq q$ is the degrees of freedom and R is a $q \times q$ symmetric non-singular matrix. To make the prior "flat", κ is set small and matrix R may be set as a diagonal matrix with diagonal elements equal to the estimates from a pilot model that assumes a diagonal D matrix.

When random effects themselves are of interest, the posteriors of $\mathbf{b}_i, i = 1, \ldots, K$, are given as follows,

$$f(\mathbf{b}_i|\mathbf{y}) = \frac{\prod_j \int_\Theta f(y_{ij}|\mathbf{b}_i, \boldsymbol{\beta}) f(\mathbf{b}_i|D) f(\boldsymbol{\theta}) d\boldsymbol{\theta}}{\int_{\mathcal{R}^q} \prod_j \int_\Theta f(y_{ij}|\mathbf{b}_i, \boldsymbol{\beta}) f(\mathbf{b}_i|D) f(\boldsymbol{\theta}) d\boldsymbol{\theta} d\mathbf{b}_i}, i = 1, \ldots, K,$$

where Θ is the parameter space.

When Monte Carlo method is applied to numerically evaluate the integrals in the the posteriors, random samples of $(\boldsymbol{\theta}, \mathbf{b}_i, i = 1, \ldots, K)$ from their joint distribution are needed. Each integral is approximated by its corresponding sample mean. However, drawing samples is a difficult task due to the curse of dimensionality. Gibbs sampling is a solution to this obstacle. The key idea of Gibbs sampling is that drawing samples from a high-dimensional joint distribution can be carried out through low dimensional conditional distributions. In the statistical literature, various sampling schemes have been developed for low dimensional (especially 1-dimensional) distributions. As a result, the implementation of Gibbs simpling method becomes feasible, but at a cost that would be very computationally intensive.

To understand the Gibbs sampling scheme, let us consider a simple case of trivariate U, V, W, and its joint distribution is denoted by $[U, V, W]$. Suppose that it is difficult or impossible to sample directly from the joint distribution $[U, V, W]$, but sampling from each of its conditional distributions, $[U|V, W]$, $[V|U, W]$, and $[W|V, U]$, is feasible. The Gibbs sampling scheme suggests that

- Step 0: Assign arbitrary starting values $U^{(0)}, V^{(0)}, W^{(0)}$.
- Step 1.1: Draw $U^{(1)} \sim [U|V^{(0)}, W^{(0)}]$,
  Step 1.2: Draw $V^{(1)} \sim [V|U^{(1)}, W^{(0)}]$,
  Step 1.3: Draw $W^{(1)} \sim [W|U^{(1)}, V^{(1)}]$;
  Complete iteration 1 and yield the first drawn sample $(U^{(1)}, V^{(1)}, W^{(1)})$.
- Step 2.1: Draw $U^{(2)} \sim [U|V^{(1)}, W^{(1)}]$,
  Step 2.2: Draw $V^{(2)} \sim [V|U^{(2)}, W^{(1)}]$,
  Step 2.3: Draw $W^{(2)} \sim [W|U^{(2)}, V^{(2)}]$;
  Complete iteration 2 and yield the second drawn sample $(U^{(2)}, V^{(2)}, W^{(2)})$.
- Continue such cycles as long as needed.

After a large number of iterations, say B, obtain sample $(U^{(B)}, V^{(B)}, W^{(B)})$. Geman and Geman (1984) showed that under some regularity conditions, the empirical distribution of the sample $(U^{(B)}, V^{(B)}, W^{(B)})$ will converge to the true underlying stationary distribution $[U, V, W]$ at an exponential rate,

$$[U^{(B)}, V^{(B)}, W^{(B)}] \stackrel{B\to\infty}{\longrightarrow} [U, V, W].$$

Therefore, after the length of *burn-in*, the empirical cumulative distribution function of the M sample values,

$$[U^{(B+k)}, V^{(B+k)}, W^{(B+k)}], \ k = 1, \ldots, M$$

converges to $[U, V, W]$. Loosely speaking, burn-in refers to such a B, after which the Markov process has no memory about the initial states that the process started at. If multiple sample paths were generated by different initial states on this Markov process, then at the point of B, these sample paths would get mixed or tangled together, and produce similar sample paths afterwards. Under such a circumstance, $[U^{(B+k)}, V^{(B+k)}, W^{(B+k)}], k = 1, \ldots, M$ may be approximately regarded as of samples generated from the joint stationary distribution $[U, V, W]$. However, since these M samples are drawn via a Markov chain, they are in general auto-correlated, which would make the use of the theory of Law of Large Number questionable. The Law of Large Number, which assumes independence, is the theoretical base for the justification of sample mean approximating integration. Gelfand and Smith (1990) suggested using a thinned chain, namely recording every b (*say*, $b = 20$) value in the sequence to reduce auto-correlation among sample values. Another method of alleviating auto-correlation is Neal's (1998) over-relaxation method that generates multiple random variates from a conditional distribution at each iteration and selects the one that is most negatively correlated with the lagged sample values. This option of over-relaxation is available in the `WinBUGS` software.

After the completion of sample generation, the next task is to estimate the marginal distribution for each of the variables, which may be given as follows:

$$\widehat{[U]} = \frac{1}{M}\sum_{k=1}^{M}[U|V^{(B+k)}, W^{(B+k)}].$$

This density estimation uses the functional form of the conditional distribution, which turns out to be more efficient than those based only on available samples. In contrast, kernel density estimation for the density of variable U only uses marginal samples, $U^{(B+k)}, k = 1, \ldots, M$. However, in many practical cases, the functional form of conditional density $[U|V, W]$ may be unavailable, so the latter one appears more convenient and hence, is often used in practice.

With the available samples, an empirical CDF may be constructed, which produces some basic statistics such as mean, standard deviation, median, 0.025 quantile, and 0.975 quantile for each of variables. Mode can also be obtained based on the estimated kernel density $\widehat{[U]}$. In cases where the estimated density $\widehat{[U]}$ is (approximately) symmetrical, mode and mean, as well as median, are approximately equal.

8.1.2 Diagnostics

The biggest challenge in the application of Bayesian inference via MCMC practically is convergence diagnosis as well as model validity. Convergence diagnosis of the MCMC algorithm must be properly conducted since it is highly relevant to the appropriateness of results given by the follow-up analysis. In other words, how many initial iterations B are needed to achieve burn-in (or convergence), so that samples after it would be usable? The length of burn-in depends on initial values chosen to start the Gibbs sampler as well as on whether the conditional distributions of values at time k, given initials, (e.g., conditional distribution $[U^{(k)}|V^{(0)}, W^{(0)}]$ of U), are close to $[U, V, W]$. This is equivalent to checking if the stationarity has achieved by time B. There are some diagnostic tools suggested in the literature that will be studied in this section. Another important decision is when the Gibbs sampler should stop, or equivalently how to choose M. The rule of stopping time may vary from case to case, depending upon a specific problem under investigation. It is difficult to find out a general answer, since it not only involves the evaluation of variance of, for example, $\widehat{[U]}$ and but also dealing with the autocorrelation among sample values.

Sensitivity analysis that aims to check whether the results are robust against certain choices of priors should also draw a critical attention. For instance, the prior distribution of the precision parameter may be specified as gamma, uniform, log-normal, or Pareto. Even if its importance has been widely accepted, the lack of suitable protocols or guidelines as well as efficient tools remains a big hurdle in the application of MCMC Bayesian inference in the GLMMs. More investigations seem to be needed in order to widen the use of MCMC further in subject-matter research areas.

There are a few relatively simple diagnostics of algorithm convergence available in the `WinBUGS` software.

(1) Use the plot of the sample paths generated from the process at several very different initial values, in which identifying the point of coalition will provide a good decision on the value of B. Also, Brooks-Gelman-Rubin's F test (R statistic) can be used further to confirm the eye-picked value of B from the plot. See Gelman and Rubin (1992) and Brooks and Gelman (1998).
(2) Simply monitor the trace plot of updates while MCMC runs. If a clear trend in the trace plot is seen, indicating a certain violation of stationarity, a longer chain has to be run till the updates become stablized.
(3) Use the plot of autocorrelation function of the generated samples to decide if an over-relaxation method is needed or if a different size of thinning bin is needed to record the updates. When every updates are recorded, the sampler generates a full sample path with bin size 1; however, this full chain can be thinned by choosing, for example, a bin size 50, which means every 50th updates will be recorded. Thinning is necessary to ensure the application of the Law of Large Number in the calculation of summary statistics, most of which are in the form of average.

In addition, there are some other convergence diagnostic methods in the literature, two of which are discussed below. One is the so-called *Geweke test* for stationarity proposed by Geweke (1992), using the idea of two-sample equal mean test. This test proceeds as follows:

Step 1: Select two segments from a sequence of M sampled values with, respectively, the first $a\%$ values and the last $b\%$ values. Usually $a = 10$ or 25, and $b = 50$.

Step 2: Divide each of the segments into 25 bins (sub-segments), *say*, and calculated sample averages and sample variances for each of the bins, denoted by,

$$\{m_i^{\text{early}}, v_i^{\text{early}}, i = 1, \ldots, 25\}, \quad \{m_i^{\text{late}}, v_i^{\text{late}}, i = 1, \ldots, 25\}.$$

If the stationarity achieves, then the mean from the early segment population should be close to that from the late segment population.

Step 3: Define a discrepancy measure Z-score

$$Z = \frac{E^{\text{early}} - E^{\text{late}}}{\sqrt{(V^{\text{early}} + V^{\text{late}})}},$$

which would follow approximately the standard normal distribution if all sample values are from the same distribution, which corresponds to the validation of stationarity (i.e., the null hypothesis). So ± 2 will be chosen as the yardstick for the test. Clearly the sample counterpart of Z-score is given by

$$Z = \frac{\bar{m}^{\text{early}} - \bar{m}^{\text{late}}}{\sqrt{(\bar{v}^{\text{early}} + \bar{v}^{\text{late}})/25}}.$$

Step 4: Display a comprehensive picture of the Geweke test through a plot of many Geweke Z-scores in terms of different extractions. A sequence of nested chains may be given as follows:

- Chain 1: Full chain with M values
- Chain 2: A sub-chain with last $M - n$ values, discarding the first n values, and its first iteration number is $n + 1$
- Chain 3: A sub-chain with last $M - 2n$ values, discarding the first $2n$ values, and its first iteration number is $2n + 1$
- Extracting procedure continues until the final sub-chain contains at least 50 values but further extraction will result in a sub-chain with less than 50 values
- Chain Q: Stop, for a suitable Q, and its first iteration number is $Qn+1$

Step 5: For sub-chain i, calculate Z_i-score by the definition, where a and b is chosen fixed over the Q chains. Plotting Z-scores, Z_i, against the first iteration numbers, $in + 1$, $i = 1, \ldots, Q$, and expect most points fall into the band $(-2, 2)$ if the stationarity achieves.

The other approach is due to Heidelberger and Welch (1983), which is developed to test the null hypothesis that sample values come from a stationary process. If the null hypothesis is rejected, the first 10% of sampled values should be discarded; if the null hypothesis is rejected again based on the 90% of sampled values, then further 10% of sampled values should be ruled out. Repeat this test until either a portion of the chain (at least 50% of sample values) passes the stationarity test, or 50% of sample values have been thrown away and the null hypothesis is still rejected. If the latter happens, the stationarity test fails and a longer run (increasing B) is deemed to achieve convergence.

Associated with the Heidelberger and Welch's test, a statistic called *halfwidth*, defined as the half length of a 95% confidence interval, is used to make a decision on whether a longer run should be considered. The halfwidth test is conducted as follows: based on the portion of the chain that passed the stationarity test, the corresponding sample mean and its asymptotic standard error can be obtained, and therefore the halfwidth of the associated 95% confidence interval for this mean is equal to $1.96 \times$ s.e. (standard error). If the halfwidth is less than $\epsilon \times$ sample mean, the test is passed and the retained sample is deemed to estimate the posterior mean with acceptable precision.

The coefficient ϵ is usually taken to be 0.1. If the test fails, a longer run is needed to achieve a satisfactory accuracy for estimates.

Both available in R software, CODA (Convergence Diagnostics and Output Analysis) and BOA (Bayesian Output Analysis) provide Geweke's test and Heidelberger and Welch's test, as well as other diagnostic methods.

8.1.3 Enhancing Burn-in

The MCMC approach is to obtain a set of "independent" draws from the joint posterior distribution of the quantities of interest after the Markov chain has reached its burn-in or stationary distribution. This stationarity has been theoretically proved achievable when the Markov chain is geometrically ergodic. The CODA or BOA packages discussed in Section 8.1.2 implement various convergence diagnostics. In practice, however, some models need to run a large number of loops in order to obtain approximately independent samples via thinning technique. This extremely slow convergence is of particular concern.

Besides the techniques of thinning and over-relaxation, an alternative way to enhance burn-in is via reparameterization based on strategies of modifying model structures or *sweeping* method suggested by Gilks et al. (1996). First, it explores the between- and within-variable correlations through a certain pilot study based on a small number of draws, which provides some basic evidence to choose a proper strategy for improvement. In the following, two common reparametrization strategies are discussed, which will be illustrated in Section 8.3 to improve the rate of burn-in and hence to considerably reduce computational time.

The first reparametrization strategy is to standardize covariates around their means, so that fixed effects parameters become orthogonal. That is, a hierarchical model takes the centered form as follows:

$$\mathrm{g}(\mu) = \bar{\mathbf{x}}^T\boldsymbol{\beta} + (\mathbf{x} - \bar{\mathbf{x}})^T\boldsymbol{\beta} + \mathbf{z}^T\mathbf{b}.$$

This reparametrization approach is, in fact, widely used in the classical setting of regression analysis to deal with the multicollinearity problem. The reparametrization is appealing here, since the resulting hierarchical model can achieve better correlation properties. The reparametrized models are equivalent to the original hierarchical models subject to a certain re-arrangement of intercept terms, but gain convergence faster than the original one at little cost.

The second reparametrization strategy takes place on the part of $\mathbf{z}^T\mathbf{b}$, which may involve nested random effects in the modeling of multi-level correlated data. *Hierarchical centering*, proposed by Gelfand et al. (1995 and 1996), aims to acquire better correlation properties in the reformulated hierarchical models, so that computational time per update is reduced. In particular, the *partial* hierarchical centering is simple and useful, which centralizes the random effects around the nested mean instead of zero. See Qiu et al. (2002) for an application of this centering approach.

8.1.4 Model Selection

The common Bayes (or Schwarz) Information Criterion (BIC/SIC) (Kass and Raftery,1995) for model selection is inappropriate because the presence of random effects in hierarchical models complicates the determination of the true number of free parameters (or degrees of freedom) in the assessment of model complexity. The *deviance information criterion* (DIC), proposed recently by Spiegelhalter et al. (2002), attempts to resolve this problem in the perspective of approximate Akaike Information Criterion (AIC), and is adopted for model selection in the framework of hierarchical models. According to Spiegelhalter et al. (2002), the Bayesian deviance is given by

$$D(\boldsymbol{\theta}|M_k) = -2\log L(\mathbf{y}|\boldsymbol{\theta}; M_k) + 2\log f(\mathbf{y})$$

for model M_k with parameter $\boldsymbol{\theta}$, where $f(\mathbf{y})$ is a certain normalizing term that is known and fully determined by data. The value of true number of free parameters is defined as

$$p_D = \mathrm{E}_{\boldsymbol{\theta}|\mathbf{y}} D(\boldsymbol{\theta} \mid M_k) - D(\mathrm{E}_{\boldsymbol{\theta}|\mathbf{y}}(\boldsymbol{\theta}) \mid M_k) \stackrel{def}{=} \bar{D} - D(\bar{\boldsymbol{\theta}}).$$

Moreover, the DIC takes the form

$$\mathrm{DIC} = \bar{D} + p_D = D(\bar{\boldsymbol{\theta}}) + 2p_D,$$

where as usual, term $\bar{D}$ explains the model fit and p_D indicates the model complexity. Spiegelhalter et al. (2002) showed asymptotically that the DIC is a generalization of the AIC.

Computing the DIC is straightforward in an MCMC implementation. Tracking both $\boldsymbol{\theta}$ and $D(\boldsymbol{\theta})$ in MCMC iterations, at the exit of sampling, one can estimate the $\bar{D}$ by the sample mean of the simulated values of D and the $D(\bar{\boldsymbol{\theta}})$ by plugging in the sample mean of the simulated values of $\boldsymbol{\theta}$. A lower value of DIC indicates a better-fit model.

One may also adopt the so-called null standardization criterion $D_0(\boldsymbol{\theta}|M_k) = -2\log L(\mathbf{y}|\boldsymbol{\theta}; M_k)$ in the model comparison. Because the normalizing term, $2\log f(\mathbf{y})$, it is independent of model M_k. For the purpose of model comparison, ignoring this term does not affect the conclusion about the ranking of candidate models. This model selection criterion allows users to assess the necessity of a certain configuration of random effects in the formulation of a GLMM, and moreover to determine a proper covariance structure for the distribution of random effects. The implementation of this model selection criterion is demonstrated in Sections 8.2 and 8.3 numerically.

8.2 An Illustration: Multiple Sclerosis Trial Data

This section illustrates the application of MCMC algorithm to fit the generalized logistic mixed effects model for multiple sclerosis trial data. The entire analysis is conducted with the utility of the `WinBUGS` software package. Note that the same data has been analyzed in Example 7.1 using `SAS` PROC NLMIXED. Refer to Section 1.3.6 for the details of data description.

In this illustration, model (b) in Example 7.1 is chosen; that is, the probability of exacerbation at the j-th visit for patient i, $\pi_{ij} = \text{prob}(Y_{ij} = 1|\mathbf{x}_{ij})$, follows the following model:

$$\begin{aligned} \text{logit}\pi_{ij} &= \beta_0 + \beta_1 \text{t}_j + \beta_2 \text{t}_j^2 + \beta_3 \text{EDSS}_i + \beta_4 \text{Ptrt}_i + \beta_5 \text{Ltrt}_i + \beta_6 \text{dur}_i \\ &\quad + b_{i0} + \text{EDSS}_i b_{i1},\ j = 1, \ldots, 17, i = 1, \ldots, 52, \end{aligned} \tag{8.2}$$

where both random effects b_{i0} and b_{i1} are assumed to be independent and normally distributed, namely

$$(b_{i0}, b_{i1}) \sim \text{MVN}_2 \left(\mathbf{0}, \begin{pmatrix} D_{11} & 0 \\ 0 & D_{22} \end{pmatrix} \right).$$

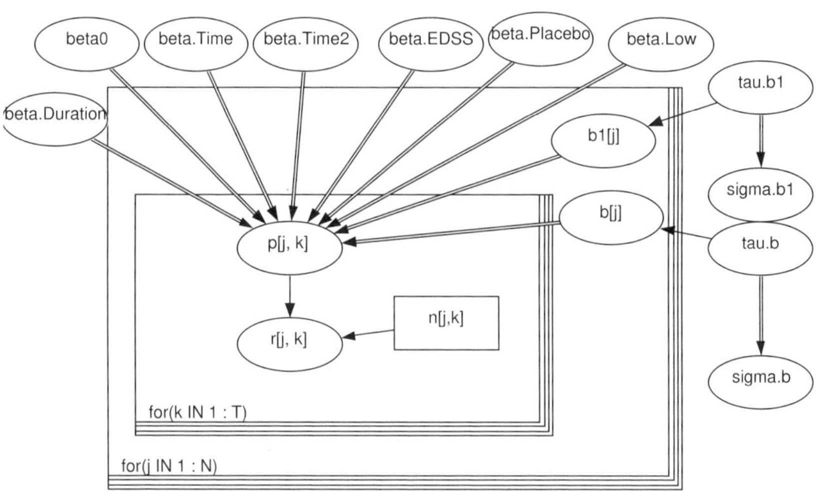

Fig. 8.1. Graphic representation of the logistic mixed effects models for the multiple sclerosis data.

The `WinBUGS` Doodle allows users to produce a graphic representation of model specification, as shown in Figure 8.1 that clearly displays relations among variables included in the GLMM 8.2. In this graph, the nodes outside of the outer plate represent the variables or parameters whose samples will be generated by the Gibbs sampler. Different plates indicate different dimensions of loops. The double-arrowed line indicates a deterministic relation, whereas the single-arrowed line indicates a stochastic relation as long as the MCMC sampling concerns. Moreover, Doodle can translate this graphic representation into a BUGS programming code, which will be discussed in detail in next section. As usual, "noninformative" priors are specified for the coefficients and precision parameters in the model.

To examine burn-in by the Brooks-Gelman-Rubin's method of multiple chains, three very different sets of initial values were chosen, as listed in Table 8.1.

Table 8.1. Three sets of initial values set to generate three chains by Gibbs sampling.

Parameter	Set 1	Set 2	Set 3
intercept	1.0	−1.0	2.0
time	0.0	0.1	−0.1
time^2	0.0	0.0	0.0
EDSS	0.0	0.0	0.0
Ptrt	0.0	0.0	0.0
Ltrt	0.0	0.0	0.0
dur	0.0	0.0	0.0
$1/D_{11}$	4.0	10.0	1.0
$1/D_{22}$	1.0	1.0	4.0

Three chains of 10,000 samples, thinned by 3, were generated with the three sets of initial values. Monitoring the trace plots suggested that the samples of the fixed effects quickly got mixed well, but the samples of the variance components took a long journey to join together. Figure 8.2 shows the worst scenarios of mixing occurred on the samples for the precision parameters $1/D_{11}$ (called `tau.b` in the top plot) and $1/D_{22}$ (called `tau.b1` in the bottom plot).

Figure 8.2 looks messy. To better examine the convergence of the Gibbs sampling algorithm, the Brooks-Gelman-Rubin diagnostic tool was applied. Discarding the first 100 (effectively the original 4,000 samples due to the thinning) simulated data, the Brooks-Gelman-Rubin statistic R can be plotted up to the 6,000 iterations in Figure 8.3.

In each plot, the red curve (or the second darkest curve) represents the statistic R; the green curve (or the least dark curve) shows the central 80% interval of the pooled runs; and the blue curve (or the darkest curve) indicates the central 80% interval of individual runs. Once convergence occurs, the blue

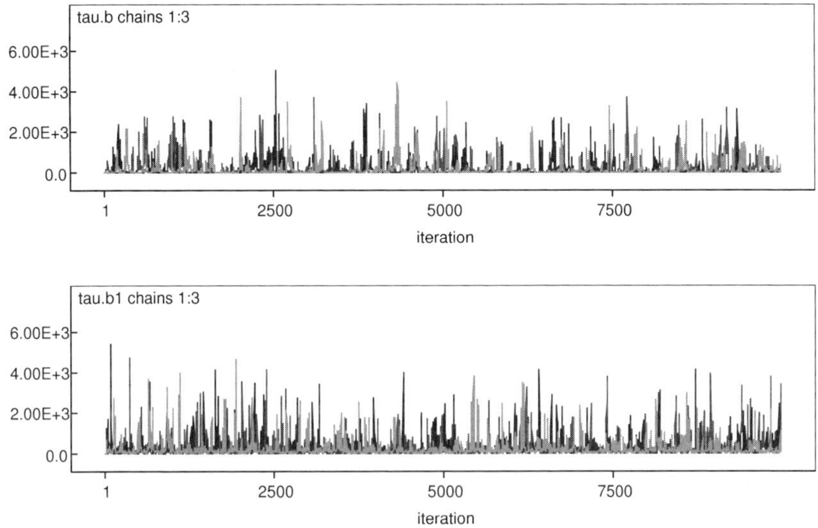

Fig. 8.2. History plots of the thinned samples for the precision parameters $1/D_{11}$ (the top panel) and $1/D_{22}$ (the bottom panel) generated by Gibbs sampler in WinBUGS.

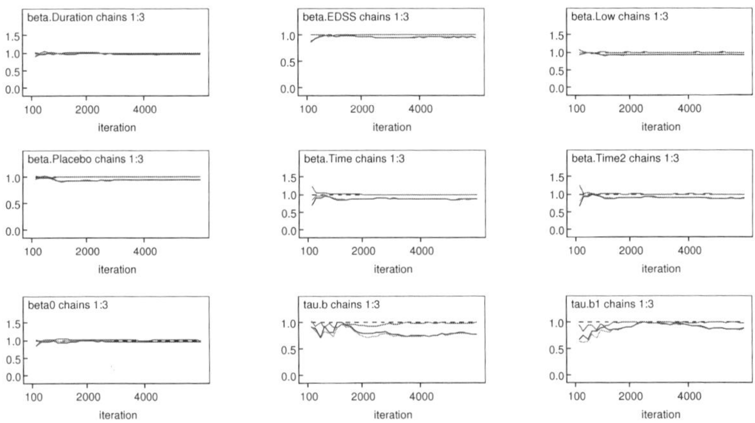

Fig. 8.3. Plots of Brooks-Gelman-Rubin statistic R for the model parameters over 6,000 iterations with the first 100 (equivalent to the original 4,000 due to the thinning) iterations being discarded.

and green curves should become almost equivalent and hence the R would be getting close to 1. A careful examination of all the plots in Figure 8.3 easily concluded the convergence after the 4,000-th iteration.

Based on the convergence diagnosis analysis, it seems reasonable to use samples generated after the 4,000-th iterations to make inference on the model parameters. In the meanwhile, the autocorrelation plots (not shown) suggested some moderate correlation between samples, which basically means longer chains needed to be generated or a larger thinning interval is required. Therefore, in the second step, 5,000 additional samples were generated and the results reported in Table 8.2 are based on a total of 9,000 samples, consisting of the last 3,000 samples of each chain. In comparison, the estimates of the GLMM obtained from `SAS` PROC NLMIXED and of the marginal GLM obtained from `SAS` PROC GENMOD under interchangeable correlation structure are also listed in the table. Estimates from `WinBUGS` are the mean and standard deviation of the estimated posterior for each of the model parameters.

Table 8.2. Estimation results of the logistic mixed effects model for the multiple sclerosis data obtained from WinBUGS, SAS PROC NLMIXED, and SAS PROC GENMOD.

Parameter	WinBUGS	NLMIXED	GENMOD
intercept	−1.7420(.5399)	−1.6348(.5241)	−1.5691(.4834)
time	−0.0318(.0150)	−0.0306(.0151)	−0.0302(.0133)
time^2	0.0003(.0001)	0.0002(.0001)	0.0002(.0001)
EDSS	0.3137(.0949)	0.2935(.0854)	0.2885(.0891)
Ptrt	0.3184(.3561)	0.2536(.3295)	0.2442(.2982)
Ltrt	0.5421(.3415)	0.4620(.3248)	0.4332(.3071)
dur	−0.0437(.0235)	−0.0433(.0215)	−0.0446(.0219)
$1/D_{11}$ or D_{11}	132.80(343.50)	0.1306(.2148)	–
$1/D_{22}$ or D_{22}	259.70(426.60)	0.0007(.0143)	–

Table 8.2 indicates that the estimates of the fixed effects are very close between the NLMIXED and `WinBUGS`, because non-informative priors were used in the `WinBUGS`. Again, the Bayesian analysis did not find the significant presence of heterogeneity among the patients. Given the ease of computation and interpretation, for the analysis of this multiple sclerosis data, the marginal GLM seems to be the best approach among the three.

8.3 Multi-Level Correlated Data

Multi-level correlated data arise from many practical studies, such as spatio-temporal data. Correlated data considered in the previous chapters, including longitudinal data, clustered data, or spatial data are special cases of multi-level data. On the line of conditional modeling approach, mixed-effects models may be extended to incorporate more random effects in a hierarchical (or nested) form, and the resulting models are often referred to as hierarchical models. In practice, hierarchical models are widely used to analyze multi-level data, even though some difficulties in statistical inference and model fitting have not completely been overcome yet. For example, because more random effects enter the model via complicated structures, the difficulty of integration in likelihood functions becomes even more challenging. MCMC algorithm appears particularly useful and handy to deal with such a high-dimensional integration problem, even if there are some caveats in the use of this algorithm.

Instead of giving a general presentation of hierarchical models, this section is intended to focus only on an example, from which readers can appreciate the utility of MCMC method, with the assistance of `WinBUGS` software package, to carry out a Bayesian inference in hierarchical models. The Tretinoin Emollient Cream (TEC) trial data, introduced in Section 1.3.7, is used to facilitate the discussion. From the data description, it is known that the response measured in this trial is in an ordinal scale. In the literature, the analysis of correlated ordinal measurements has been investigated by many researchers, such as Heagerty and Zeger (1996) for marginal models, Hederker and Mermelstein (2000) for random effects models, and Kosorok and Chao (1996) for ordinal longitudinal data in continuous time.

Let Y_{ijl} be the response variable for the l-th observation of the j-th location ($j = 1$ indicating arm; $j = 2$ indicating face; $l = 1, \cdots, \kappa_j$, where $\kappa_1 = 4$ and $\kappa_2 = 1$) in the i-th patient ($i = 1, 2, \cdots, K$, where $K = 32$). Because the response variable is ordinal with 4 levels, a hierarchical proportional odds model is considered for the data analysis. To analyze correlated ordinal measurements, readers may refer to Heagerty and Zeger (1996) for marginal models and Hederker and Mermelstein (2000) for random effects models.

Let the cumulative probability be $\pi_{ijl,m} = \text{Prob}(Y_{ijl} \leq m)$ with

$$m = \begin{cases} 1, & \text{no change or worsening} \\ 2, & \text{slight improvement} \\ 3, & \text{improvement} \\ 4, & \text{great improvement.} \end{cases}$$

The cumulative probability $\pi_{ijl,m}$ represents the probability that the symptom score for patient i at the j-th location during the l-th visit is not better than category m.

Then, the point mass probabilities are

$$\mu_{ijl,m} = \pi_{ijl,m} - \pi_{ijl,m-1}, \quad m = 1, 2, 3, 4.$$

The proportional odds model (McCullagh and Nelder, 1989) usually assumes that these probabilities arise, via the threshold method, from an underlying continuum Z_{ijl} that takes values on $\mathcal{R}$ according to the following one-to-one relation:

$$Y_{ijl} = m \quad \text{if and only if} \quad Z_{ijl} \in (a_{m-1}, a_m], \; m = 1, 2, 3, 4,$$

where the threshold (or cut) points are constrained with $-\infty = a_0 < a_1 < a_2 < a_3 < a_4 = \infty$. The hierarchical model takes the form

$$g(\pi_{ijl,m}) = a_m + \mathbf{x}_{ij}^T \boldsymbol{\beta} + b_i + d_{ij}, \tag{8.3}$$

where the three threshold points are $a_m, m = 1, 2, 3$ with $a_1 < a_2 < a_3$, b_i are the patient-specific random effects and d_{ij} are the location-specific random effect. In effect, model (8.3) with the logit link g is resulted from the assumption that each latent variable Z_{ijl} follows marginally a standardized logistic distribution. The three-element vector of covariates, $\mathbf{x}_{ij}$, includes $x_{1ijl} = 1$ if the active drug TEC is applied to the j-th location of the i-th patient, and $x_{1ijl} = 0$ if placebo is applied; x_{2ijl} is the number of weeks for the treatment, which is either 24 or 48; and x_{3ijl} is the interaction of treatment and time.

The complication in statistical analysis is the presence of the correlation among multiple locations on a patient (spatial correlation at level 2) and the serial correlation across multiple time points (autocorrelation at level 3). This hierarchical model is a natural extension of the generalized linear mixed effects model with nested random effects. To invoke MCMC based inference, priors on both regression coefficients and parameters in the distribution of random effects have to be specified. As discussed in the preceding section, if flat (or diffuse) priors are used, this model essentially gives results numerically similar to those from the frequentist maximum likelihood inference. This is because the resulting posterior is equal or approximately equal to the normalized likelihood function.

To satisfy the order restriction on the cut-points $a_1 < a_2 < a_3$, one may introduce two parameters $\theta_1 > 0$, $\theta_2 > 0$ such that $a_2 = a_1 + \theta_1$ and $a_3 = a_2 + \theta_2$. Under this reparametrization, the assumed priors are listed in Table 8.3.

Table 8.3. Priors specified in the hierarchical proportional odds model.

Parameters	Prior
Each $\beta_k, k = 1, \ldots, p$	$N(0, 1000^2)$
Cut-point a_2	$N(0, 1000^2)$
Each $\theta_k, k = 1, 2$	Truncated $N(0, 1000^2)$ on $(0, \infty)$

Four candidate models may be resulted from the above model specification and hence will be compared in the following analysis. They are:

(a) the naive model, with no random effects (that is, both b_i and d_{ij} are degenerated at zero);
(b) the level-1 hierarchical model that just incorporates location-specific random effects b_i, but with no time-dependent random effects d_{ij};
(c) the level-2 (variance component) hierarchical model that includes both random effects b_i and d_{ij}, but the variance matrix of random effects D of the random effects vector $\gamma_i = (b_i, d_{i1}, d_{i2})^T$ is assumed to be a diagonal structure, denoted by D =diag(τ_1, τ_2, τ_2), where τ_1 and τ_2 are the respective variances of b_i and d_{ij};
(d) the level-2 (full) hierarchical model as specified in (c) with a general variance matrix $D = (\tau_{ij})_{3\times 3}$.

Obviously Model (d) specification refers to the case that all random effects are correlated while Model (c) implies that the random effects at the patient level and those at the location level are uncorrelated. In Model (c), it is possible to further reduce the number of variance parameters to 2 by assuming two arms to share common random effects d_{i1}. This seems to be a reasonable reduction because the skin tissue on the left arm is anatomically almost identical with that on the right arm.

The Wishart prior distribution for the inverse of the fully unspecified variance matrix D is adopted in Model (d), where the hyper-parameters are set as follows: the scale matrix $R = \text{diag}(1, 1, 1)$ and the degrees of freedom equal to 3. In Model (c), priors for the variance components are specified as the conjugate inverse gamma distribution IG(a_0, b_0) with hyper-parameter values $a_0 = b_0 = 0.0001$. The conjugate prior is commonly used in the literature as prior distributions for variance parameters. The variance of each prior for β_k or the variance components τ_k is set so large that the resulting prior would be (approximately) non-informative. Thus the choice of priors would have little effect on the analysis (Sun et al., 2001).

Table 8.4. Comparison among candidate hierarchical proportional odds models for the TEC data analysis.

Model	$\bar{D}_0$	$D_0(\bar{\boldsymbol{\theta}})$	p_D	DIC
(a) Naive	254.8	248.804	5.996	260.796
(b) Level-1	199.4	169.614	29.759	229.133
(c) Level-2 VC	91.2	39.663	51.514	142.692
(d) Level-2 FM	107.3	58.224	49.076	156.376

Table 8.4 reports the values of the DIC model selection criterion based on 5,000 records after the burn-in of 5, 000 iterations generated by the `WinBUGS`

package with the option of Neal's over-relaxation. It is noticeable that Model (c) with the variance components structure appears to be superior over the all other three models, including the model with a fully unspecified variance matrix, as Model (c) receives the smallest DIC value, as well as the minimum BIC. Therefore, Model (c) is used as the final model to draw conclusions.

Three chains of 15,000 iterations were first run for Model (c) in `WinBUGS`, with three different sets of initial values and a thinning interval of 2. All parameters have indicated reasonably fast rates of mixing, except for the variance parameters $\tau_k, k = 1, 2$ that appeared to have a very slow pace of mixing. With first 2,500 iterations discarded, panel (a) in Figure 8.4 shows the Brooks-Gelman-Rubin diagnostic plots for $\tau_k, k = 1$ (left) and $k = 2$ (right). The red lines (the second darkest) in both plots seem to suggest that iteration 5,000 would be a safer point to draw the line of burn-in, judged by the closeness to the horizontal line at 1.

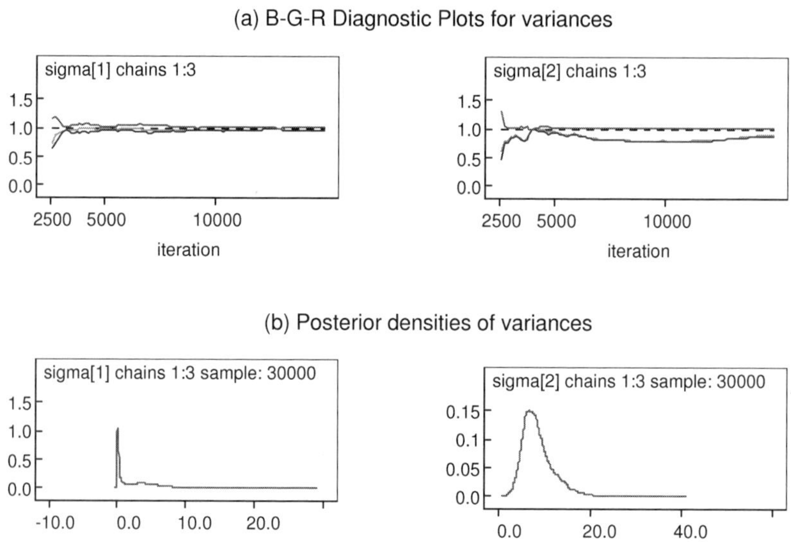

Fig. 8.4. Brooks-Gelman-Rubin diagnostic plots and posterior densities for the variance component parameters $\tau_k, k = 1, 2$ based on three chains of 15,000 iterations in the analysis of TEC data.

Also in Figure 8.4, the posterior densities of the two variance parameters are displayed in panel (b). The density of τ_1 has a sharp peak near zero as well as a very long tail, which means that this parameter was poorly estimated. This is further confirmed by Figure 8.5 in that the history plots of the last 1,000 iterations clearly show a low quality of mixing, especially for parameter τ_1. This is due largely to the limited size of observed data–only one measurement on face and two repeated measurements on arms, as opposed to a large number of random effects involved in the model. Therefore, in order to calculate summary statistics for the model parameters, a long chain seems to be necessary.

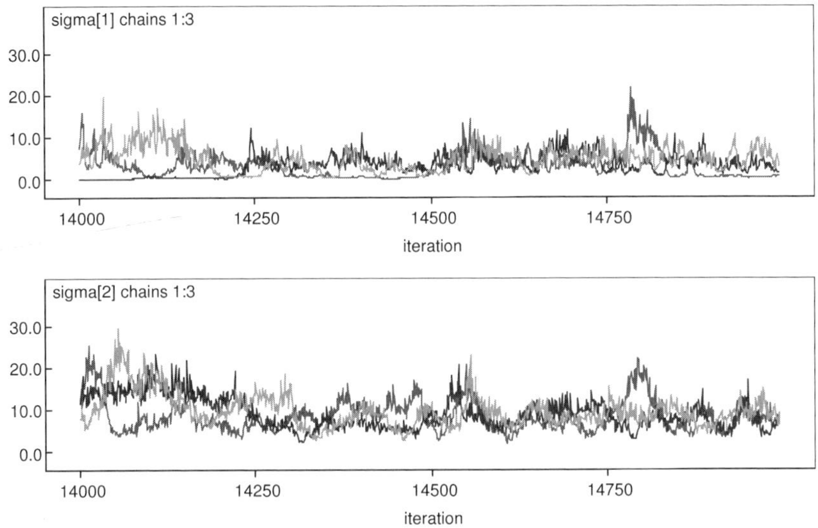

Fig. 8.5. History plots for the variance component parameters $\tau_k, k = 1, 2$ based on three chains of 15,000 iterations in the analysis of TEC data.

Table 8.5 lists the summary statistics on the basis of a single chain of 40,000 iterations, with deletion of the first 10,000 iterations. For this case, the burn-in was further confirmed by both Geweke's Z-scores and Heidelberger & Welch test using `R` package CODA.

Table 8.5 also reports the estimated log odds ratios for TEC treatment versus placebo at 24 weeks, $\log(\mathrm{OR}_{24}) = \beta_1 + 24 \times \beta_3$, and at 48 weeks, $\log(\mathrm{OR}_{48}) = \beta_1 + 48 \times \beta_3$. Both of them are highly significant. This indicates strong evidence for the efficacy of TEC drug in treating photo-age skin.

In this data analysis, because the convergence for the chains associated with the variance parameters appeared very slow, drawing the final conclusion on the treatment effectiveness requires extra caution. Section 8.4.2 provides the `WinBUGS` code used to run the above analysis.

Table 8.5. Summary statistics from the respective posteriors of the model parameters in the TEC drug treatment analysis, using WinBUGS.

Parameter	Mean	Std Dev	MC error	2.5%	Median	97.5%
a_1	10.290	4.658	0.233	2.815	9.718	20.870
a_2	19.710	7.593	0.437	8.278	18.510	37.520
a_3	28.360	10.380	0.597	13.170	26.570	52.940
β_1	−15.730	6.490	0.347	−31.030	−14.700	−5.744
β_2	0.029	0.075	0.002	−0.116	0.027	0.186
β_3	−0.180	0.089	0.003	−0.376	−0.174	−0.019
$\log(\mathrm{OR}_{24})$	−20.050	6.795	0.390	−36.240	−18.820	−10.280
$\log(\mathrm{OR}_{48})$	−24.370	7.716	0.438	−42.700	−23.100	−13.120
τ_1	2.462	2.865	0.171	0.015	1.286	9.547
τ_2	9.296	3.784	0.215	3.893	8.570	18.500

8.4 WinBUGS Software

`WinBUGS` is a free software package that is useful to generate random samples from multivariate distributions via Markov chain Monte Carlo (MCMC) algorithm. This software can be downloaded from

`http://www.mrc-bsu.cam.ac.uk/bugs/welcome.shtml`

Users need to apply for a key to unlock and upgrade the package from an educational version to an advanced version in order to run the codes given in this section.

Beginners of the software may visit a webpage

`http://www.statslab.cam.ac.uk/~krice/winbugsthemovie.html`

to watch a movie about *Getting Started with* `WinBUGS`. Also, it is beneficial to join in a discussion mailing list, `jscmail@jiscmail.ac.uk`, to receive or ask questions regarding `WinBUGs` problems. Most importantly, always get help from an online users' manual or a hard copy manual that can be found on the home page of `WinBUGS` at

`http://www.mrc-bsu.cam.ac.uk/bugs/winbugs/manual14.pdf.`

In the manual, the tutorial section is always worth reading.

This section presents `WinBUGS` codes, respectively, used in the fit of the multiple sclerosis data to a logistic mixed effects model and in the fit of the TEC drug data to a hierarchical proportional odds model.

8.4.1 WinBUGS Code in Multiple Sclerosis Trial Data Analysis

The WinBUGS program for the analysis of multiple sclerosis trial data produced by the WinBUGS' Doodle in Figure 8.1 is given as follows. The diagnostic analysis of the burn-in is based on the Brooks-Gelman-Rubin's R statistic via multiple chains (available in WinBUGS), where three very different initial values are specified; see Table 8.1.

```
model
{
 for(j in 1 : N) {
    for(k in 1 : T) {
      logit(p[j,k]) <- beta0+beta.Time*Time[k]
             +beta.Time2*Time2[k]+beta.EDSS*EDSS[j,k]
             +beta.Placebo*Placebo[j]+beta.Low*Low[j]
             +beta.Duration*Duration[j]
             +b1[j]+b[j]*EDSS[j,k]
        y[j,k] ~ dbin(p[j, k],1)
            }
        # subject random effects
        b1[j] ~ dnorm(0.0, tau.b)
        b[j] ~ dnorm(0.0, tau.b1)
      }
    # priors:
        beta0 ~ dnorm(0.0,1.0E-4)
        beta.Time ~ dnorm(0.0,1.0E-4)
                beta.Time2 ~ dnorm(0.0,1.0E-4)
        beta.EDSS ~ dnorm(0.0,1.0E-4)
        beta.Placebo ~ dnorm(0.0,1.0E-4)
        beta.Low ~ dnorm(0.0,1.0E-4)
        beta.Duration ~ dnorm(0.0,1.0E-4)
        tau.b1 ~ dgamma(1.0E-3,1.0E-3);
        sigma.b1 <- 1.0 /sqrt(tau.b1)
        tau.b ~ dgamma(1.0E-3,1.0E-3);
        sigma.b  <- 1.0/sqrt(tau.b)
}
```

In this program, the priors for the coefficients β_k are specified as normal with a variance of 10^4 (or precision 10^{-4}), and the priors for the precisions of the random effects are specified as gamma with both small rate and scale parameters 10^{-3}.

For the convergence diagnostics, WinBUGS provides trace plots that monitor updates while programs are running, autocorrelation plot, and Brooks-Gelman-Rubin diagnostic with multiple chains. In addition, CODA (Convergence Diagnostics and Output Analysis) and BOA (Bayesian Output

Analysis), both available in R packages, provide Geweke test and Heidelberger and Welch's test.

8.4.2 WinBUGS Code for the TEC Drug Analysis

Below lists the WinBUGS code used in the implementation of the hierarchical proportional odds model (the selected level-2 VC model) in the TEC drug data analysis. Denote $N = 160$ as the number of observations, $NS = 32$ as the number of subjects, $NJ = 2$ as as the number of locations, and $Ncut = 3$ as the number of cut-points. Also, let h be the index of overall observations, so $i = U(h), j = V(h)$, where U and V are the index variables for level i and j. In addition, x1.bar, x2.bar, and x3.bar are the mean values (given in the dataset) of covariates x1, x2, and x3, respectively. In the code, variable sigma denotes the vector of variance component parameters τ_k in the covariance matrix D, and cut denotes the threshold points a_m.

```
Inits
       list(beta=c(0, 0, 0), a=c(1, 0, 1), tau=c(1,1) )
       list(beta=c(0, 0, 0), a=c(1, 0, 1), tau=c(0.1, 0.1) )
model {
   for (h in 1:N) {
             covc[h] <- beta[1] * (x1[h] - x1.bar)
                        + beta[2] * (x2[h] - x2.bar)
                        + beta[3] * (x3[h] - x3.bar)
             logit(Q[h,1]) <- -a[1] + d[U[h], V[h]] + covc[h]
             logit(Q[h,2]) <- d[U[h], V[h]] + covc[h]
             logit(Q[h,3]) <- a[3] + d[U[h], V[h]] + covc[h]
  # probability of response = m
         mu[h,1] <- Q[h,1]
         for (m in 2:Ncut) {
                     mu[h,m] <- Q[h,m] - Q[h,m - 1]
         }
         mu[h,(Ncut+1)] <- 1 - Q[h,Ncut]
         y[h] ~ dcat(mu[h, 1: (Ncut + 1)])
  }
# Patient-specific/Patient*location-specific random effects
    for (i in 1: NS) {
         b[i] ~ dnorm(a[2], tau[1])
         for (j in 1: NJ ) {
              d[i, j] ~ dnorm(b[i], tau[2])
         }
    }
# Priors
     for(k in 1:3){
           beta[k] ~ dnorm(0, 1.0E-06)
```

```
     }
    a[2] ~ dnorm(0, 1.0E-06)
    a[1] ~ dnorm(0, 1.0E-06)I(0, );
    a[3] ~ dnorm(0, 1.0E-06)I(0, )
     for(k in 1:2){
          tau[k] ~ dgamma(0.0001, 0.0001)
          sigma[k] <- 1 / sqrt(tau[k])
     }
# log odds ratios and intercepts on original scale:
    LOR[1] <- beta[1] + 24*beta[3]
    LOR[2] <- beta[1] + 48*beta[3]
    x.bar <- beta[1]*x1.bar + beta[2]*x2.bar + beta[3]*x3.bar
    cut[1] <- a[2] - a[1] - x.bar
    cut[2] <- a[2] - x.bar
    cut[3] <- a[2] + a[3] - x.bar
}
```

9

Linear Predictors

This chapter is devoted to the best linear unbiased predictor (BLUP). Starting with the definition of BLUP, Section 9.1 investigates some basic properties of BLUP. Then, Section 9.2 derives the estimates of random effects in linear or generalized linear mixed models. Finally, Section 9.3 presents the Kalman filter and smoother, originally proposed by Kalman (1960) and Kalman and Bucy (1961), under a more general hierarchical structure than the classic linear hierarchical model.

9.1 General Results

Throughout this section, we consider random vectors $\mathbf{X}$, $\mathbf{Y}$, and $\mathbf{Z}$ with finite second moments whose means, variances, and covariances are denoted by

$$\mathrm{E}\begin{pmatrix}\mathbf{X}\\ \mathbf{Y}\\ \mathbf{Z}\end{pmatrix}=\begin{pmatrix}\boldsymbol{\mu}_X\\ \boldsymbol{\mu}_Y\\ \boldsymbol{\mu}_Z\end{pmatrix} \quad \text{and} \quad \mathrm{Var}\begin{pmatrix}\mathbf{X}\\ \mathbf{Y}\\ \mathbf{Z}\end{pmatrix}=\begin{pmatrix}\boldsymbol{\Sigma}_X & \boldsymbol{\Sigma}_{XY} & \boldsymbol{\Sigma}_{XZ}\\ \boldsymbol{\Sigma}_{YX} & \boldsymbol{\Sigma}_Y & \boldsymbol{\Sigma}_{YZ}\\ \boldsymbol{\Sigma}_{ZX} & \boldsymbol{\Sigma}_{ZY} & \boldsymbol{\Sigma}_Z\end{pmatrix}.$$

The *linear predictor* (or the least squares prediction) of $\mathbf{X}$ given $\mathbf{Y}$ is defined by

$$\mathbf{X}|\mathbf{Y} \sim [\mathbf{m}_{X|Y}; \mathbf{C}_{X|Y}],$$

where $\mathbf{m}_{X|Y}$ and $\mathbf{C}_{X|Y}$, called respectively the *predictor* or BLUP and the *mean square error* (MSE), are given by

$$\mathbf{m}_{X|Y} = \boldsymbol{\mu}_X + \boldsymbol{\Sigma}_{XY}\boldsymbol{\Sigma}_Y^{-1}(\mathbf{Y}-\boldsymbol{\mu}_Y) \tag{9.1}$$

$$\mathbf{C}_{X|Y} = \boldsymbol{\Sigma}_X - \boldsymbol{\Sigma}_{XY}\boldsymbol{\Sigma}_Y^{-1}\boldsymbol{\Sigma}_{YX}. \tag{9.2}$$

Here Greek letters $\boldsymbol{\mu}$ and $\boldsymbol{\Sigma}$ denote the true moments, while the Roman letters $\mathbf{m}$ and $\mathbf{C}$ denote the linear predictor and the MSE, with appropriate subscripts to index particular variables in expression.

When no conditioning is involved, the linear predictor gives the true mean and variance. If $\mathbf{X}$ and $\mathbf{Y}$ are jointly multivariate normally distributed, then $\mathbf{m}_{X|Y}$ and $\mathbf{C}_{X|Y}$ coincide with the conditional mean and variance, which is otherwise not generally the case. Furthermore, $\mathbf{m}_{X|Y}$ is the linear combination of the elements of $\mathbf{Y}$ that minimizes the MSE for prediction of $\mathbf{X}$, i.e., the best linear unbiased predictor (BLUP) (e.g., Brockwell and Davis, 1991, p. 64).

In general, the linear predictor behaves much like the conditional mean in the context of the multivariate normal distribution, with some basic properties listed below.

(1) The MSE may be expressed in the form

$$\mathbf{C}_{X|Y} = \mathrm{E}\left\{(\mathbf{X} - \mathbf{m}_{X|Y})(\mathbf{X} - \mathbf{m}_{X|Y})^T\right\} = \mathrm{Var}\left(\mathbf{X} - \mathbf{m}_{X|Y}\right).$$

Thus, $\mathbf{C}_{X|Y}$ resembles $\mathrm{Var}\left\{\mathbf{X} - \mathrm{E}(\mathbf{X}|\mathbf{Y})\right\} = \mathrm{E}\{\mathrm{Var}(\mathbf{X}|\mathbf{Y})\}$.

(2) For suitable matrices or scalars $\boldsymbol{\alpha}$ and $\boldsymbol{\beta}$,

$$(\boldsymbol{\alpha}\mathbf{X} + \boldsymbol{\beta})|\mathbf{Y} \sim [\boldsymbol{\alpha}\mathbf{m}_{X|Y} + \boldsymbol{\beta}; \boldsymbol{\alpha}\mathbf{C}_{X|Y}\boldsymbol{\alpha}^T].$$

(3) For the multivariate normal distribution, the conditional mean is a linear function of $\mathbf{Y}$ and the conditional variance is not functionally dependent on the value of $\mathbf{Y}$. Similarly for linear predictors, if $\mathrm{E}(\mathbf{X}|\mathbf{Y})$ is linear in $\mathbf{Y}$, then

$$\mathbf{X}|\mathbf{Y} \sim [\mathrm{E}(\mathbf{X}|\mathbf{Y}); \mathrm{E}\{\mathrm{Var}(\mathbf{X}|\mathbf{Y})\}]. \tag{9.3}$$

(4) The prediction error $\mathbf{X} - \mathbf{m}_{X|Y}$ is uncorrelated with $\mathbf{Y}$; that is,

$$\mathrm{cov}(\mathbf{X} - \mathbf{m}_{X|Y}, \mathbf{Y}) = \mathrm{cov}(\mathbf{X}, \mathbf{Y}) - \boldsymbol{\Sigma}_{XY}\boldsymbol{\Sigma}_Y^{-1}\mathrm{Var}(\mathbf{Y}) = 0.$$

This is an important property used later.

(5) If $\mathbf{Y}$ and $\mathbf{Z}$ are uncorrelated, then $\mathbf{X}|(\mathbf{Y}, \mathbf{Z}) \sim [\mathbf{m}_{X|Y,Z}; \mathbf{C}_{X|Y,Z}]$, with

$$\mathbf{m}_{X|Y,Z} = \boldsymbol{\mu}_X + \boldsymbol{\Sigma}_{XY}\boldsymbol{\Sigma}_Y^{-1}(\mathbf{Y} - \boldsymbol{\mu}_Y) + \boldsymbol{\Sigma}_{XZ}\boldsymbol{\Sigma}_Z^{-1}(\mathbf{Z} - \boldsymbol{\mu}_Z) \tag{9.4}$$

$$\mathbf{C}_{X|Y,Z} = \boldsymbol{\Sigma}_X - \boldsymbol{\Sigma}_{XY}\boldsymbol{\Sigma}_Y^{-1}\boldsymbol{\Sigma}_{YX} - \boldsymbol{\Sigma}_{XZ}\boldsymbol{\Sigma}_Z^{-1}\boldsymbol{\Sigma}_{ZX}. \tag{9.5}$$

Theorem 9.1 shows that the joint, marginal, and conditional linear predictors behave much like the joint, marginal, and conditional means and variances for the multivariate normal distribution.

Theorem 9.1. *Let* $\mathbf{X}$, $\mathbf{Y}$, *and* $\mathbf{Z}$ *be random vectors with finite second moments. Then the joint predictor of* $\mathbf{X}, \mathbf{Y}$ *given* $\mathbf{Z}$ *is given by*

$$\begin{pmatrix} \mathbf{X} \\ \mathbf{Y} \end{pmatrix}\Bigg| \mathbf{Z} \sim \left[\begin{pmatrix} \mathbf{m}_{X|Z} \\ \mathbf{m}_{Y|Z} \end{pmatrix}; \begin{pmatrix} \mathbf{C}_{X|Z} & \mathbf{C}_{XY|Z} \\ \mathbf{C}_{YX|Z} & \mathbf{C}_{Y|Z} \end{pmatrix}\right] \tag{9.6}$$

where $\mathbf{C}_{XY|Z} = \boldsymbol{\Sigma}_{XY} - \boldsymbol{\Sigma}_{XZ}\boldsymbol{\Sigma}_Z^{-1}\boldsymbol{\Sigma}_{ZY}$. *The linear predictor of* $\mathbf{X}$ *given both* $\mathbf{Y}$ *and* $\mathbf{Z}$ *is given by*

$$\mathbf{X}\left|\begin{pmatrix}\mathbf{Y}\\\mathbf{Z}\end{pmatrix}\right. \sim \Big[\mathbf{m}_{X|Z} + \mathbf{C}_{XY|Z}\mathbf{C}_{Y|Z}^{-1}\left(\mathbf{Y} - \mathbf{m}_{Y|Z}\right); \\ \mathbf{C}_{X|Z} - \mathbf{C}_{XY|Z}\mathbf{C}_{Y|Z}^{-1}\mathbf{C}_{YX|Z}\Big]. \tag{9.7}$$

Proof. The joint predictor (9.6) is obtained directly from the definition. To show (9.7), the facts that the vector $(\mathbf{Y}, \mathbf{Z})$ has the same span as $(\mathbf{Y} - \mathbf{m}_{Y|Z}, \mathbf{Z})$ and that the two components of the latter vector are uncorrelated imply that the linear predictor of $\mathbf{X}$ given $(\mathbf{Y}, \mathbf{Z})$ is the same as that given $\{(\mathbf{Y} - \mathbf{m}_{Y|Z}), \mathbf{Z}\}$. Also, note that $\text{cov}(\mathbf{X}, \mathbf{Y} - \mathbf{m}_{Y|Z}) = \mathbf{C}_{XY|Z}$ and $\text{Var}(\mathbf{Y} - \mathbf{m}_{Y|Z}) = \mathbf{C}_{Y|Z}$. Then, (9.4) leads to

$$\begin{aligned}\mathbf{m}_{X|YZ} &= \boldsymbol{\mu}_X + \left(\text{cov}(\mathbf{X}, \mathbf{Y} - \mathbf{m}_{Y|Z}), \text{cov}(\mathbf{X}, \mathbf{Z})\right)\begin{pmatrix}\mathbf{C}_{Y|Z} & \mathbf{0}\\ \mathbf{0} & \boldsymbol{\Sigma}_Z\end{pmatrix}\begin{pmatrix}\mathbf{Y} - \mathbf{m}_{Y|Z}\\ \mathbf{Z} - \boldsymbol{\mu}_Z\end{pmatrix}\\ &= \boldsymbol{\mu}_X + \left(\mathbf{C}_{XY|Z}\mathbf{C}_{Y|Z}^{-1}, \boldsymbol{\Sigma}_{XZ}\boldsymbol{\Sigma}_Z^{-1}\right)\begin{pmatrix}\mathbf{Y} - \mathbf{m}_{Y|Z}\\ \mathbf{Z} - \boldsymbol{\mu}_Z\end{pmatrix}\\ &= \mathbf{m}_{X|Z} + \mathbf{C}_{XY|Z}\mathbf{C}_{Y|Z}^{-1}\left(\mathbf{Y} - \mathbf{m}_{Y|Z}\right).\end{aligned}$$

And (9.5) results in

$$\begin{aligned}\mathbf{C}_{X|YZ} &= \boldsymbol{\Sigma}_X - \left(\mathbf{C}_{XY|Z}\mathbf{C}_{Y|Z}^{-1}, \boldsymbol{\Sigma}_{XZ}\boldsymbol{\Sigma}_Z^{-1}\right)\begin{pmatrix}\mathbf{C}_{YX|Z}\\ \boldsymbol{\Sigma}_{ZX}\end{pmatrix}\\ &= \boldsymbol{\Sigma}_X - \mathbf{C}_{XY|Z}\mathbf{C}_{Y|Z}^{-1}\mathbf{C}_{YX|Z} - \boldsymbol{\Sigma}_{XZ}\boldsymbol{\Sigma}_Z^{-1}\boldsymbol{\Sigma}_{ZX}\\ &= \mathbf{C}_{X|Z} - \mathbf{C}_{XY|Z}\mathbf{C}_{Y|Z}^{-1}\mathbf{C}_{YX|Z}.\end{aligned}$$

This completes the proof.

Corollary 9.2. *If* $\mathbf{X}$ *and* $\mathbf{Y}$ *are conditionally uncorrelated given* $\mathbf{Z}$, *and if either* $E(\mathbf{X}|\mathbf{Z})$ *or* $E(\mathbf{Y}|\mathbf{Z})$ *is linear in* $\mathbf{Z}$, *then*

$$\mathbf{X}\left|\begin{pmatrix}\mathbf{Y}\\\mathbf{Z}\end{pmatrix}\right. \equiv \mathbf{X}|\mathbf{Z} \sim \left[\mathbf{m}_{X|Z}; \mathbf{C}_{X|Z}\right]. \tag{9.8}$$

Proof. From (9.7), if $\mathbf{C}_{XY|Z} = \mathbf{0}$, then (9.8) is true. Hence it suffices to show $\mathbf{C}_{XY|Z} = \mathbf{0}$ under the given conditions. By symmetry, it is equivalent to proving the result in the case $\text{E}(\mathbf{X}|\mathbf{Z}) = \boldsymbol{\alpha}\mathbf{Z} + \boldsymbol{\beta}$. Since $\text{cov}(\mathbf{X}, \mathbf{Y}|\mathbf{Z}) = \mathbf{0}$,

$$\begin{aligned}\boldsymbol{\Sigma}_{XY} &= \text{cov}(\mathbf{X}, \mathbf{Y})\\ &= \text{E}\{\text{cov}(\mathbf{X}, \mathbf{Y}|\mathbf{Z})\} + \text{cov}\{\text{E}(\mathbf{X}|\mathbf{Z}), \text{E}(\mathbf{Y}|\mathbf{Z})\}\\ &= \text{cov}\{\text{E}(\mathbf{X}|\mathbf{Z}), \text{E}(\mathbf{Y}|\mathbf{Z})\}\\ &= \text{cov}(\boldsymbol{\alpha}\mathbf{Z} + \boldsymbol{\beta}, \mathbf{Y})\\ &= \boldsymbol{\alpha}\boldsymbol{\Sigma}_{ZY}\end{aligned}$$

and

$$\begin{aligned}\mathbf{\Sigma}_{XZ} &= \mathrm{cov}(\mathbf{X},\mathbf{Z}) = \mathrm{cov}[\mathrm{E}(\mathbf{X}|\mathbf{Z}),\mathbf{Z}] \\ &= \mathrm{cov}(\boldsymbol{\alpha}\mathbf{Z}+\boldsymbol{\beta},\mathbf{Z}) = \boldsymbol{\alpha}\mathbf{\Sigma}_Z.\end{aligned}$$

This implies that

$$\mathbf{C}_{XY|Z} = \mathbf{\Sigma}_{XY} - \mathbf{\Sigma}_{XZ}\mathbf{\Sigma}_Z^{-1}\mathbf{\Sigma}_{ZY} = \boldsymbol{\alpha}\mathbf{\Sigma}_{ZY} - \boldsymbol{\alpha}\mathbf{\Sigma}_Z\mathbf{\Sigma}_Z^{-1}\mathbf{\Sigma}_{ZY} = \mathbf{0}.$$

This proves the corollary.

Theorem 9.3. *Assume* $\mathbf{Y}|\mathbf{Z} \sim [\mathbf{m}_{Y|Z};\mathbf{C}_{Y|Z}]$ *and* $\mathbf{X}|\mathbf{Y},\mathbf{Z} \sim [\boldsymbol{\alpha}\mathbf{Y}+\boldsymbol{\beta}\mathbf{Z}+\boldsymbol{\gamma};\mathbf{C}_{X|YZ}]$. *The joint predictor of* $\mathbf{X},\mathbf{Y}$ *given* $\mathbf{Z}$ *is*

$$\left.\begin{pmatrix}\mathbf{X}\\ \mathbf{Y}\end{pmatrix}\right|\mathbf{Z} \sim \left[\begin{pmatrix}\boldsymbol{\alpha}\mathbf{m}_{Y|Z}+\boldsymbol{\beta}\mathbf{Z}+\boldsymbol{\gamma}\\ \mathbf{m}_{Y|Z}\end{pmatrix};\begin{pmatrix}\mathbf{C}_{X|YZ}+\boldsymbol{\alpha}\mathbf{C}_{Y|Z}\boldsymbol{\alpha}^T & \boldsymbol{\alpha}\mathbf{C}_{Y|Z}\\ \mathbf{C}_{Y|Z}\boldsymbol{\alpha}^T & \mathbf{C}_{Y|Z}\end{pmatrix}\right]. \tag{9.9}$$

In particular, $\mathbf{X}|\mathbf{Z} \sim [\boldsymbol{\alpha}\mathbf{m}_{Y|Z}+\boldsymbol{\beta}\mathbf{Z}+\boldsymbol{\gamma};\mathbf{C}_{X|YZ}+\boldsymbol{\alpha}\mathbf{C}_{Y|Z}\boldsymbol{\alpha}^T]$.

Proof. First note that

$$\begin{aligned}(\boldsymbol{\alpha},\boldsymbol{\beta}) &= (\mathbf{\Sigma}_{XY},\mathbf{\Sigma}_{XZ})\begin{pmatrix}\mathbf{\Sigma}_Y & \mathbf{\Sigma}_{YZ}\\ \mathbf{\Sigma}_{ZY} & \mathbf{\Sigma}_Z\end{pmatrix}^{-1}\\ \boldsymbol{\gamma} &= \boldsymbol{\mu}_X - \boldsymbol{\alpha}\boldsymbol{\mu}_Y - \boldsymbol{\beta}\boldsymbol{\mu}_Z.\end{aligned}$$

This implies that

$$\begin{aligned}\mathbf{\Sigma}_{XY} &= \boldsymbol{\alpha}\mathbf{\Sigma}_Y + \boldsymbol{\beta}\mathbf{\Sigma}_{ZY}\\ \mathbf{\Sigma}_{XZ} &= \boldsymbol{\alpha}\mathbf{\Sigma}_{YZ} + \boldsymbol{\beta}\mathbf{\Sigma}_Z.\end{aligned}$$

It follows from the definition (9.1) that

$$\begin{aligned}\mathbf{m}_{X|Z} &= \boldsymbol{\mu}_X + (\boldsymbol{\alpha}\mathbf{\Sigma}_{YZ}+\boldsymbol{\beta}\mathbf{\Sigma}_Z)\,\mathbf{\Sigma}_Z^{-1}(\mathbf{Z}-\mathbf{g}_Z)\\ &= \boldsymbol{\mu}_X + \boldsymbol{\alpha}\left(\mathbf{m}_{Y|Z}-\boldsymbol{\mu}_Y\right) + \boldsymbol{\beta}(\mathbf{Z}-\boldsymbol{\mu}_Z)\\ &= \boldsymbol{\alpha}\mathbf{m}_{Y|Z} + \boldsymbol{\beta}\mathbf{Z} + \boldsymbol{\gamma}.\end{aligned}$$

Similarly it is seen from (9.2) that

$$\begin{aligned}\mathbf{C}_{XY|Z} &= \mathbf{\Sigma}_{XY} - \mathbf{\Sigma}_{XZ}\mathbf{\Sigma}_Z^{-1}\mathbf{\Sigma}_{ZY}\\ &= \boldsymbol{\alpha}\mathbf{\Sigma}_Y + \boldsymbol{\beta}\mathbf{\Sigma}_{ZY} - (\boldsymbol{\alpha}\mathbf{\Sigma}_{YZ}+\boldsymbol{\beta}\mathbf{\Sigma}_Z)\,\mathbf{\Sigma}_Z^{-1}\mathbf{\Sigma}_{ZY}\\ &= \boldsymbol{\alpha}\mathbf{C}_{Y|Z}.\end{aligned} \tag{9.10}$$

To derive $\mathbf{C}_{X|Z}$, note that $\boldsymbol{\beta} = (\mathbf{\Sigma}_{XZ}-\boldsymbol{\alpha}\mathbf{\Sigma}_{YZ})\,\mathbf{\Sigma}_Z^{-1}$, which gives

$$\begin{aligned}\mathbf{C}_{X|YZ}+\boldsymbol{\alpha}\mathbf{C}_{Y|Z}\boldsymbol{\alpha}^T &= \mathbf{\Sigma}_X - \boldsymbol{\alpha}\mathbf{\Sigma}_{YX} - \boldsymbol{\beta}\mathbf{\Sigma}_{ZX} + \boldsymbol{\alpha}\mathbf{C}_{Y|Z}\boldsymbol{\alpha}^T\\ &= \mathbf{\Sigma}_X - \mathbf{\Sigma}_{XZ}\mathbf{\Sigma}_Z^{-1}\mathbf{\Sigma}_{ZX} - \boldsymbol{\alpha}\left(\mathbf{C}_{YX|Z}-\mathbf{C}_{Y|Z}\boldsymbol{\alpha}^T\right).\end{aligned}$$

According to (9.10) the last term of this expression vanishes, and hence

$$\begin{aligned}\mathbf{C}_{X|Z} &= \mathbf{\Sigma}_X - \mathbf{\Sigma}_{XZ}\mathbf{\Sigma}_Z^{-1}\mathbf{\Sigma}_{ZX}\\ &= \mathbf{C}_{X|YZ} + \boldsymbol{\alpha}\mathbf{C}_{Y|Z}\boldsymbol{\alpha}^T.\end{aligned}$$

9.2 Estimation of Random Effects in GLMMs

This section discusses the estimation of random effects in the linear mixed models (LMMs) and in the generalized linear mixed models (GLMMs), respectively.

9.2.1 Estimation in LMMs

For subject/cluster i, let $\mathbf{Y}_i = (Y_{i1}, \ldots, Y_{in_i})^T$ be an n_i-dimensional vector of responses, $X_i^T = (\mathbf{x}_{i1}, \ldots, \mathbf{x}_{in_i})$ be a $p \times n_i$ matrix, and $Z_i^T = (\mathbf{z}_{i1}, \ldots, \mathbf{z}_{in_i})$ be a $q \times n_i$ matrix. A LMM can be rewritten as of the matrix form as follows:

$$\mathbf{y}_i = X_i\boldsymbol{\beta} + Z_i\mathbf{b}_i + \boldsymbol{\varepsilon}_i$$

where $\boldsymbol{\varepsilon}_i = (\varepsilon_{i1}, \ldots, \varepsilon_{in_i})^T \sim \mathrm{MVN}_{n_i}(0, \Sigma_\varepsilon)$, $\mathbf{b}_i \sim \mathrm{MVN}_q(0, D)$, and $\boldsymbol{\varepsilon}_i$ and $\mathbf{b}_i$ are independent. Note that since random effects are cluster-specific, only data from cluster i will be relevant in the estimation of $\mathbf{b}_i$.

Under the multivariate normality assumption, according to (9.1) the BLUP of $\mathbf{b}_i$ is then the conditional expectation, $\mathrm{E}(\mathbf{b}_i|\tilde{\mathbf{Y}}_i)$, with $\tilde{\mathbf{Y}}_i = \mathbf{Y}_i - X_i\boldsymbol{\beta}$, which is given as follows:

$$\begin{aligned}\mathrm{E}(\mathbf{b}_i|\tilde{\mathbf{Y}}_i) &= \mathrm{E}(\mathbf{b}_i) + \mathrm{cov}(\mathbf{b}_i, \tilde{\mathbf{Y}}_i)\mathrm{Var}(\tilde{\mathbf{Y}}_i)^{-1}\{\tilde{\mathbf{Y}}_i - \mathrm{E}(\tilde{\mathbf{Y}}_i)\} \\ &= DZ_i^T\{Z_iDZ_i^T + \Sigma_\varepsilon\}^{-1}(\mathbf{Y}_i - X_i\boldsymbol{\beta}).\end{aligned}$$

In particular, when ε_{ij}'s are *i.i.d.* $N(0, \sigma^2)$, $\Sigma_\varepsilon = \sigma^2 I_{n_i}$. In addition, by (9.2) the MSE of the BLUP is

$$\begin{aligned}\mathrm{E}\{\mathrm{Var}(\mathbf{b}_i|\tilde{\mathbf{Y}}_i)\} &= \mathrm{Var}(\mathbf{b}_i) - \mathrm{cov}(\mathbf{b}_i, \tilde{\mathbf{Y}}_i)\mathrm{Var}(\tilde{\mathbf{Y}}_i)^{-1}\mathrm{cov}(\tilde{\mathbf{Y}}_i, \mathbf{b}_i) \\ &= D - DZ_i^T\{Z_iDZ_i^T + \Sigma_\varepsilon\}^{-1}Z_iD.\end{aligned}$$

When estimates $\widehat{\boldsymbol{\beta}}$ and $\widehat{D}$ as well as $\widehat{\sigma}^2$ are available, the estimated BLUP of the $\mathbf{b}_i$ and the corresponding MSE are given by

$$\begin{aligned}\widehat{\mathbf{b}}_i &= \widehat{D}Z_i^T\{Z_i\widehat{D}Z_i^T + \widehat{\sigma}^2 I_{n_i}\}^{-1}(\mathbf{Y}_i - X_i\widehat{\boldsymbol{\beta}}), \\ \widehat{\mathrm{MSE}}(\widehat{\mathbf{b}}_i) &= \widehat{D} - \widehat{D}Z_i^T\{Z_i\widehat{D}Z_i^T + \widehat{\sigma}^2 I_{n_i}\}^{-1}Z_i\widehat{D}.\end{aligned}$$

When the cluster size n_i is small, the performance of the above estimates may be unstable.

9.2.2 Estimation in GLMMs

In general, the theory of BLUP is not directly applicable for the estimation of the random effects in the GLMMs. An approximate inference method has been suggested in the literature to estimate the random effects iteratively.

Let $\boldsymbol{\beta}^{(1)}, \mathbf{b}_i^{(1)}, D^{(1)}, (\sigma^2)^{(1)}$ be the updates of the model parameters obtained at the previous iteration 1. At the current iteration 2, define surrogate responses

$$Y_{ij}^* = g(\mu_{ij}^b) + (Y_{ij} - \mu_{ij}^b)\dot{g}(\mu_{ij}^b), \; j = 1, \ldots, n_i, \; i = 1, \ldots, K,$$

and set errors

$$\varepsilon_{ij}^* = (Y_{ij} - \mu_{ij}^b)\dot{g}(\mu_{ij}^b).$$

In matrix notation, the above equations can be rewritten into the following form:

$$\mathbf{Y}_i^* = X_i\boldsymbol{\beta} + Z_i\mathbf{b}_i + \boldsymbol{\varepsilon}_i^*,$$

where $\mathbf{Y}_i^* = (Y_{i1}^*, \ldots, Y_{in_i}^*)^T$ and the variance matrix of $\boldsymbol{\varepsilon}_i^*$ is

$$\Sigma_{\varepsilon^*,i} = \operatorname{diag}[\sigma^2 V(\mu_{ij}^b)\{\dot{g}(\mu_{ij}^b)\}^2], \; i = 1, \ldots, K,$$

which are computed on the basis of the previous updates. Then, after $\boldsymbol{\beta}$ and D being updated first at iteration 2, the random effects are then updated by

$$\mathbf{b}_i^{(2)} = D^{(2)} Z_i^T \left\{ Z_i D^{(2)} Z_i^T + \Sigma_{\varepsilon^*,i}^{(1)} \right\}^{-1} \left(\mathbf{Y}_i^{*(1)} - X_i\boldsymbol{\beta}^{(2)} \right),$$

with the corresponding MSE given by

$$\operatorname{MSE}(\widehat{\mathbf{b}}_i^{(2)}) = D^{(2)} - D^{(2)} Z_i^T \left\{ Z_i D^{(2)} Z_i^T + \Sigma_{\varepsilon^*,i}^{(1)} \right\}^{-1} Z_i D^{(2)}.$$

9.3 Kalman Filter and Smoother

This section presents both Kalman filter and smoother under a general framework, where only the mean structure is assume to be linear, extending the classical theory given by, for example, Harvey (1981). In other words, the classical Kalman filtering and smoothing recursions would be special cases of the formulas given in this section.

9.3.1 General Forms

Assume that two vector-valued stochastic processes $\{\mathbf{Y}_t\}_{t=1}^n \in \mathcal{R}^p$ and $\{\boldsymbol{\theta}_t\}_{t=0}^n \in \mathcal{R}^q$ follow the *comb structure* defined as follows:

$(A1)$ Given $\boldsymbol{\theta}_t$, $\mathbf{Y}_t$ is uncorrelated with the rest of the $\mathbf{Y}_t$'s;
$(A2)$ $\{\boldsymbol{\theta}_t\}$ is a first-order Markov process.

Figure 9.1 gives a graphic illustration of the comb structure at time $t-1, t, t+1$. In addition, the two processes are assumed to satisfy the following first and second moment structures:

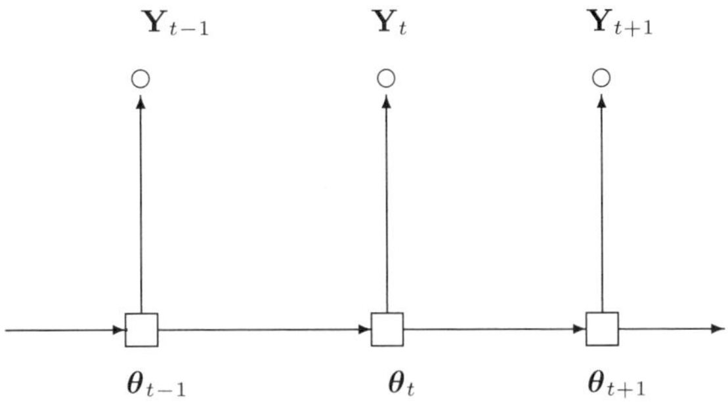

Fig. 9.1. The comb structure.

$(A3)$ $\mathrm{E}\left(\mathbf{Y}_t|\boldsymbol{\theta}_t\right) = \mathbf{A}_t\boldsymbol{\theta}_t + \mathbf{a}_t$, and $\mathrm{Var}\left(\mathbf{Y}_t|\boldsymbol{\theta}_t\right) = \mathbf{W}_t(\boldsymbol{\theta}_t) + \mathbf{W}_t^0$;

and

$(A4)$ $\mathrm{E}\left(\boldsymbol{\theta}_t|\boldsymbol{\theta}_{t-1}\right) = \mathbf{B}_t\boldsymbol{\theta}_{t-1} + \mathbf{b}_t$, and $\mathrm{Var}\left(\boldsymbol{\theta}_t|\boldsymbol{\theta}_{t-1}\right) = \mathbf{D}_t(\boldsymbol{\theta}_{t-1}) + \mathbf{D}_t^0$.

Here the first moment structures are assumed to be linear, but the second moment structures may be nonlinear. In the second moments, both $\mathbf{W}_t(\boldsymbol{\theta}_t)$ and $\mathbf{D}_t(\boldsymbol{\theta}_t)$ are matrices of functions in $\boldsymbol{\theta}_t$ entrywise, with constant terms $\mathbf{W}_t^0$ and $\mathbf{D}_t^0$, respectively. The inclusion of these constant terms explicitly is just for the mathematical convenience in the development of theories later in Chapters 11 and 12. Denote $\mathrm{E}\left\{\mathbf{W}_t(\boldsymbol{\theta}_t)\right\} = \bar{\mathbf{W}}_t$, $\mathrm{E}\left\{\mathbf{D}_t(\boldsymbol{\theta}_t)\right\} = \bar{\mathbf{D}}_t$ and $\mathrm{E}(\boldsymbol{\theta}_t) = \boldsymbol{\tau}_t$.

Let $\mathbf{Y}^t$ be the set of the first t vectors $\mathbf{Y}_1, \ldots, \mathbf{Y}_t$. In general, the Kalman filter is defined as the BLUP of $\boldsymbol{\theta}_t$ given $\mathbf{Y}^t$, and the Kalman smoother is the BLUP of $\boldsymbol{\theta}_t$ based on the all observations $\mathbf{Y}^n$. Both predictions are calculated via recursive procedures given in the following theorems, respectively.

Theorem 9.4. *Under assumptions (A1)–(A4), for the given prediction at time* $t-1$, $\boldsymbol{\theta}_{t-1}|\mathbf{Y}^{t-1} \sim [\mathbf{m}_{t-1}; \mathbf{C}_{t-1}]$, *the Kalman filter proceeds recursively as follows:*

Step 1: Compute two predictions

$$\boldsymbol{\theta}_t|\mathbf{Y}^{t-1} \sim [\mathbf{B}_t\mathbf{m}_{t-1} + \mathbf{b}_t;\ \mathbf{H}_t] \text{ and } \mathbf{Y}_t|\mathbf{Y}^{t-1} \sim [\mathbf{f}_t;\ \mathbf{Q}_t]$$

where

$$\begin{aligned} H_t &= \bar{\mathbf{D}}_{t-1} + \mathbf{D}_t^0 + \mathbf{B}_t\mathbf{C}_{t-1}\mathbf{B}_t^T \\ \mathbf{f}_t &= \mathbf{A}_t\left(\mathbf{B}_t\mathbf{m}_{t-1} + \mathbf{b}_t\right) + \mathbf{a}_t,\ \mathbf{Q}_t = \bar{\mathbf{F}}_{t-1} + \mathbf{F}_t^0 + \mathbf{A}_t\mathbf{B}_t\mathbf{C}_{t-1}\mathbf{B}_t^T\mathbf{A}_t^T. \end{aligned}$$

Step 2: Update the prediction of $\boldsymbol{\theta}_t$ *given* $\mathbf{Y}^t$*,*

$$\boldsymbol{\theta}_t|\mathbf{Y}^t \sim [\mathbf{m}_t;\ \mathbf{C}_t]$$

where

$$\mathbf{m}_t = \mathbf{B}_t\mathbf{m}_{t-1} + \mathbf{b}_t + \mathbf{H}_t^T\mathbf{A}_t^T\mathbf{Q}_t^{-1}(\mathbf{Y}_t - \mathbf{f}_t), \text{ and } \mathbf{C}_t = \mathbf{H}_t - \mathbf{H}_t^T\mathbf{A}_t^T\mathbf{Q}_t^{-1}\mathbf{A}_t\mathbf{H}_t.$$

Note that the Kalman filtering recursions move forward from time 1 to n.

Theorem 9.5. *Suppose that the Kalman filtering has been complete. Under assumptions (A1)-(A4), the Kalman smoother proceeds backwards as follows: Given the prediction at time* $t+1$, $\boldsymbol{\theta}_{t+1}|\mathbf{Y}^n \sim [\mathbf{m}^*_{t+1};\ \mathbf{C}^*_{t+1}]$, *the Kalman smoother and the corresponding error are given by*

$$\mathbf{m}^*_t = \mathbf{m}_t + \mathbf{C}_t\mathbf{B}^T_{t+1}\mathbf{H}^{-1}_{t+1}(\mathbf{m}^*_{t+1} - \mathbf{B}_{t+1}\mathbf{m}_t - \mathbf{b}_{t+1})$$

and

$$\mathbf{C}^*_t = \mathbf{C}_t - \mathbf{C}_t\mathbf{B}^T_{t+1}\mathbf{H}^{-1}_{t+1}\mathbf{B}_{t+1}\mathbf{C}_t + \mathbf{C}_t\mathbf{B}^T_{t+1}\mathbf{H}^{-1}_{t+1}\mathbf{C}^*_{t+1}\mathbf{H}^{-T}_{t+1}\mathbf{B}_{t+1}\mathbf{C}_t.$$

The recursion starts with $t = n$ *by taking* $\mathbf{m}^*_n = \mathbf{m}_n$ *and* $\mathbf{C}^*_n = \mathbf{C}_n$.

These two theorems are proved below.

Proof. (Theorem 9.4) It follows from the assumptions (A1)-(A4) that

$$\begin{aligned}
\mathrm{E}(\mathbf{Y}_t|\boldsymbol{\theta}_{t-1}) &= \mathbf{A}_t\mathbf{B}_t\boldsymbol{\theta}_{t-1} + \mathbf{A}_t\mathbf{b}_t + \mathbf{a}_t, \qquad (9.11)\\
\mathrm{Var}(\mathbf{Y}_t|\boldsymbol{\theta}_{t-1}) &= \mathrm{E}\{\mathrm{Var}(\mathbf{Y}_t|\boldsymbol{\theta}_t)|\boldsymbol{\theta}_{t-1}\} + \mathrm{Var}\{\mathrm{E}(\mathbf{Y}_t|\boldsymbol{\theta}_t)|\boldsymbol{\theta}_{t-1}\}\\
&= \mathbf{F}_t(\boldsymbol{\theta}_{t-1}) + \mathbf{F}^0_t\\
\mathrm{cov}(\mathbf{Y}_t, \boldsymbol{\theta}_t|\boldsymbol{\theta}_{t-1}) &= \mathrm{cov}\{\mathrm{E}(\mathbf{Y}_t|\boldsymbol{\theta}_t), \boldsymbol{\theta}_t|\boldsymbol{\theta}_{t-1}\} = \mathbf{A}_t\mathrm{cov}(\boldsymbol{\theta}_t, \boldsymbol{\theta}_t|\boldsymbol{\theta}_{t-1})\\
&= \mathbf{A}_t\mathbf{D}_t(\boldsymbol{\theta}_{t-1}) + \mathbf{A}_t\mathbf{D}^0_t,
\end{aligned}$$

where

$$\mathbf{F}_t(\boldsymbol{\theta}_{t-1}) = \mathrm{E}\{\mathbf{W}_t(\boldsymbol{\theta}_t)|\boldsymbol{\theta}_{t-1}\} + \mathbf{A}_t\mathbf{D}_t(\boldsymbol{\theta}_{t-1})\mathbf{A}^T_t, \text{ and } \mathbf{F}^0_t = \mathbf{W}^0_t + \mathbf{A}_t\mathbf{D}^0_t\mathbf{A}^T_t.$$

Since both conditional expectations $\mathrm{E}(\mathbf{Y}_t|\boldsymbol{\theta}_{t-1})$ and $\mathrm{E}(\boldsymbol{\theta}_t|\boldsymbol{\theta}_{t-1})$ are linear in $\boldsymbol{\theta}_{t-1}$, by the property (9.3),

$$\begin{pmatrix}\mathbf{Y}_t\\ \boldsymbol{\theta}_t\end{pmatrix}\Bigg|\,\boldsymbol{\theta}_{t-1} \sim \left[\begin{pmatrix}\mathbf{A}_t\mathbf{B}_t\boldsymbol{\theta}_{t-1} + \mathbf{A}_t\mathbf{b}_t + \mathbf{a}_t\\ \mathbf{B}_t\boldsymbol{\theta}_{t-1} + \mathbf{b}_t\end{pmatrix};\right.$$
$$\left.\begin{pmatrix}\bar{\mathbf{F}}_{t-1} + \mathbf{F}^0_t & \mathbf{A}_t\bar{\mathbf{D}}_{t-1} + \mathbf{A}_t\mathbf{D}^0_t\\ \bar{\mathbf{D}}^T_{t-1}\mathbf{A}^T_t + \mathbf{D}^{0T}_t\mathbf{A}^T_t & \bar{\mathbf{D}}_{t-1} + \mathbf{D}^0_t\end{pmatrix}\right] \qquad (9.12)$$

where

$$\begin{aligned}
\bar{\mathbf{F}}_{t-1} &= \mathrm{E}\{\mathbf{F}_t(\boldsymbol{\theta}_{t-1})\}\\
&= \mathrm{E}[\mathrm{E}\{\mathbf{W}_t(\boldsymbol{\theta}_t)|\boldsymbol{\theta}_{t-1}\}] + \mathbf{A}_t\mathrm{E}\{\mathbf{D}_t(\boldsymbol{\theta}_{t-1})\}\mathbf{A}^T_t\\
&= \bar{\mathbf{W}}_t + \mathbf{A}_t\bar{\mathbf{D}}_{t-1}\mathbf{A}^T_t.
\end{aligned}$$

With the availability of the filter at $t-1$, at the present step, assume that one currently knows

$$\boldsymbol{\theta}_{t-1}|\mathbf{Y}^{t-1} \sim [\mathbf{m}_{t-1};\ \mathbf{C}_{t-1}].$$

Theorem 9.3 leads to the predictions of $(\mathbf{Y}_t, \boldsymbol{\theta}_t)$ given $\mathbf{Y}^{t-1}$ as follows:

$$\left.\begin{pmatrix} \mathbf{Y}_t \\ \boldsymbol{\theta}_t \end{pmatrix}\right| \mathbf{Y}^{t-1} \sim \left[\widetilde{\mathbf{m}}_{t|t-1};\ \widetilde{\mathbf{C}}_{t|t-1}\right] \tag{9.13}$$

where the predictor is

$$\widetilde{\mathbf{m}}_{t|t-1} = \begin{pmatrix} \mathbf{A}_t \\ \mathbf{I} \end{pmatrix} \mathbf{B}_t \mathbf{m}_{t-1} + \begin{pmatrix} \mathbf{A}_t \\ \mathbf{I} \end{pmatrix} \mathbf{b}_t + \begin{pmatrix} \mathbf{a}_t \\ \mathbf{0} \end{pmatrix},$$

with the prediction error

$$\begin{aligned} \widetilde{\mathbf{C}}_{t|t-1} = &\begin{pmatrix} \bar{\mathbf{F}}_{t-1} + \mathbf{F}_t^0 & \mathbf{A}_t \bar{\mathbf{D}}_{t-1} + \mathbf{A}_t \mathbf{D}_t^0 \\ \bar{\mathbf{D}}_{t-1}^T \mathbf{A}_t^T + \mathbf{D}_t^{0T} \mathbf{A}_t^T & \bar{\mathbf{D}}_{t-1} + \mathbf{D}_t^0 \end{pmatrix} \\ &+ \begin{pmatrix} \mathbf{A}_t \mathbf{B}_t \mathbf{C}_{t-1} \mathbf{B}_t^T \mathbf{A}_t^T & \mathbf{A}_t \mathbf{B}_t \mathbf{C}_{t-1} \mathbf{B}_t^T \\ \mathbf{B}_t \mathbf{C}_{t-1} \mathbf{B}_t^T \mathbf{A}_t^T & \mathbf{B}_t \mathbf{C}_{t-1} \mathbf{B}_t^T \end{pmatrix}. \end{aligned}$$

In particular,

$$\boldsymbol{\theta}_t|\mathbf{Y}^{t-1} \sim [\mathbf{B}_t \mathbf{m}_{t-1} + \mathbf{b}_t;\ \mathbf{H}_t]$$

with

$$\mathbf{H}_t = \bar{\mathbf{D}}_{t-1} + \mathbf{D}_t^0 + \mathbf{B}_t \mathbf{C}_{t-1} \mathbf{B}_t^T.$$

In addition,

$$\mathbf{Y}_t|\mathbf{Y}^{t-1} \sim [\mathbf{f}_t;\ \mathbf{Q}_t]$$

with

$$\mathbf{f}_t = \mathbf{A}_t \left(\mathbf{B}_t \mathbf{m}_{t-1} + \mathbf{b}_t\right) + \mathbf{a}_t,\ \ \mathbf{Q}_t = \bar{\mathbf{F}}_{t-1} + \mathbf{F}_t^0 + \mathbf{A}_t \mathbf{B}_t \mathbf{C}_{t-1} \mathbf{B}_t^T \mathbf{A}_t^T.$$

Now update the prediction of $\boldsymbol{\theta}_t$ given a new vector $\mathbf{Y}_t$ and $\mathbf{Y}^{t-1}$ from the expression (9.13). It follows from Theorem 9.1 that

$$\boldsymbol{\theta}_t|\mathbf{Y}^t \sim [\mathbf{m}_t;\ \mathbf{C}_t]$$

where

$$\begin{aligned} \mathbf{m}_t &= \mathbf{B}_t \mathbf{m}_{t-1} + \mathbf{b}_t + \left(\bar{\mathbf{D}}_{t-1} + \mathbf{D}_t^0 + \mathbf{B}_t \mathbf{C}_{t-1} \mathbf{B}_t^T\right)^T \mathbf{A}^T \\ &\quad \left(\bar{\mathbf{F}}_{t-1} + \mathbf{F}_t^0 + \mathbf{A}_t \mathbf{B}_t \mathbf{C}_{t-1} \mathbf{B}_t^T \mathbf{A}_t^T\right)^{-1} (\mathbf{Y}_t - \mathbf{f}_t) \\ &= \mathbf{B}_t \mathbf{m}_{t-1} + \mathbf{b}_t + \mathbf{H}_t^T \mathbf{A}_t^T \mathbf{Q}_t^{-1} (\mathbf{Y}_t - \mathbf{f}_t) \\ \mathbf{C}_t &= \mathbf{H}_t - \mathbf{H}_t^T \mathbf{A}_t^T \mathbf{Q}_t^{-1} \mathbf{A}_t \mathbf{H}_t. \end{aligned}$$

The proof of Theorem 9.4 is complete.

Proof. (Theorem 9.5) Suppose that the current Kalman smoother is available

$$\boldsymbol{\theta}_{t+1}|\mathbf{Y}^n \sim \left[\mathbf{m}_{t+1}^*;\ \mathbf{C}_{t+1}^*\right] \tag{9.14}$$

and the entire sequence of the Kalman filters have been completed

$$\boldsymbol{\theta}_t|\mathbf{Y}^t \sim \left[\mathbf{m}_t;\ \mathbf{C}_t\right],\ t = 1, \ldots, n. \tag{9.15}$$

Note that for given $\boldsymbol{\theta}_t$, $\boldsymbol{\theta}_{t+1}$ is independent of $\mathbf{Y}^t$ by the assumptions (A1)–(A4), then from (9.12),

$$\boldsymbol{\theta}_{t+1}|\mathbf{Y}^t, \boldsymbol{\theta}_t \stackrel{d}{\equiv} \boldsymbol{\theta}_{t+1}|\boldsymbol{\theta}_t \sim \left[\mathbf{B}_{t+1}\boldsymbol{\theta}_t + \mathbf{b}_{t+1};\ \bar{\mathbf{D}}_t + \mathbf{D}_{t+1}^0\right], \tag{9.16}$$

where $U \stackrel{d}{\equiv} V$ means that random variables U and V are identically distributed. By Theorem 9.3, (9.15) and (9.16),

$$\begin{pmatrix} \boldsymbol{\theta}_t \\ \boldsymbol{\theta}_{t+1} \end{pmatrix} \Bigg| \mathbf{Y}^t \sim \left[\begin{pmatrix} \mathbf{m}_t \\ \mathbf{B}_{t+1}\mathbf{m}_t + \mathbf{b}_{t+1} \end{pmatrix};\begin{pmatrix} \mathbf{C}_t & \mathbf{C}_t\mathbf{B}_{t+1}^T \\ \mathbf{B}_{t+1}\mathbf{C}_t & \mathbf{H}_{t+1} \end{pmatrix}\right]. \tag{9.17}$$

Furthermore, it follows from Theorem 9.1 and (9.17) that

$$\begin{aligned} \boldsymbol{\theta}_t|\boldsymbol{\theta}_{t+1}, \mathbf{Y}^t \sim \big[&\mathbf{m}_t + \mathbf{C}_t\mathbf{B}_{t+1}^T\mathbf{H}_{t+1}^{-1}(\boldsymbol{\theta}_{t+1} - \mathbf{B}_{t+1}\mathbf{m}_t - \mathbf{b}_{t+1}); \\ &\mathbf{C}_t - \mathbf{C}_t\mathbf{B}_{t+1}^T\mathbf{H}_{t+1}^{-1}\mathbf{B}_{t+1}\mathbf{C}_t\big]. \end{aligned} \tag{9.18}$$

Similar to the calculation of the equation (9.11), it is easy to prove that for each $s \geq 1$, $\mathrm{E}(\mathbf{Y}_{t+s}|\boldsymbol{\theta}_{t+1})$ is linear in $\boldsymbol{\theta}_{t+1}$. From the Corollary 9.2,

$$\boldsymbol{\theta}_t|\boldsymbol{\theta}_{t+1}, \mathbf{Y}^n \stackrel{d}{\equiv} \boldsymbol{\theta}_t|\boldsymbol{\theta}_{t+1}, \mathbf{Y}^t$$

and again from Theorem 9.3 and (9.14),

$$\begin{pmatrix} \boldsymbol{\theta}_t \\ \boldsymbol{\theta}_{t+1} \end{pmatrix} \Bigg| \mathbf{Y}^n \sim \left[\begin{pmatrix} \mathbf{m}_t^* \\ \mathbf{m}_{t+1}^* \end{pmatrix};\begin{pmatrix} \mathbf{C}_t^* & \mathbf{C}_t\mathbf{B}_{t+1}^T\mathbf{H}_{t+1}^T\mathbf{C}_{t+1}^* \\ \mathbf{C}_{t+1}^*\mathbf{H}_{t+1}\mathbf{B}_{t+1}\mathbf{C}_t & \mathbf{C}_{t+1}^* \end{pmatrix}\right].$$

Here $\mathbf{m}_t^*$ and $\mathbf{C}_t^*$ are the Kalman smoother and the corresponding prediction error given by

$$\begin{aligned} \mathbf{m}_t^* &= \mathbf{m}_t + \mathbf{C}_t\mathbf{B}_{t+1}^T\mathbf{H}_{t+1}^{-1}(\mathbf{m}_{t+1}^* - \mathbf{B}_{t+1}\mathbf{m}_t - \mathbf{b}_{t+1}) \\ \mathbf{C}_t^* &= \mathbf{C}_t - \mathbf{C}_t\mathbf{B}_{t+1}^T\mathbf{H}_{t+1}^{-1}\mathbf{B}_{t+1}\mathbf{C}_t + \mathbf{C}_t\mathbf{B}_{t+1}^T\mathbf{H}_{t+1}^{-1}\mathbf{C}_{t+1}^*\mathbf{H}_{t+1}^{-T}\mathbf{B}_{t+1}\mathbf{C}_t. \end{aligned}$$

The proof of Theorem 9.5 is complete.

10

Generalized State Space Models

10.1 Introduction

Generalized state space models (GSSM) refer to a class of nonlinear state space models with parametric error distributions, which are possibly non-normal. The class of models attempts to model the dynamic feature of time series of, for example, counts or binary observations. Such models and the associated Kalman filtering technique have had a profound impact on time series analysis and longitudinal data analysis when the number of repeated observations is large. First introduced by Kalman (1960) in connection to the theory of controls in linear systems, state space models appear very flexible in the modeling of certain stochastic systems and include Box and Jenkins' linear ARIMA models as a special case. In practice, many types of models may be formulated in the form of state space models; for example, the structural time series models by Harvey (1990) and mean-drifting time series models considered by Kitagawa (1987).

Although the ordinary linear Gaussian state space model has been well studied in the literature, in many practical studies, time series are primarily not measurements from normal distributions. In biomedical and health sciences, time series of counts or binary observations are often collected. For instance, in Section 1.3.9 readers have seen a time series of binomial observations, and in Section 1.3.10, a 4-dimensional time series of counts. In comparison to the conventional time series analysis, longitudinal data analysis focuses on modeling systematic trends and estimating effects of time-varying covariates at both first two moments and transition structures, with the incorporation of stochastic mechanism governed by a certain time series model.

Generalized state space models are flexible and suitable for a variety of longitudinal data, which has been demonstrated in the literature such as West and Harrison (1997), Kitagawa (1987), Carlin et al. (1992), and Jørgensen et al. (1999), among others. A crucial step in the application of all these models to data analysis is to estimate the conditional distribution of state variables given all or part of the data as well as the other unknown model parameters. In an abstract fashion, there is no difficulty in deriving MLE, but the actual implementation is extremely challenging due to the lack of closed form expressions for typically very high-dimensional integrals in the likelihood function. Thus, one has to rely on various approximations that essentially fall into three categories: analytic, Monte Carlo, and numerical.

An analytic solution concerning the approximation may be given by the extended Kalman filtering recursions based on BLUP, discussed in Section 9.3. Monte Carlo approximation in the form of MCMC has received most attention in recent years; see, for example, Carlin et al. (1992) and de Jong and Shephard (1995). Another version of Monte Carlo method is proposed by Durbin and Koopman (1997) in that they derived a direct approximation to the log-likelihood for generalized state space models and invoked a Monte Carlo simulation via importance sampling scheme. The numerical approximation proposed by Kitagawa (1987) uses a crude numerical evaluation of integration over the state space via the piece-wise linear approximation. One shortcoming of this method is that it does not utilize related distributional properties in the approximation, and hence the resulting approximation is not satisfactorily accurate, unless a very large number of nodes are used.

This book concentrates on two methods: the BLUP based frequentist inference and MCMC based Bayesian inference. The selection of the topics is made purely according to the author's familiarity with them, given that there are some alternatives also working well in inference. Durbin and Koopman's simulated MLE will be discussed briefly at the end of this Chapter. The BLUP based inference essentially resembles the EM algorithm, in which the E-step is obtained approximately by BLUP and the M-step proceeds to maximizing an augmented likelihood as usual. This method is termed as *Kalman Estimating Equation* (KEE) in this book. As discussed in the context of generalized linear mixed-effects models in Chapter 8, MCMC is appealing to overcome the difficulty of high-dimensional integration when priors and convergence diagnostics are carefully handled. In particular, Chapter 11 illustrates that de Jong and Shephard's (1995) simulation smoother is an efficient sampler, which considerably enhances the computational speed in the analysis of binomial longitudinal data.

Let us begin with the model. A generalized state space model consists of two stochastic processes: an d-dimensional observation process $\{\mathbf{Y}_t\}$ and a q-dimensional state process $\{\boldsymbol{\theta}_t\}$ given as follows.

$\mathbf{M}_1$: The state process $\boldsymbol{\theta}_0, \boldsymbol{\theta}_1, \ldots,$ is a Markov chain with initial condition $\boldsymbol{\theta}_0 \sim p_0(\boldsymbol{\theta})d\boldsymbol{\theta}$ and transition (conditional) distribution is given by

$$\boldsymbol{\theta}_t|\boldsymbol{\theta}_{t-1} \sim g_t(\boldsymbol{\theta}|\boldsymbol{\theta}_{t-1})d\boldsymbol{\theta}. \tag{10.1}$$

$\mathbf{M}_2$: The observation process $\{\mathbf{Y}_t\}$ are conditionally independent given the state process $\{\boldsymbol{\theta}_t, t \geq 0\}$ and each $\mathbf{Y}_t$ is conditionally independent of $\boldsymbol{\theta}_s, s \neq t$; Given $\boldsymbol{\theta}_t$, the conditional distribution is

$$\mathbf{Y}_t|\boldsymbol{\theta}_t \sim f_t(\mathbf{y}|\boldsymbol{\theta}_t)d\mathbf{y}. \tag{10.2}$$

This model can be graphically presented by a comb structure shown in Figure 10.1.

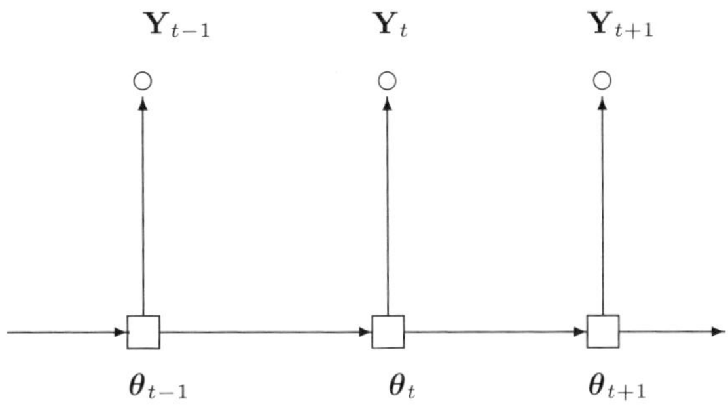

Fig. 10.1. Graphic representation of a generalized state space model.

Let $\mathbf{Y}^s$ be the collection of all observations up to time s, namely $\mathbf{Y}^s = (\mathbf{Y}_1, \ldots, \mathbf{Y}_s)$. Denote the conditional density of $\boldsymbol{\theta}_t$, given $\mathbf{Y}^s = \mathbf{y}^s$, by $f_{t|s}(\boldsymbol{\theta}|\mathbf{y}^s)$. Then, the prediction, filter, or smoother density is defined, respectively, according to whether $t > s$, $t = s$ or $t < s$. This conditional density $f_{t|s}(\boldsymbol{\theta}|\mathbf{y}^s)$ is the key component of statistical inference in GSSMs.

In particular, one-step prediction densities, $f_{t|t-1}$, and filter densities, $f_{t|t}$, can be given by the Kalman recursions, respectively:

$$f_{t|t-1}\left(\boldsymbol{\theta}_t|\mathbf{y}^{t-1}\right) = \int_{\mathcal{R}^q} f_{t-1|t-1}\left(\boldsymbol{\theta}_{t-1}|\mathbf{y}^{t-1}\right) g_t(\boldsymbol{\theta}_t|\boldsymbol{\theta}_{t-1}) d\boldsymbol{\theta}_{t-1}, \quad (10.3)$$

$$f_{t|t}\left(\boldsymbol{\theta}_t|\mathbf{y}^t\right) = \frac{f_{t|t-1}\left(\boldsymbol{\theta}_t|\mathbf{y}^{t-1}\right) f_t(\mathbf{y}_t|\boldsymbol{\theta}_t)}{\int_{\mathcal{R}^q} f_{t|t-1}\left(\boldsymbol{\theta}_t|\mathbf{y}^{t-1}\right) f_t(\mathbf{y}_t|\boldsymbol{\theta}_t) d\boldsymbol{\theta}_t}, \quad (10.4)$$

with the recursion starting with $f_{0|0}(\boldsymbol{\theta}) = p_0(\boldsymbol{\theta})$. In general, exact evaluation of the integrals in (10.3) and (10.4) is analytically unavailable, unless in some simple situations, such as both processes $\mathbf{M}_1$ and $\mathbf{M}_2$ being linear and normally distributed. For the linear Gaussian state space model, all $f_{t|s}$ are Gaussian, so the first two moments of (10.3) and (10.4) can be easily derived from the conventional Kalman filtering procedure, as discussed in Section 9.3.

To develop the maximum likelihood inference for model parameters in GSSMs, the one-step prediction densities $f_{t|t-1}$ are the key components for the computation of the likelihood function. Given a time series data $\{\mathbf{Y}_t, t = 1, \ldots, n\}$, the likelihood of $\mathbf{Y}^n$ is

$$
\begin{aligned}
f(\mathbf{Y}^n) &= \int_{\mathcal{R}^q} f(\mathbf{Y}_1, \ldots, \mathbf{Y}_{n-1}|\boldsymbol{\theta}_n) f_n(\mathbf{Y}_n|\boldsymbol{\theta}_n) g_n(\boldsymbol{\theta}_n) d\boldsymbol{\theta}_n \\
&= \int_{\mathcal{R}^q} f(\mathbf{Y}_1, \ldots, \mathbf{Y}_{n-1}) f(\boldsymbol{\theta}_n|\mathbf{Y}_1, \ldots, \mathbf{Y}_{n-1}) f_n(\mathbf{Y}_n|\boldsymbol{\theta}_n) d\boldsymbol{\theta}_n \\
&= \cdots\cdots \text{(applying the same trick repeatedly)} \\
&= \prod_{t=2}^{n} f(\mathbf{Y}_1) \int_{\mathcal{R}^q} f_{t|t-1}\left(\boldsymbol{\theta}_t|\mathbf{Y}^{t-1}\right) f_t(\mathbf{Y}_t|\boldsymbol{\theta}_t) d\boldsymbol{\theta}_t \\
&= \prod_{t=1}^{n} \int_{\mathcal{R}^q} f_{t|t-1}\left(\boldsymbol{\theta}_t|\mathbf{Y}^{t-1}\right) f_t(\mathbf{Y}_t|\boldsymbol{\theta}_t) d\boldsymbol{\theta}_t,
\end{aligned}
$$

where $f_1(\mathbf{Y}_1)$ is expressed as follows:

$$
f_1(\mathbf{Y}_1) = \int_{\mathcal{R}^q} f_1(\mathbf{Y}_1|\boldsymbol{\theta}_1) g_1(\boldsymbol{\theta}_1) d\boldsymbol{\theta}_1 = \int_{\mathcal{R}^q} f_1(\mathbf{Y}_1|\boldsymbol{\theta}_1) f_{1|0}(\boldsymbol{\theta}_1|\mathbf{Y}^0) d\boldsymbol{\theta}_1
$$

where by convention $g_1(\boldsymbol{\theta}_1) = f_{1|0}(\boldsymbol{\theta}_1|\mathbf{Y}_0)$, conditional on an imaginary observation $\mathbf{Y}_0$ at time 0. The challenge arises from the fact that, in general, all the integrals in the likelihood have no closed form expressions. Numerical evaluation of related integrals is possible via quadrature numerical evaluation, only when the dimension of $\boldsymbol{\theta}_t$ is low. The difficulty in the implementation of MLE is really rooted in the fact that the assignment of quadrature points and weights varies over time t, because the probability distribution is different at a different time point.

When evaluating the integrals is not analytically feasible, a certain approximation seems inevitable. This book focuses on BLUP and Markov chain Monte Carlo, both of which are generally applicable for a variety of models and data types.

On the other hand, in the computation of smoother densities $f_{t|n}, t < n$, namely the conditional densities of state variable $\boldsymbol{\theta}_t$ given all observations $\mathbf{Y}^n$, a backward recursion procedure is performed:

$$
\begin{aligned}
\boldsymbol{\theta}_n|\mathbf{Y}^n = \mathbf{y}^n &\sim f_{n|n}(\boldsymbol{\theta}|\mathbf{y}^n) d\boldsymbol{\theta}, \\
\boldsymbol{\theta}_t|(\boldsymbol{\theta}_{t+1}, \mathbf{y}^n) &\sim \frac{g_{t+1}(\boldsymbol{\theta}_{t+1}|\boldsymbol{\theta}) f_{t|t}(\boldsymbol{\theta}|\mathbf{y}^t)}{f_{t+1|t}(\boldsymbol{\theta}_{t+1}|\mathbf{y}^t)} d\boldsymbol{\theta}.
\end{aligned}
$$

It follows that the smoother density at time $t, t < n$, is

$$
\begin{aligned}
f_{t|n}(\boldsymbol{\theta}|\mathbf{y}^n) &= \int \frac{g_{t+1}(\boldsymbol{\theta}_{t+1}|\boldsymbol{\theta}) f_{t|t}(\boldsymbol{\theta}|\mathbf{y}^t)}{f_{t+1|t}(\boldsymbol{\theta}_{t+1}|\mathbf{y}^t)} f_{t+1|n}(\boldsymbol{\theta}_{t+1}|\mathbf{y}^n) d\boldsymbol{\theta}_{t+1} \\
&= f_{t|t}(\boldsymbol{\theta}|\mathbf{y}^t) \int \frac{g_{t+1}(\boldsymbol{\theta}_{t+1}|\boldsymbol{\theta})}{f_{t+1|t}(\boldsymbol{\theta}_{t+1}|\mathbf{y}^t)} f_{t+1|n}(\boldsymbol{\theta}_{t+1}|\mathbf{y}^n) d\boldsymbol{\theta}_{t+1}. \quad (10.5)
\end{aligned}
$$

Generalized state space models, ($\mathbf{M}_1$ and $\mathbf{M}_2$), can accommodate a variety of discrete and continuous longitudinal data. This book is devoted to the analysis of longitudinal discrete data, Chapter 11 for longitudinal binary data and Chapter 12 for longitudinal count data. Statistical inference is discussed selectively based on some specific settings of $\mathbf{M}_1$ and $\mathbf{M}_2$. For example, MCMC based inference is illustrated in GSSMs for binary or binomial data, and BLUP based inference is demonstrated in GSSMs for count data. In effect, the two methods are general and suitable for many other data types. It is suggested that readers pay attention to ideas and procedures of developing these inference methods.

10.2 Linear State Space Models

The classical state space models refer to a class of linear Gaussian state models in that the observation process $\{\mathbf{Y}_t\}$ is driven by a latent state process $\{\boldsymbol{\theta}_t\}$ by a linear observation equation, as described below,

$$\mathbf{Y}_t = \mathbf{A}_t \boldsymbol{\theta}_t + \boldsymbol{\varepsilon}_t, \text{ with } \boldsymbol{\varepsilon}_t \stackrel{iid}{\sim} \mathrm{MVN}_d(\mathbf{0}, \mathbf{W}_t)$$

and the state process is governed by a linear transition equation,

$$\boldsymbol{\theta}_t = \mathbf{B}_t \boldsymbol{\theta}_{t-1} + \boldsymbol{\xi}_t, \text{ with } \boldsymbol{\xi}_t \stackrel{iid}{\sim} \mathrm{MVN}_q(\mathbf{0}, \mathbf{D}_t),$$

where the design matrix $\mathbf{A}_t$ and the transition matrix $\mathbf{B}_t$ are known, and the covariance matrices $\mathbf{W}_t$ and $\mathbf{D}_t$ may be known or unknown. The initial state $\boldsymbol{\theta}_0$ is assumed to follow $\boldsymbol{\theta}_0 \sim \mathrm{MVN}_q(\mathbf{g}_0, \mathbf{D}_0)$.

Assume that the two processes satisfy the conditions of a comb structure in Section 9.3; that is, $\{\boldsymbol{\varepsilon}_t\}$, $\{\boldsymbol{\xi}_t\}$ and $\boldsymbol{\theta}_0$ are mutually independent. Note that Box and Jenkins' ARMA(p,q) models are a special case of the linear state space model (Brockwell and Davis, 1996, Section 8.3). Under the Gaussian errors in both observation and state processes, the integration can be carried out analytically, and the resulting recursions for filtering and smoothing are exactly the same as those given in Theorem 9.4 and 9.5 with $\mathbf{W}_t(\boldsymbol{\theta}_t) = \mathbf{W}_t$, $\mathbf{W}_t^0 = 0$, $\mathbf{D}_t(\boldsymbol{\theta}_{t-1}) = \mathbf{D}_t$, and $\mathbf{D}_t^0 = 0$. The normality assumption can be relaxed to requiring only the existence of first two moments. The resulting model is referred to as the *linear state space model.* According to the results in Section 9.3, Kalman filter and smoother, in the form of BLUP, are still available in the linear state space model.

However, the linear Gaussian state space model is challenged by many real world time series data. For example, in many financial time series data, distributions of processes are often highly positively skewed with heavy tails, which essentially impairs the normality assumption. When data are discrete, such as time series of binomial observations or time series of counts, normal distributed errors are no longer suitable. Chapters 11 and 12 will discuss some extensions of the linear state space model to handle time series of discrete observations.

10.3 Shift-Mean Model

Modeling structural change is of great interest in time series data analysis. Mean shift in a stochastic process presents a challenge to the linear state space model, when a systematic trend in the mean is broken at multiple times by strong change points. To elucidate, consider a time series data simulated by Kitagawa (1987) from the following shift-mean model:

$$
\begin{aligned}
Y_t &\sim N(\mu_t, 1) \\
\mu_t &= 0, \quad t = 1, \cdots, 100, \\
&= -1, \quad t = 101, \cdots, 250, \\
&= 1, \quad t = 251, \cdots, 350, \\
&= 0, \quad t = 351, \cdots, 500.
\end{aligned}
$$

Figure 10.2 displays a simulated sample path. Clearly, the mean function is not continuous, with jumps occurring at time $t = 101, 251$, and 351. The objective is to estimate the shifting mean value function, μ_t, and especially to identify the jump locations.

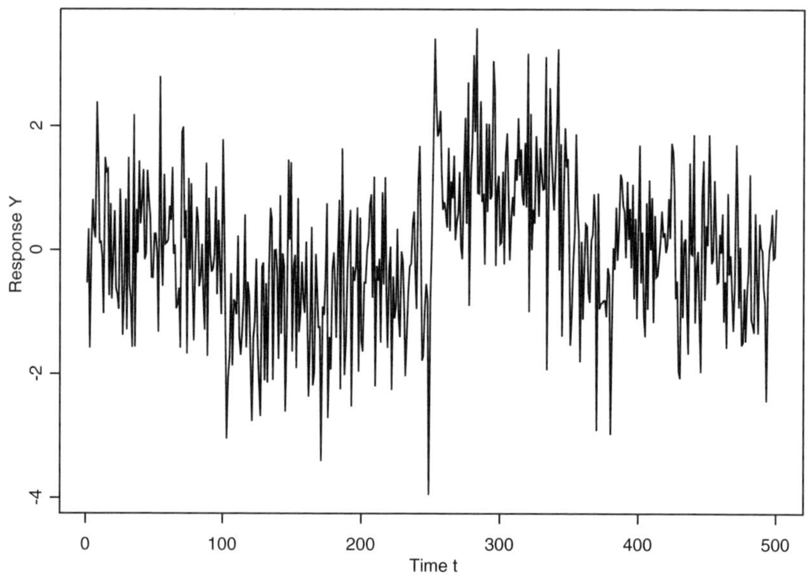

Fig. 10.2. Simulated sample path with the shifting mean function.

Kitagawa (1987) proposed a state space model to analyze the data, by borrowing a latent state process that is presumably able to capture the jumps at proper times. The model takes the form:

$$\begin{aligned} Y_t &= \theta_t + \varepsilon_t, \quad \text{with } \varepsilon_t \sim \mathrm{N}(0, \sigma^2), \\ \theta_t &= \theta_{t-1} + \xi_t, \quad \text{with } \xi_t \sim Q(b, \tau^2), \end{aligned}$$

where $b = 0.75$ according to Kitagawa (1987), and $Q(b, \tau^2)$ denotes the distribution of the Pearson system, which has the density

$$q(x; b, \tau) = C(\tau^2 + x^2)^{-b}$$

with $C = \tau^{2b-1}\Gamma(b)/\{\Gamma(b - \frac{1}{2})\Gamma(\frac{1}{2})\}$. Note that this density function does not produce finite second moments, which in fact is a desirable property to respond to jumps of any size. Intuitively, when the process arrives at the time at which a jump occurs, the Q distribution will react to the discontinuity by an extraordinarily large second moment. Kitagawa used a crude numerical integration over the state space by a piecewise linear approximation and obtained ML estimates of τ^2 and σ^2, $\hat{\tau}^2 = 2.2 \times 10^{-7}$, and $\hat{\sigma}^2 = 1.022$, respectively. See Table 10.1.

Table 10.1. Maximum likelihood estimates of the variance parameters τ^2 and σ^2 in the shift-mean state space model.

M	$\hat{\tau}^2_{SBQF}$	$\hat{\sigma}^2_{SBQF}$	Log-likelihood
25	9.966×10^{-10}	1.026	−742.6331
50	1.448×10^{-9}	1.028	−742.7902
100	1.603×10^{-9}	1.029	−742.7885
200	1.668×10^{-9}	1.029	−742.7946
400	1.692×10^{-9}	1.030	−742.7982
800	1.700×10^{-9}	1.030	−742.7998
	$\hat{\tau}^2_{Kitagawa}$	$\hat{\sigma}^2_{Kitagawa}$	Log-likelihood
400	2.2×10^{-7}	1.022	−741.944

Since the estimate of τ^2 is extremely small and corresponding error density is very sharply peaked, it appears very difficult to evaluate related integrals in filter densities or smoother densities accurately by only a piecewise linear approximation. To mitigate this problem, some numerical techniques are proposed to improve Kitagawa's crude numerical evaluation. One solution proposed in the same paper by Kitagawa was a method of variable mesh, which essentially divides the intervals around the mode of a filter or smoother density further into a finer mesh. This second step effort on a finer scale increases tremendously the computational intensity in all related procedures of estimation and integral evaluation.

A more efficient method was proposed recently by Xing (2004) based on high-order smoothed best quadrature formulas (SBQF). Xing's quadrature numerical evaluation of integration controls a global approximation precision on the nq-multiple integral. A noticeable merit of this method is that it provides a universal allocation of quadrature points and weights for all integrals, regardless of probability measures at different time points. It is shown that the optimal quadrature points should be placed at respective quantiles. So, the SBQF ends up with a global assessment of approximation accuracy and greatly improves the computational efficiency. Table 10.1 reports the maximum likelihood estimates of the model parameters, $\hat{\tau}^2$ and $\hat{\sigma}^2$, with the utility of Xing's SBQF and under different numbers (M) of quadrature points equally spaced on interval $(-4, 4)$. Kitagawa's estimates are obtained with $M = 400$ and are similar to Xing's estimates.

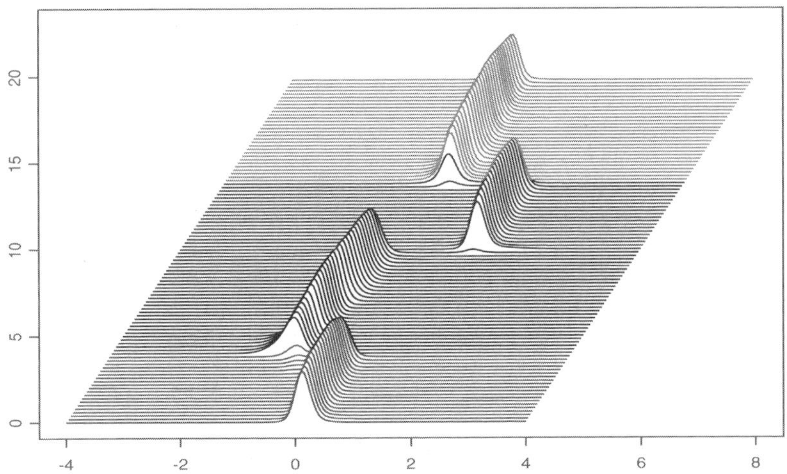

Fig. 10.3. Estimated smoother densities $f_{t|n}(x|\mathbf{Y}^n)(t = 5, 10, \cdots, 500)$ by the shift-mean model with $\hat{\tau}^2 = 1.667863 \times 10^{-9}$ and $\hat{\sigma}^2 = 1.029391$.

Figure 10.3 displays a collection of smoother densities at time $t = 5, 10, \cdots, 500$, including those time points where jumps occur. They are computed from the formula (10.5) with parameter estimates $\hat{\tau}^2 = 1.667863 \times 10^{-9}$

and $\hat{\sigma}^2 = 1.029391$ using $M = 200$. It is easy to see from this figure that the means of the smoother densities are shifted at the jump points. Figure 10.4 shows the median (bold curve) and $0.13, 2.27, 15.87, 84.13, 97.73$, 99.87 percentiles of the smoother density functions that correspond to $\pm 1, \pm 2, \pm 3$ sigma points of Gaussian density, respectively.

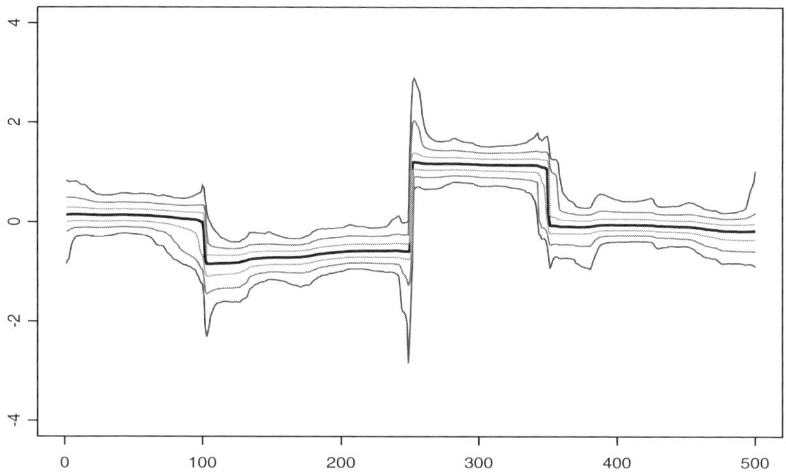

Fig. 10.4. Estimated median indicated by the bold line and $0.13, 2.27, 15.87, 84.13$, $97.73, 99.87$ percentiles in the shifted-mean model.

10.4 Monte Carlo Maximum Likelihood Estimation

Monte Carlo maximum likelihood estimation (MCMLE) proposed by Durbin and Koopman (1997) is a general likelihood-based inference in that related integrals are evaluated by Monte Carlo simulation. In comparison to the simulated maximum likelihood method that uses Monte Carlo simulation to directly evaluate integrals in likelihood function, this MCMLE approach explores some structures of the proposed model by borrowing strength from a similar but well-studied existing model. For example, when the errors in the model of an observation process follow a heavy-tailed distribution, a direct MLE would be generally hard to carry out. However, the linear state space model with normally distributed errors has been well studied, which can be therefore utilized to assist the search for the MLE in the model with nonnormal errors. Durbin and Koopman (1997) proposed a way to embed the likelihood of a chosen working model into that of the model under investigation.

To elucidate, suppose $f_t(\mathbf{y}|\boldsymbol{\theta}_t)$ for the observation process (10.2) is non-normal, but $g_t(\boldsymbol{\theta}|\boldsymbol{\theta}_{t-1})$ in the state process (10.1) is normal. Let $L(\boldsymbol{\eta})$ denote the likelihood function of this proposed state space model, where $\boldsymbol{\eta}$ represents the vector of all model parameters. The working model would be the linear Gaussian state space model, in which the $\tilde{f}_t(\mathbf{y}|\boldsymbol{\theta}_t)$ is assumed to be normal, while the $g_t(\boldsymbol{\theta}|\boldsymbol{\theta}_{t-1})$ remains the same. The resulting likelihood of the working model is denoted by $L_w(\boldsymbol{\eta})$. Note that this working model has been well studied, where Kalman filter and smoother are available and easily calculated. Then, the likelihood for the proposed model is

$$\begin{aligned} L(\boldsymbol{\eta}) &= \int f(\mathbf{y}^n, \boldsymbol{\theta}^n) d\boldsymbol{\theta}^n \\ &= \int f_w(\mathbf{y}^n) \frac{f(\mathbf{y}^n, \boldsymbol{\theta}^n)}{f_w(\mathbf{y}^n, \boldsymbol{\theta}^n)} f_w(\boldsymbol{\theta}^n \mid \mathbf{y}^n) d\boldsymbol{\theta}^n \\ &= L_w(\boldsymbol{\eta}) \mathrm{E}_w \left\{ \frac{f(\mathbf{y}^n, \boldsymbol{\theta}^n)}{f_w(\mathbf{y}^n, \boldsymbol{\theta}^n)} \right\} \\ &= L_w(\boldsymbol{\eta}) L_e(\boldsymbol{\eta}), \end{aligned}$$

where $L_e(\boldsymbol{\eta})$ is the remainder likelihood that bridges between the wanted likelihood $L(\boldsymbol{\eta})$ and the working likelihood $L_w(\boldsymbol{\eta})$. The $L_e(\boldsymbol{\eta})$ takes the form

$$\begin{aligned} L_e(\boldsymbol{\eta}) &= \mathrm{E}_w \left\{ \frac{f(\mathbf{y}^n, \boldsymbol{\theta}^n)}{f_w(\mathbf{y}^n, \boldsymbol{\theta}^n)} \right\} \\ &= \mathrm{E}_w C(\boldsymbol{\eta}; \boldsymbol{\theta}^n, \mathbf{y}^n), \end{aligned}$$

where the expectation is taken under the conditional distribution, $f_w(\boldsymbol{\theta}^n \mid \mathbf{y}^n)$, in the setup of the working model. It follows that the log likelihood functions satisfy an additive relation:

$$\ell(\boldsymbol{\eta}) = \ell_w(\boldsymbol{\eta}) + \ell_e(\boldsymbol{\eta}),$$

where the piece $\ell_w(\boldsymbol{\eta})$ is analytically easy with no involvement of integration, and the other piece $\ell_e(\boldsymbol{\eta})$ is hard to be dealt with both analytically and numerically. Durbin and Koopman (1997) suggested the use of Monte Carlo simulation to evaluate the second piece $\ell_e(\boldsymbol{\eta})$. Suppose, $\boldsymbol{\theta}^{n(k)}, k = 1, \ldots, M$, are M *i.i.d.* samples drawn from $f_w(\boldsymbol{\theta}^n \mid \mathbf{Y}^n)$. Then, the resulting Monte Carlo likelihood is

$$\ell_{mc}(\boldsymbol{\eta}) = \ell_w(\boldsymbol{\eta}) + \frac{1}{M} \sum_{k=1}^{M} C(\boldsymbol{\eta}; \boldsymbol{\theta}^{n(k)}, \mathbf{Y}^n),$$

where

$$C^{(k)}(\boldsymbol{\eta}; \boldsymbol{\theta}^{n(k)}, \mathbf{Y}^n)) = \frac{f(\mathbf{Y}^n, \boldsymbol{\theta}^{n(k)})}{f_w(\mathbf{Y}^n, \boldsymbol{\theta}^{n(k)})}.$$

The MCML estimation of $\boldsymbol{\eta}$ is obtained by maximizing the ℓ_{mc} *w.r.t.* $\boldsymbol{\eta}$. The maximization is carried out iteratively by, *say*, a Newton algorithm and the sampling from $f_w(\boldsymbol{\theta}^n \mid \mathbf{Y}^n)$ may be implemented via de Jong and Shephard's (1995) simulation smoother. Alternatively, one may consider the maximization by parts (MBP) algorithm of Song et al. (2005) discussed in Section 6.5.1 to directly maximize the $\ell(\boldsymbol{\eta})$, in which Monte Carlo may be used to evaluate integrals in the first order derivatives $\dot{\ell}_e(\boldsymbol{\eta})$.

An issue in the use of MCML is the choice of the working model or working likelihood function. In the analysis of time series of continuous observations, the linear Gaussian state space model seems to be a sensible candidate for a working model, but in the analysis of time series of discrete observations, it is not so clear in general to opt a proper working model. This is simply because none of models for discrete-valued time series has been recognized as being central, simple, and handy, in comparison to the linear Gaussian state space model for continuous-valued time series. Chapters 11 and 12 will discuss the modeling and inference in the generalized state space models for time series of binomial observations and counts, respectively.

11

Generalized State Space Models for Longitudinal Binomial Data

11.1 Introduction

Assume $\{(\mathbf{Y}_t, \mathbf{x}_t), t = 1, \ldots, n\}$ is a collection of time series observations, where $\mathbf{Y}_t$ is a d-dimensional binomial response and $\mathbf{x}_t$ is a p-dimensional covariate vector $\mathbf{x}_t$. According to models $\mathbf{M}_1$ 10.1 and $\mathbf{M}_2$ 10.2, a class of GSSMs for such binomial longitudinal data is formulated by specifying the following equations, respectively:

$$\begin{aligned} \boldsymbol{\theta}_t | \boldsymbol{\theta}_{t-1} &\sim \mathrm{MVN}_q(\boldsymbol{\mu}(\boldsymbol{\theta}_{t-1}), \Sigma(\boldsymbol{\theta}_{t-1})) \\ Y_{it} | \boldsymbol{\theta}_t &\sim \mathrm{Bi}(k_{it}, \pi_{it}(\boldsymbol{\theta}_t)). \end{aligned}$$

For the state process $\boldsymbol{\theta}_t$, a widely used model in the literature is a linear transition model given by

$$\boldsymbol{\theta}_t = \mathbf{B}_t \boldsymbol{\theta}_{t-1} + \boldsymbol{\xi}_t, \tag{11.1}$$

where $\mathbf{B}_t$ is a $q \times q$-dimensional transition matrix and $\boldsymbol{\xi}_t$ is the q-variate Gaussian white noise with zero mean and covariance matrix Q_t, i.e., $\boldsymbol{\xi}_t \sim \mathrm{MVN}_q(0, Q_t)$. For the special case of one-dimensional state process with $q = 1$, the following two models are popular. One is the random walk process, with $\mathbf{B}_t \equiv 1$,

$$\theta_t = \theta_{t-1} + \xi_t, \tag{11.2}$$

where $\xi_t \sim N(0, \sigma^2)$. This model assumes essentially the increments $\theta_t - \theta_{t-1}$ are *i.i.d.* normal random variates.

The other one is Box and Jenkins' stationary AR(1) process, with $\mathbf{B}_t \equiv \rho, |\rho| < 1$,

$$\theta_t = \rho \theta_{t-1} + \xi_t, \tag{11.3}$$

where $\xi_t \sim N(0, \sigma^2)$, and ρ represents the autocorrelation coefficient. This implies that the autocorrelation function (ACF) of this process is $\rho^{|h|}, h = 0, 1, \ldots,$ with lag h.

One primary difference between these two types of processes is rooted in their variances: the AR(1) process has a bounded variance, $\sigma^2/(1-\rho^2)$, but the random walk has an unbounded variance, $t\sigma^2$, which effectively increases in time and hence is unbounded.

For the observation process $\mathbf{Y}_t$, each probability component of the vector $\boldsymbol{\pi}_t = (\pi_{1t}, \ldots, \pi_{dt})^T$ is assumed to follow a GLM of the form

$$g(\pi_{it}) = \eta_{it} + G_{it}^T \boldsymbol{\theta}_t, \ i = 1, \ldots, d,$$

where the η_{it} is the componentwise deterministic predictor that may be specified in a similar way as one of those given in (i)–(v) for model (4.1). In particular, both trend and seasonality would be modeled via the η_{it} term. For example, if one wants to fit the data by a linear predictor, then specifying $\eta_{it} = \mathbf{x}_{it}^T \boldsymbol{\alpha}, i = 1, \ldots, d$, in which $\boldsymbol{\alpha}$ is a vector of regression coefficients to be estimated. Other parameters in the model to be estimated include the state variables $\boldsymbol{\theta}_t$, the variance parameters σ^2, and/or the autocorrelation parameter ρ.

11.2 Monte Carlo Kalman Filter and Smoother

Following Song (2000b), let us consider a simple case with no deterministic predictors, namely $\eta_{it} = 0, i = 1, \ldots, d$. When the link function g is chosen to be the probit link, $g(\pi) = \Phi^{-1}(\pi)$, the above GSSM for binomial time series may be rewritten via the latent variable representation, so that the classical Kalman filter and smoothing available in the Gaussian linear state space models can be transplanted to this probit GSSM. Let the initial state $\boldsymbol{\theta}_0 \sim \mathrm{MVN}_q(\mathbf{a}_0, Q_0)$ where both $\mathbf{a}_0$ and Q_0 are known.

The probit state space model for binary time series may be regarded as being merged from a linear Gaussian state space model. To proceed, let $\{\mathbf{Z}_t = (Z_{1t}, \ldots, Z_{dt})^T\}$ be a d-dimensional latent process satisfying

$$\mathbf{Z}_t = \mathbf{G}_t^T \boldsymbol{\theta}_t + \boldsymbol{\varepsilon}_t, \ t = 1, \ldots, n, \tag{11.4}$$

where $\mathbf{G}_t = (G_{1t}, \ldots, G_{dt})$ is a $q \times d$ matrix and $\boldsymbol{\varepsilon}_t$ are *i.i.d.* $\mathrm{MVN}_d(\mathbf{0}, I)$ Gaussian innovations. For the i-th component, define a one-to-one correspondence

$$Y_{it} = 1 \ \text{ if and only if } \ Z_{it} \geq 0, \ i = 1, \ldots, d. \tag{11.5}$$

Therefore, as desired, the latent variable representation results in

$$\pi_{it} = \mathrm{P}(Z_{it} \geq 0 | \boldsymbol{\theta}_t) = \Phi(G_{it}^T \boldsymbol{\theta}_t), \ i = 1, \ldots, d.$$

It is noted that models (11.4) and (11.1) together form a linear and Gaussian state space model, for which the optimal linear Kalman filter and smoothing are available in terms of the latent process $\{\mathbf{Z}_t\}$. They are the minimum mean square error estimates of the states and could be computed using the standard recursive procedures given in Theorems 9.4 and 9.5 in Chapter 9, if these latent vectors $\mathbf{Z}_t$ were known. For convenience, such a state space model, consisting of (11.4) and (11.1), is called *the interim model.*

Clearly,

$$\mathbf{Z}^n = (\mathbf{Z}_1^T, \ldots, \mathbf{Z}_n^T)^T \sim \text{MVN}_{dn}(\boldsymbol{\mu}, \Sigma),$$

where $\boldsymbol{\mu} = (\boldsymbol{\mu}_1^T, \ldots, \boldsymbol{\mu}_n^T)^T$ with

$$\begin{aligned} \mu_{it} &= G_{it}^T \text{E}(\boldsymbol{\theta}_t) \\ &= G_{it}^T \mathbf{B}_t \cdots \mathbf{B}_1 \mathbf{a}_0, \end{aligned} \tag{11.6}$$

and covariance matrix is $\Sigma = (\Sigma_{st})$ with the (s,t)-th element $[\Sigma]_{st}$ being a $d \times d$ variance-covariance matrix of $\text{cov}(\mathbf{Z}_s, \mathbf{Z}_t), s, t = 1, \ldots, n$. It follows immediately from the model specification that the block-diagonals of the Σ are equal to

$$\Sigma_{tt} = \text{Var}(\mathbf{Z}_t) = I + \mathbf{G}_t^T \text{Var}(\boldsymbol{\theta}_t) \mathbf{G}_t, \ t = 1, \ldots, n, \tag{11.7}$$

where

$$\begin{aligned} &\text{Var}(\boldsymbol{\theta}_t) = Q_t + \mathbf{B}_t Q_{t-1} \mathbf{B}_t^T + (\mathbf{B}_t \mathbf{B}_{t-1}) Q_{t-2} (\mathbf{B}_t \mathbf{B}_{t-1})^T + \cdots + \\ &\quad (\mathbf{B}_t \mathbf{B}_{t-1} \cdots \mathbf{B}_2) Q_1 (\mathbf{B}_t \mathbf{B}_{t-1} \cdots B_2)^T + (\mathbf{B}_t \mathbf{B}_{t-1} \cdots \mathbf{B}_1) Q_0 (\mathbf{B}_t \mathbf{B}_{t-1} \cdots \mathbf{B}_1)^T. \end{aligned}$$

The off-block-diagonals of the Σ are given by

$$\Sigma_{t,t+s} = \mathbf{G}_t^T \text{cov}(\boldsymbol{\theta}_t, \boldsymbol{\theta}_{t+s}) \mathbf{G}_{t+s}, \tag{11.8}$$

where

$$\text{cov}(\boldsymbol{\theta}_t, \boldsymbol{\theta}_{t+s}) = \text{Var}(\boldsymbol{\theta}_t)(\mathbf{B}_{t+s} \cdots \mathbf{B}_{t+1})^T.$$

Clearly, the Gaussianity of the latent process $\{\mathbf{Z}_t\}$ and that of state variables $\{\boldsymbol{\theta}_t\}$ imply that the joint distribution of $(\mathbf{Z}^n, \boldsymbol{\theta})$ is Gaussian, and so is the marginal distribution of $\mathbf{Z}^n$, with its mean vector $\boldsymbol{\mu}$ and covariance matrix Σ given by (11.6), (11.7), and (11.8).

To obtain the Kalman filter and smoother, the central task is to compute two conditional mean estimates, $\text{E}(\boldsymbol{\theta}_t | \mathbf{Y}^t)$ and $\text{E}(\boldsymbol{\theta}_t | \mathbf{Y}^n)$, respectively, given the information available up to time t and all information.

It follows from (11.5) that

$$\text{E}(\boldsymbol{\theta}_t | \mathbf{Y}^s, \mathbf{Z}^s) = \text{E}(\boldsymbol{\theta}_t | \mathbf{Z}^s), \ t, s = 1, \ldots, n.$$

Hence,

$$\begin{aligned} \text{E}\left(\boldsymbol{\theta}_t | \mathbf{Y}^t\right) &= \text{E}\left\{\text{E}\left(\boldsymbol{\theta}_t | \mathbf{Y}^t, \mathbf{Z}^t\right) | \mathbf{Y}^t\right\} \\ &= \text{E}\left\{\text{E}\left(\boldsymbol{\theta}_t | \mathbf{Z}^t\right) | \mathbf{Y}^t\right\} \\ &= \text{E}\left(\Theta_t | \mathbf{Y}^t\right), \end{aligned} \tag{11.9}$$

and similarly,

$$\begin{aligned} \mathrm{E}\left(\boldsymbol{\theta}_t|\mathbf{Y}^n\right) &= \mathrm{E}\left\{\mathrm{E}\left(\boldsymbol{\theta}_t|\mathbf{Z}^n\right)|\mathbf{Y}^n\right\} \\ &= \mathrm{E}\left(\Theta_t^*|\mathbf{Y}^n\right), \end{aligned} \tag{11.10}$$

where $\Theta_t = \mathrm{E}\left(\boldsymbol{\theta}_t|\mathbf{Z}^t\right)$ and $\Theta_t^* = \mathrm{E}\left(\boldsymbol{\theta}_t|\mathbf{Z}^n\right)$ are the conditional expectations of $\boldsymbol{\theta}_t$ with respect to the latent process $\{\mathbf{Z}_t\}$, respectively.

If $\mathbf{Z}_t$'s were observed, both Θ_t and Θ_t^* would be computed recursively according to the following standard Kalman filter and smoother recursions:

1. Filter Prediction Step

$$\begin{aligned} \Theta_{t|t-1} &= \mathrm{E}(\boldsymbol{\theta}_t|\mathbf{Z}^{t-1}) = \mathbf{B}_t\Theta_{t-1}, \quad \text{with } \Theta_0 = \mathbf{a}_0, \\ \Lambda_{t|t-1} &= \mathbf{B}_t\Lambda_{t-1}\mathbf{B}_t^T + Q_t, \quad \text{with } \Lambda_0 = Q_0. \end{aligned}$$

2. Filter Correction Step

$$\begin{aligned} \Theta_t &= \Theta_{t|t-1} + \Lambda_{t|t-1}\mathbf{G}_t\Delta_t^{-1}\left(\mathbf{Z}_t - \mathbf{G}_t^T\Theta_{t|t-1}\right), \\ \Lambda_t &= \Lambda_{t|t-1} - \Lambda_{t|t-1}\mathbf{G}_t\Delta_t^{-1}\mathbf{G}_t^T\Lambda_{t|t-1}, \end{aligned}$$

where

$$\Delta_t = \mathbf{G}_t^T\Lambda_{t|t-1}\mathbf{G}_t + I.$$

3. Smoothing Step

$$\begin{aligned} \Theta_t^* &= \mathrm{E}(\boldsymbol{\theta}_t|\mathbf{Z}^n) = \Theta_t + P_t\left(\Theta_{t+1}^* - B_{t+1}\Theta_t\right), \\ \Lambda_t^* &= \Lambda_t + P_t\left(\Lambda_{t+1}^* - \Lambda_{t+1|t}\right)P_t^T \end{aligned}$$

where $P_t = \Lambda_t\mathbf{B}_{t+1}^T\left(\Lambda_{t+1|t}\right)^{-1}$, $t = n-1, \ldots, 1$. At time n, $\Theta_T^* = \Theta_n$ and $\Lambda_T^* = \Lambda_T$.

Based on equations (11.9) and (11.10), Song (2000b) suggested applying the Monte Carlo technique to approximate both conditional mean estimates of the states based on $\{\mathbf{Y}_t\}$. For convenience, let $[\mathbf{u}|\mathbf{w}]$ denote the conditional distribution of $\mathbf{u}$ given $\mathbf{w}$.

Suppose $\mathbf{Z}^{n(1)}, \ldots, \mathbf{Z}^{n(M)}$ are M *i.i.d.* samples generated from $[\mathbf{Z}^n|\mathbf{Y}^n]$. For each sample $\mathbf{Z}^{n(i)} = (\mathbf{Z}_1^{T(i)}, \ldots, \mathbf{Z}_n^{T(i)})^T$, the Kalman filter and smoother given by above steps 1–3 produce $\left\{\Theta_t^{(i)}, \Lambda_t^{(i)}\right\}$ and $\left\{\Theta_t^{*(i)}, \Lambda_t^{*(i)}\right\}$, respectively, $i = 1, \ldots, M$. By the Law of Large Number, $\mathrm{E}(\boldsymbol{\theta}_t|\mathbf{Y}^n)$ is then approximated by the average of M smoothers of the form

$$\mathbf{m}_t^* = \frac{1}{M}\sum_{i=1}^{M}\Theta_t^{*(i)}, \quad t = 1, \ldots, n. \tag{11.11}$$

According to Song (2000b), the $\mathbf{m}_t^*$ in (11.11) is called *the Monte Carlo Kalman smoother* (MCKS). The Monte Carlo approximation of $\mathrm{E}(\boldsymbol{\theta}_t|\mathbf{Y}^t)$ needs to draw samples from $[\mathbf{Z}^t|\mathbf{Y}^t]$ for each t and the corresponding estimator may be defined in a way similar to that of the MCKS. Given the availability of full samples $\mathbf{Z}^{n(1)}, \ldots, \mathbf{Z}^{n(M)}$, however, a better estimator of $\mathrm{E}(\boldsymbol{\theta}_t|\mathbf{Y}^t)$, which has smaller mean square error than the ordinary one, may be obtained by

$$\mathbf{m}_t = \frac{1}{M}\sum_{i=1}^{M}\Theta_t^{(i)}, \ t = 1, \ldots, n. \tag{11.12}$$

The $\mathbf{m}_t$ in (11.12) is referred to as *the Monte Carlo Kalman filter* (MCKF).

It is easy to show that the conditional distribution $[\mathbf{Z}^n|\mathbf{Y}^n]$ is a truncated multivariate normal distribution with mean vector $\boldsymbol{\mu}$ in (11.6) and covariance matrix Σ in (11.7) and (11.8), where the truncated region is a rectangular area specified by $a_j \le z_j \le b_j$, with $a_j = \log(y_j)$ and $b_j = -\log(1-y_j)$, $j = 1, \ldots, dn$. Generating random variates from truncated multivariate normal distributions has been discussed extensively in the literature; see, for example, Robert (1995).

To assess the accuracy of the given Kalman filter or smoother, the respective mean square errors (MSE) are used. For the MC smoother $\mathbf{m}_t^*$, the MSE is $\mathrm{E}\left(\boldsymbol{\theta}_t - \mathbf{m}_t^*\right)\left(\boldsymbol{\theta}_t - \mathbf{m}_t^*\right)^T$. Song (2000b) found that for large M, this MSE can be approximated by

$$\mathrm{E}(\boldsymbol{\theta}_t - \mathbf{m}_t^*)(\boldsymbol{\theta}_t - \mathbf{m}_t^*)^T \approx \frac{1}{M}\sum_{i=1}^{M}\Lambda_t^{*(i)} \tag{11.13}$$

where $\Lambda_t^{*(i)}$ are the mean square errors corresponding to the Kalman smoother $\Theta_t^{*(i)}$ available in terms of M samples $\mathbf{Z}^{n(i)}$, $i = 1, \ldots, M$.

In the Kalman filter and smoother, both covariance Q_t and autocorrelation coefficient ρ (if present) are usually unknown, and a method of moments estimation is suggested by Song (2000b). Assume $Q_t = Q$, independent of t. A consistent estimator of Q may be obtained by iteratively applying the following formula (11.14) until convergence, with initializing matrix being specified by, for example, $Q = I$,

$$\begin{aligned}\widehat{Q} &= \frac{1}{n}\sum_{t=1}^{n}\left(\mathbf{m}_t^* - \mathbf{B}_t\mathbf{m}_{t-1}^*\right)\left(\mathbf{m}_t^* - \mathbf{B}_t\mathbf{m}_{t-1}^*\right)^T \\ &+ \frac{1}{n}\sum_{t=1}^{n}\left(\bar{\Lambda}_t^* + \mathbf{B}_t\bar{\Lambda}_{t-1}^*\mathbf{B}_t^T - 2\bar{\Lambda}_{t,t-1}^*\mathbf{B}_t^T\right),\end{aligned} \tag{11.14}$$

where $\bar{\Lambda}_t^* = M^{-1}\sum_{i=1}^{M}\Lambda_t^{*(i)}$ and $\bar{\Lambda}_{t,t-1}^* = M^{-1}\sum_{i=1}^{M}\Lambda_{t-1}^{(i)}\mathbf{B}_t^T\left(\Lambda_{t|t-1}^{(i)}\right)^{-1}\Lambda_t^{*(i)}$.

If the state process follows the 1-dimensional stationary AR(1) model, the autocorrelation coefficient ρ may be consistently estimated as follows. Let $\phi = \rho/(1-\rho^2)$. Then a consistent estimate of ϕ is given by

$$\widehat{\phi} = \frac{1}{n\widehat{\sigma}^2}\sum_{t=1}^{n-1} m_t^* m_{t+1}^* + \frac{1}{n\widehat{\sigma}^2}\sum_{t=1}^{n-1}\bar{\Lambda}_{t,t+1}^*. \tag{11.15}$$

This leads to a consistent estimate of ρ as

$$\hat{\rho} = \frac{-1 + \sqrt{1 + 4\hat{\phi}}}{2\hat{\phi}}. \tag{11.16}$$

Example 11.1 (Infant Sleep Data).

The binary time series of infant sleep status, reported by Stoffer et al. (1988), were recorded in a 120 minute EEG study where the response $y = 1$ if the infant was judged to be in REM sleep during minute t, $y = 0$ otherwise. The two horizontal lines of dots in Figure 11.1 represent the time series of binary observations. The data were previously analyzed by Carlin and Polson (1992) using MCMC algorithm and now is re-analyzed by applying the Monte Carlo Kalman filter and smoother approach. According to Carlin and Polson (1992), the probit state space model is comprised of the following two models:

$$\begin{aligned} Y_t|\theta_t &\sim \text{Bi}(1, \pi_t), \text{ with } \pi_t = \Phi(\theta_t) \\ \theta_t &= \rho\theta_{t-1} + \varepsilon_t, \end{aligned}$$

$t = 1, \ldots, 120$, with the initializing state $\theta_0 \sim \mathcal{N}(0, 1)$. Here θ_t may be thought essentially of as an underlying continuous "sleep state" following a stationary Markov process of order 1 with mean zero and variance $\sigma^2/(1 - \rho^2)$. The objective is to estimate the process θ_t and hence the probability π_t of being in REM sleep status. The application of the MCKS algorithm, with $d = k = q = 1$, $G_{it} = 1$, $\mathbf{B}_t = \rho$, and $Q_t = \sigma^2$ leads to the MCKS estimate of the state process θ_t shown in Figure 11.1, with the 95% upper and lower confidence bounds determined by the estimated MSE.

Figure 11.2 shows the patterns for updates in the estimation of ϕ (the lower curve) and σ^2 (the upper curve) over 70 iterations of the MCKS for the state variables, initialized with $\phi = 0$ and $\sigma^2 = 1$. The figure clearly indicates that both procedures of updates for ϕ and σ^2 got stablized after iteration 20, which is hence taken as the convergence cutoff point. The corresponding estimated values at this iteration are $\hat{\phi} = 0.2252$ and $\hat{\sigma}^2 = 0.9921$, leading to, by (11.16), $\hat{\rho} = 0.8408$. The bumps over the zero-line indicate that the probabilities of being in REM are bigger than 0.50, and their patterns closely follow the observed time series of 0's and 1's.

Notice that Figure 11.1 shows a great deal of similarity to Figure 1 of Carlin and Polson's (1992) analysis based on the MCMC algorithm. This indicates that the MCKS estimate approximates to the state variables, at least in this example, as competitively well as the MCMC estimate, but the MCKS method is much simpler conceptually and much less burdensome computationally.

The above development can be extended to analyze time series of binomial observations with little effort. That is, let $\{\mathbf{Y}_t\}$ be d-dimensional binomial responses with the i-th component following as 1-dimensional binomial distribution $\text{Bi}(k_{it}, \pi_{it})$ and the marginal probability π_{it} following the probit model

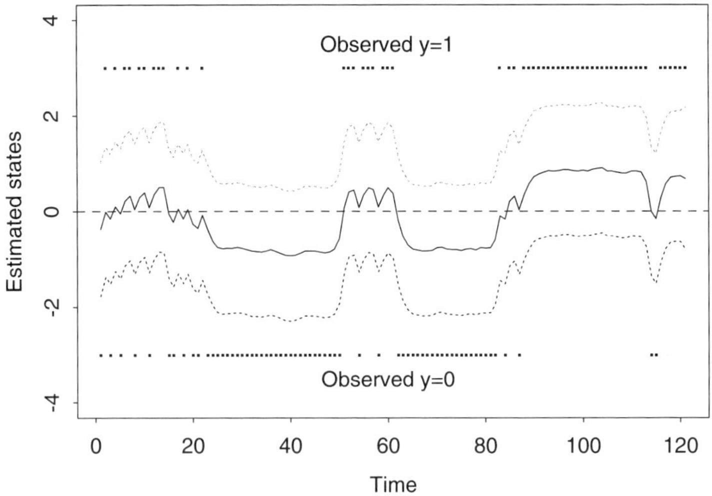

Fig. 11.1. MC Kalman smoothing estimates of state process with 95% confidence bounds. The raw time series of binary observations are indicated by two horizontal lines of dots.

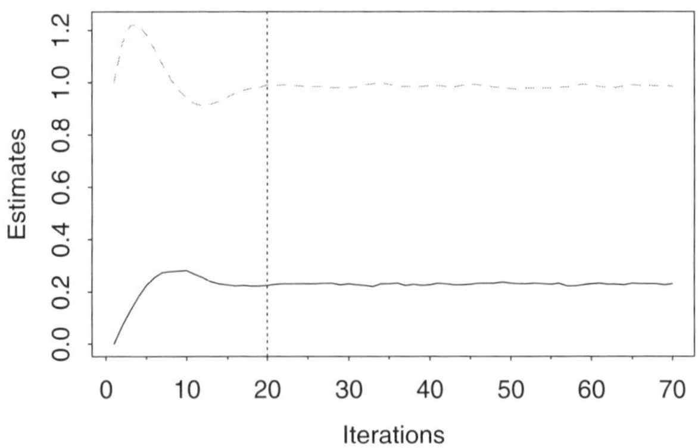

Fig. 11.2. Patterns for updates of hyperparameters over 70 iterations. (—) represents the estimation for ϕ; (- - -) represents the estimation for σ^2.

$\pi_{it} = \Phi(G_{it}^T \boldsymbol{\theta}_t)$. For simplicity, assume $k_{it} = k$ for all i and t. Now decompose Y_{it} into an independent sum of binary variables, namely $Y_{it} = Y_{it1} + \cdots + Y_{itk}$ where $Y_{it1}, \ldots, Y_{itk}$ are *i.i.d.* binary variables with probability π_{it}. Following Tanner and Wong (1987), draw n *i.i.d.* d-variates $\mathbf{Z}_{tj} = (Z_{1tj}, \ldots, Z_{dtj})^T$,

$j = 1, \dots, k$ from the normal distribution $\text{MVN}_d(\mathbf{G}_t^T \boldsymbol{\theta}_t, I)$ conditional on $\boldsymbol{\theta}_t$, implying $Z_{itj}|\boldsymbol{\theta}_t \sim \text{N}(G_{it}^T \boldsymbol{\theta}_t, 1)$. Define the one-to-one correspondence:

$$Y_{itj} = 1 \text{ if and only if } Z_{itj} \geq 0, \; j = 1, \dots, k.$$

It follows that

$$\pi_{it} = \text{P}(Y_{itj} = 1|\boldsymbol{\theta}_t) = \text{P}(Z_{itj} \geq 0|\boldsymbol{\theta}_t) = \Phi(G_{it}^T \boldsymbol{\theta}_t), \; j = 1, \dots, k.$$

Let $\mathbf{Z}_t = (\mathbf{Z}_{t1}^T, \dots, \mathbf{Z}_{tk}^T)^T$ with $\mathbf{Z}_{tj} = (Z_{1tj}, \dots, Z_{dtj})^T$.

$$\mathbf{Z}_t = \mathbf{G}_t^T \boldsymbol{\theta}_t + \boldsymbol{\varepsilon}_t \tag{11.17}$$

where $\mathbf{G}_t^* = (\mathbf{G}_t, \dots, \mathbf{G}_t)$ is a $q \times dk$ matrix and $\boldsymbol{\varepsilon}_t$ are independent innovations with $\text{MVN}_{dk}(\mathbf{0}, I)$. As a result, the interim state space model consisting of (11.17) and (11.1) is apparently linear and Gaussian, similar to those obtained in the probit model for time series of binary observation.

Example 11.2 (Tokyo Rainfall Data).

This extension is now applied to the Tokyo rainfall data introduced in Section 1.3.9, which consist of the daily number of occurrences of rainfall over 1 mm in Tokyo for years 1983-1984. The data were previously analyzed by many authors, for example, Kitagawa (1987) and Fahrmeir (1992), in which the probit state space model takes the form

$$\begin{aligned} Y_t|\theta_t &\sim \text{Bi}(2, \pi_t), \text{ with } \pi_t = \Phi(\theta_t), \\ \theta_t &= \theta_{t-1} + \xi_t, \end{aligned}$$

for $t = 1, \dots, 366$, where $\xi_t \sim \mathcal{N}(0, \sigma^2)$, with repetition number $k = 2$. This state process is assumed to be a random walk, with the initial state $\theta_0 \sim \mathcal{N}(-1.5, 0.002)$ known from Fahrmeir and Tutz (1994, Page 282). Figure 11.3 shows the estimated probabilities $\hat{\pi}_t$ of rain using MCKS estimate for the state process θ_t with $\hat{\sigma}^2 = 0.028$. Figure 11.3 shows a similar pattern to both Kitagawa's (1987) Figure 11 and Fahrmeir's (1992) Figure 9. Once again, this simple MC Kalman filter and smoother is fairly reliable compared to other rather sophisticated methods of estimating state space variables in the generalized state space models for time series of binary or binomial observations.

This figure clearly indicates some seasonal patterns of rainfall over a year in Tokyo. The period of July-August seems to be the most wet season of a year. However, this model does not explicitly address the seasonality, but simply invokes a random walk process to deal with the nonstationarity. This data will be analyzed again in the next section 11.3 by explicitly incorporating seasonal covariates in the probit state space model.

11.3 Bayesian Inference Based on MCMC

This section concerns a Bayesian inference in the generalized linear state space models for time series of binary or binomial observations, based on MCMC

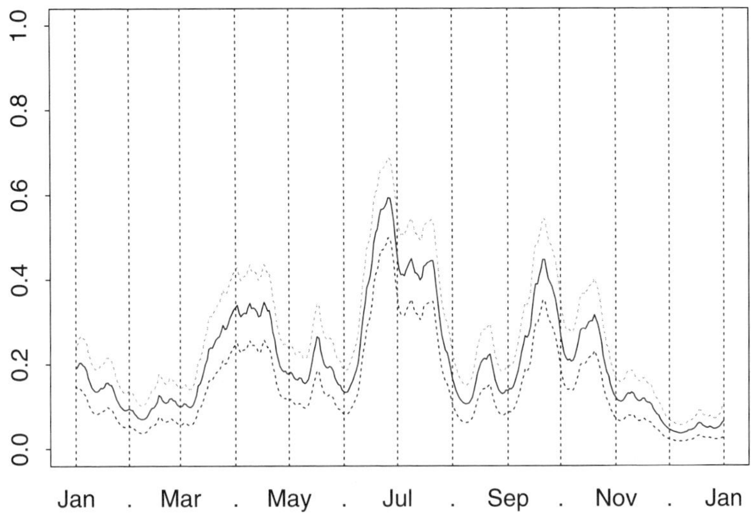

Fig. 11.3. Monte Carlo predicted probability process with 95% confidence bounds based on Kalman smoother.

algorithm. Because MCMC gives a great deal of flexibility in inference, one may add deterministic linear predictors η_{it} in a probit state space model, and the resulting model is useful to explicitly address the effects of trend, seasonal, or other time-varying covariates. In the context of state space models, de Jong and Shephard's (1995) simulation smoother is known as an efficient sampling procedure for the state variables, which hence is adopted in the development of MCMC. For the ease of exposition, fix $d = 1$ in the following presentation.

A probit state space model with the inclusion of linear predictor $\mathbf{x}_t^T \boldsymbol{\alpha}$ is given by

$$Y_t|\theta_t \sim \text{Bi}(1, \pi_t), \text{ with } \pi_t = \Phi(-\mathbf{x}_t^T \boldsymbol{\alpha} - \theta_t),$$
$$\theta_t = \rho\theta_{t-1} + \xi_t,$$

where $\xi_t \sim N(0, \sigma^2)$. Here the latent process follows a univariate stationary AR(1) process, $|\rho| < 1$. Note that θ_t represents a time-specific effect on the observed process, similar to the subject-specific in the generalized linear mixed models considered in Chapter 7. Czado and Song (2007) refers this class of state space models to as *the state space mixed models.*

Again the latent variable representation is applied in that Y_t is supposedly generated through dichotomization of an underlying continuous process Z_t on the basis of the one-to-one correspondence

$$Y_t = 1 \text{ if and only if } Z_t \le 0.$$

Consequently, with the latent variable vector $\mathbf{Z}^n = (Z_1, \cdots, Z_n)^T$, the interim state space model is

$$Z_t = -\mathbf{x}_t^T \boldsymbol{\alpha} - \theta_t + \varepsilon_t, \; t = 1, \cdots, n, \tag{11.18}$$

$$\theta_{t+1} = \rho\theta_t + \xi_t, \; t = 0, 1, \cdots, n. \tag{11.19}$$

Further both error terms are assumed to be independent and normally distributed,

$$\varepsilon_t \overset{iid}{\sim} N(0, 1) \text{ and } \xi_t \overset{iid}{\sim} N(0, \sigma^2),$$

and hence the expressions (11.18) and (11.19) together represent a linear Gaussian state space model. Note that the mutual independence between ε_t's and ξ_t's implies that given θ_t, Z_t is conditionally independent of the other Z_t's and θ_t's. In addition, the initial condition is set as $\theta_0 = 0$ and $\theta_1 \sim N(0, \frac{\sigma^2}{1-\rho^2})$, which is the unconditional distribution of the AR(1) process.

The parameters of primary interest include $\boldsymbol{\alpha}$ and ρ. In order to make forecasting or to conduct model diagnostics, estimates of the state variables θ_t and the variance σ^2 are also needed.

To implement MCMC, Czado and Song (2007) suggested the following prior distributions: $[\boldsymbol{\alpha}, \sigma^2, \rho] = [\boldsymbol{\alpha}] \times [\sigma^2|\rho] \times [\rho]$, where

$$\boldsymbol{\alpha} \sim \text{MVN}_p(\boldsymbol{\alpha}_0, \Sigma_0), \tag{11.20}$$

$$\sigma^2|\rho \sim IG(a, b(\rho)) \text{ with } b(\rho) = c^2[(1-\rho^2)(a-1)]^{-1}, \tag{11.21}$$

$$\rho \sim \text{Uniform}(-1, 1). \tag{11.22}$$

The hyper-parameters $\boldsymbol{\alpha}_0, \Sigma_0, a$, and c are pre-specified. Here the $IG(a, b)$ denotes the inverse gamma distribution with density given by

$$f(\sigma^2) = \frac{1}{b^a \Gamma(a)} (\sigma^2)^{a+1} \exp(-\frac{1}{\sigma^2 b}), \; \sigma^2 > 0,$$

with $\mathrm{E}(\sigma^2) = [b(a-1)]^{-1}$ and $\mathrm{Var}(\sigma^2) = [(a-1)^2(a-2)b^2]^{-1}$ if $a > 2$.

The motivation behind the choice of the hyper-parameter b in (11.21), as a function of c, ρ, and a, is given as follows. First, note that this prior implies

$$\mathrm{E}(\sigma^2|\rho) = \frac{1-\rho^2}{c^2}. \tag{11.23}$$

Second, to balance the effect between ε_t and θ_t, similar to the familiar signal-to-noise ratio, it is useful to compare the standard deviation of ε_t (equal to 1) to the unconditional standard deviation of the AR(1) process, which equals to $\sqrt{\frac{\sigma^2}{1-\rho^2}}$. Hence, the resulting ratio between these two random sources is defined by

$$c^2 = \frac{\mathrm{Var}(\varepsilon_t)}{\mathrm{Var}(\theta_t)} = \frac{1-\rho^2}{\sigma^2},$$

which leads to the restriction $\sigma^2 = \frac{1-\rho^2}{c^2}$. Finally, by comparing this restriction to (11.23), it is easy to see that the prior mean for σ^2 may be chosen in such a way that the parameter c balances the relative variability between the ε_t and the θ_t.

A major gain from balancing the two sources of variations, as presented in (11.21)-(11.22), is to reduce the sensitivity of the posterior distributions on the hyper-parameters in the prior distributions, especially parameter b in the IG prior. When the prior for σ^2 is independent of ρ, Czado and Song (2007) found that the posteriors are very sensitive to the hyper-parameter b in the IG prior; in particular, if the b is not appropriately chosen, the MCMC runs would not even converge. Therefore, the conditional prior choice for σ^2 given in (11.21) is crucial to alleviate this unpleasant sensitivity.

Based on the specification of the probit state space model, the marginal likelihood function for the parameters $(\boldsymbol{\alpha}, \rho, \sigma^2)$ is,

$$L(\boldsymbol{\alpha}, \rho, \sigma^2 | \mathbf{Y}^n) = \int [Y_t|\theta_t; \boldsymbol{\alpha}][\theta_t|\theta_{t-1}; \rho, \sigma^2][\theta_0] d\theta_0 d\theta_1 \cdots d\theta_n$$

where the integral is clearly $(n+1)$-dimensional, with $[Y_t|\theta_t; \boldsymbol{\alpha}]$ being a Bernoulli distribution and $[\theta_t|\theta_{t-1}; \rho, \sigma^2]$ a conditional normal. MCMC is invoked to evaluate this high-dimensional integral, which basically draws sufficiently large number of joint samples from the posterior distribution $[\boldsymbol{\alpha}, \boldsymbol{\theta}^n, \sigma^2, \rho, \mathbf{Z}^n | \mathbf{Y}^n]$. Then, these samples can be utilized to infer these parameters via, for example, their marginal posteriors. Refer to Chapter 8 for a general introduction to the MCMC.

As suggested by Czado and Song (2007), the simulation smoother of de Jong and Shephard (1995) is efficient to sample the state variables, and this method is feasible since the probit state space model of (11.18)-(11.19) is a special case of the general state space models considered by de Jong (1991).

Similar to Section 8.1.4, the DIC can be calculated for model comparison and selection. In the present setting, the DIC is given by

$$\text{DIC} = \bar{D} + p_D = D(\bar{\boldsymbol{\alpha}}, \bar{\boldsymbol{\theta}}^n) + 2p_D,$$

with the Bayesian deviance

$$D(\boldsymbol{\alpha}, \boldsymbol{\theta}^n) = -2 \log L(\boldsymbol{\alpha}, \boldsymbol{\theta}^n) = -2 \sum_{t=1}^{T} [y_t \log(\pi_t) + (1 - y_t) \log(1 - \pi_t)]$$

and the effective number of parameters

$$\begin{aligned} p_D &= \text{E}_{\boldsymbol{\alpha}, \boldsymbol{\theta}^n, \rho, \sigma^2 | \mathbf{Y}^n} D(\boldsymbol{\alpha}, \boldsymbol{\theta}^n) - D(\text{E}_{\boldsymbol{\alpha}|\mathbf{Y}^n}(\boldsymbol{\alpha}), \text{E}_{\boldsymbol{\theta}^n|\mathbf{Y}^n}(\boldsymbol{\theta}^n)) \\ &= \bar{D} - D(\bar{\boldsymbol{\alpha}}, \bar{\boldsymbol{\theta}}^n). \end{aligned}$$

As usual, $\bar{D}$ explains the model fit and p_D indicates the model complexity. Computing the DIC is straightforward in an MCMC implementation. Monitoring both $(\boldsymbol{\alpha}, \boldsymbol{\theta}^n)$ and $D(\boldsymbol{\alpha}, \boldsymbol{\theta}^n)$ in MCMC updates, at the end one can estimate the $\bar{D}$ by the sample mean of the simulated values of D and the $D(\bar{\boldsymbol{\alpha}}, \bar{\boldsymbol{\theta}}^n)$ by plugging in the sample means of the simulated posterior values of $\boldsymbol{\alpha}$ and $\boldsymbol{\theta}^n$. A lower value of DIC indicates a better-fitting model.

The posterior distributions required in the MCMC updating of unknown variates are listed below.

Updating Latent Variable: Since the latent variables Z_t are conditionally independent given $\boldsymbol{\theta}^n$, one can reduce the update of $[\mathbf{Z}^n|\mathbf{Y}^n, \boldsymbol{\alpha}, \boldsymbol{\theta}^n, \sigma^2, \rho]$ to the individual updates of $[Z_t|\mathbf{Y}^n, \boldsymbol{\alpha}, \boldsymbol{\theta}^n, \sigma^2, \rho]$ for $t = 1, \cdots, n$. Each of these univariate conditional distribution is equivalent to $[Z_t|\mathbf{Y}^n, \boldsymbol{\alpha}, \boldsymbol{\theta}^n]$, since given $\boldsymbol{\theta}^n$ the information contained in σ^2 and ρ has no influence on the $\mathbf{Z}^n$. Moreover, $[Z_t|\mathbf{Y}^n, \boldsymbol{\alpha}, \boldsymbol{\theta}^n] = [Z_t|Y_t, \boldsymbol{\alpha}, \theta_t]$ holds for $t = 1, \ldots, n$, due again to the conditional independence. It is easy to see that these distributions are univariate truncated normal with mean $\mathbf{x}_t^T \boldsymbol{\alpha} + \theta_t$ and variance 1. Truncation interval is $(-\infty, 0]$(or $[0, \infty)$) when $Y_t = 1$(or $Y_t = 0$). The inversion method for the generation of truncated univariate normal random variables, proposed by Robert (1995), is easily implemented.

Updating State Variables and Regression Coefficients: The fact that $\mathbf{Y}^n$ is completely determined with given $\mathbf{Z}^n$ produces the following reduction, $[\boldsymbol{\alpha}|\mathbf{Y}^n, \mathbf{Z}^n, \boldsymbol{\theta}^n, \sigma^2, \rho] = [\boldsymbol{\alpha}|\mathbf{Z}^n, \boldsymbol{\theta}^n]$, which uses the simulation smoother of de Jong and Shephard (1995). To update the state variables and the regression parameters jointly, the simulation smoother applied to the probit state space model for (11.18)-(11.19) suggests the following steps:

Step 1: Perform the Kalman filter recursions for $t = 1, \cdots, n$,

$$
\begin{aligned}
E_t &= (\mathbf{x}_t^T \Sigma_0, Z_t + \mathbf{x}_t^T \boldsymbol{\alpha}_0) + A_t \\
A_{t+1} &= \rho A_t - \frac{\rho P_t}{P_t + 1} E_t \\
Q_{t+1} &= Q_t + \frac{1}{P_t + 1} E_t^T E_t \\
P_{t+1} &= \frac{\rho^2 P_t}{P_t + 1} + \sigma^2 \,, P_1 = \sigma^2 .
\end{aligned}
$$

In the meanwhile, identify matrix S and vector s from the partition of $Q_{n+1} = \begin{pmatrix} S & -s \\ -s^T & * \end{pmatrix}$ and then

$$\text{draw} \quad \delta \sim \text{MVN}_p((S + I_p)^{-1} s, (S + I_p)^{-1}).$$

Step 2: Perform the smoothing recursions for $t = n, n-1, \cdots, 1$:

$$
\begin{aligned}
&\text{set } r_n = 0,\ U_n = 0 \\
&\text{draw } \epsilon_t \sim N(0, \sigma^2(1-\sigma^2 U_t)) \\
&\text{with } e_t = E_t \begin{pmatrix} \delta \\ 1 \end{pmatrix} \\
&r_{t-1} = \frac{1}{P_t+1}\left[\rho r_t - e_t - \frac{\rho U_t e_t}{1-\sigma^2 U_t}\right] \\
&U_{t-1} = \frac{1}{P_t+1}\left[1 + \frac{\rho^2 U_t}{(P_t+1)(1-\sigma^2 U_t)}\right].
\end{aligned}
$$

Step 3: At the end of the filtering and smoothing recursions, set for $t = 0, 1, \cdots, n$

$$\chi_t = \sigma^2 r_t + \epsilon_t,\ t = 0, 1, \ldots, n,$$

and update the state variables by

$$\theta_{t+1} = \rho\theta_t + \chi_t,\ t = 0, 1, \cdots, n,$$

with $\theta_0 = 0$. According to de Jong and Shephard (1995), this gives a draw from the conditional distribution $[\boldsymbol{\theta}^n|\boldsymbol{\alpha}, \mathbf{Z}^n, \rho, \sigma^2]$.
Step 4: Furthermore, to make a draw from the conditional distribution $\boldsymbol{\alpha}|\mathbf{Z}^n, \boldsymbol{\theta}^n, \rho, \sigma^2 \sim \text{MVN}_p(\boldsymbol{\alpha}_0 + \Sigma_0(S+I_p)^{-1}\Sigma_0^T)$, simply set

$$\boldsymbol{\alpha} = \boldsymbol{\alpha}_0 + \Sigma_0\delta.$$

Updating State Variance: The conditional prior $IG(a, b(\rho))$ given in (11.21) is used for σ^2. A straightforward calculation gives that the density of $[\boldsymbol{\theta}^n|\sigma^2, \rho]$ is

$$f(\boldsymbol{\theta}^n|\sigma^2, \rho) = \frac{1}{(2\pi\sigma^2)^{\frac{n+1}{2}}} \exp\left\{-\frac{1}{2\sigma^2}\sum_{t=1}^{n}(\theta_{t+1} - \rho\theta_t)^2 - \frac{1-\rho^2}{2\sigma^2}\theta_1^2\right\}. \tag{11.24}$$

It follows that

$$
\begin{aligned}
f(\sigma^2|\boldsymbol{\theta}^n, \rho) &\propto f(\boldsymbol{\theta}^n|\sigma^2, \rho)\pi(\sigma^2) \\
&\propto \frac{1}{(\sigma^2)^{\frac{n+1}{2}+a+1}} \times \\
&\quad \exp\left\{-\frac{1}{\sigma^2}\left[\frac{1}{b(\rho} + \frac{1}{2}\left\{\sum_{t=1}^{n}(\theta_{t+1} - \rho\theta_t)^2 + (1-\rho^2)\theta_1^2\right\}\right]\right\},
\end{aligned}
$$

which shows that the conditional distribution of $[\sigma^2|\boldsymbol{\alpha}, \rho]$ is an inverse gamma $IG(a^*, b^*)$ with

$$
\begin{aligned}
a^* &= \frac{n+1}{2} + a \\
b^* &= \left[\frac{1}{b(\rho)} + \frac{1}{2}\left\{\sum_{t=1}^{n}(\theta_{t+1} - \rho\theta_t)^2 + (1-\rho^2)\theta_1^2\right\}\right]^{-1}.
\end{aligned}
$$

Updating Autocorrelation Coefficient: A causal state process given by (11.19) requires that $\rho \in (-1, 1)$. So a uniform prior distribution on $(-1, 1)$ is assumed for ρ. First writing the exponent of $[\rho|\boldsymbol{\theta}^n, \sigma^2]$ and then turning it into a quadratic form, it is easy to find that $[\rho|\boldsymbol{\theta}^n, \sigma^2]$ is truncated univariate normal on $(-1, 1)$ with mean μ_ρ and variance σ_ρ^2 given by, respectively,

$$\mu_\rho = \frac{\sum_{t=1}^n \theta_t \theta_{t+1}}{\sum_{t=2}^n \theta_t^2} \text{ and } \sigma_\rho^2 = \frac{\sigma^2}{\sum_{t=2}^n \theta_t^2}.$$

The above development can be extended to deal with the binomial state space model by the method of aggregation of *i.i.d.* Bernoulli variates, in a similar fashion to that used in Example 11.2. For more details, refer to Czado and Song (2007).

Example 11.3 (Tokyo Rainfall Data).

Now the Tokyo rainfall data is analyzed again using the binomial state space mixed model. As pointed in the previous section, a binomial variate can be treated as an aggregation of *i.i.d.* Bernoulli variates, so the inference established for the Bernoulli model can be extended to deal with binomial data with little effort.

The analysis given by Example 11.2 assumes a (non-stationary) random walk model for the state process. The state variable θ_t may be thought of as a certain underlying meteorological variate, such as moisture, most directly responsible for rainfall. One obvious limitation of the random walk formulation is that it effectively increases the variability of moisture over time, which does not seem to be realistic. In addition, it does not allow us to examine directly the seasonal rainfall cycle.

The new model directly addresses seasonal and monthly effects through covariates $\mathbf{x}_t = (\cos 1_t, \sin 1_t, \cos 4_t, \sin 4_t, \cos 12_t, \sin 12_t)^T$, where

$$\cos m_t = \cos\left(\frac{2\pi m t}{n}\right) \text{ and } \sin m_t = \sin\left(\frac{2\pi m t}{n}\right), \quad m = 1, \ldots, n.$$

So the latent variables $\{Z_{it}\}$ follow

$$\begin{aligned} Z_{it} = & -\alpha_0 \\ & -\alpha_1 \cos 1_t - \alpha_2 \sin 1_t - \alpha_3 \cos 4_t - \alpha_4 \sin 4_t - \alpha_5 \cos 12_t - \alpha_6 \sin 12_t \\ & -\theta_t + \varepsilon_{it}, \ i = 1, 2; t = 1, \ldots, 366, \end{aligned} \tag{11.25}$$

and the state variables $\{\theta_t\}$ follow the stationary AR(1) model. In each of all the cases described in Table 11.1, a total of 20,000 iterations of the MCMC algorithm were run with every 10th iteration recorded. A burn-in of 100 recorded iterations (equivalent to 1000 unthinned iterations) was used for the posterior density estimation.

To compare several possible combinations of seasonal covariates as well as the effect of the ratio parameter c, the DIC information criterion is used.

Table 11.1. Model Fit $\overline{D}$, effective number of parameters p_D and DIC for the rainfall data.

Covariate $\mathbf{x}_t$	Ratio c	$\overline{D}$	p_D	DIC
$(\cos 4_t, \sin 4_t, \cos 12_t, \sin 12_t)^T$	1.0	702.31	79.83	782.14
	2.0	745.67	33.76	779.42
	5.0	763.21	11.53	774.74
$(\cos 1_t, \sin 1_t, \cos 4_t, \sin 4_t)'$	1.0	705.01	82.02	787.03
	2.0	737.09	50.98	788.07
	5.0	774.18	22.27	796.45
$(\cos 1_t, \sin 1_t, \cos 4_t, \sin 4_t, \cos 12_t, \sin 12_t)^T$	1.0	696.50	78.57	775.07
	2.0	727.19	47.25	774.43
	5.0	755.25	18.39	773.63

Totally, the MCMC is run in 9 different settings, each with a set of chosen covariates and a value of the c. The results are summarized in Table 11.1.

From this table it is easy to see that the DIC values of the second submodel are steadily higher than those of the two other models, so the second combination $\mathbf{x}_t = (\cos 1_t, \sin 1_t, \cos 4_t, \sin 4_t)^T$ should not be considered further. By comparing the DIC values between the first submodel and the full model (11.25), the full model appeared to have a more stable performance over different levels of the c and reached the minimum at $c = 5$. It is important to replicate the above exercise a few more times and to ensure the observed evidence occurs by the chance of MCMC sampling. Over a few replications, the DIC always appears to favor the full model. Thus, the full model is selected for the further analysis of the data.

Figure 11.4 displays the estimated posterior densities for the regression parameters $\alpha_l, l = 0, \cdots, 6$, the standard error parameter σ, and the autocorrelation parameter ρ, at different values $c = 1, 2$ and 5. It is clear that the densities of the regression coefficients $\alpha_l, l = 0, 1, \ldots, 6$ appeared to be less affected by the c than the densities of the σ and γ. For comparison, the naive point estimates obtained by a cross-sectional GLM fit (under the independence correlation), are indicated by the vertical lines in the plots. It is evident that the state space modeling of auto-correlated data does produce different estimations from the naive GLM analysis that assumes independent observations. Furthermore, the 90% credible intervals of the regression parameters based on the full model suggest that a yearly covariate $\cos(1_t)$, a seasonal covariate $\cos(4_t)$, and two monthly covariates $\sin(12_t)$ and $\cos(12_t)$ turn out to be important explanatory variables, at all of the c levels considered. In the meanwhile, for the modeling of the state variables, the average estimate of the autocorrelation coefficient ρ over the three c levels is around .41, and such a medium sized ρ is supportive to the AR model, rather than the random walk

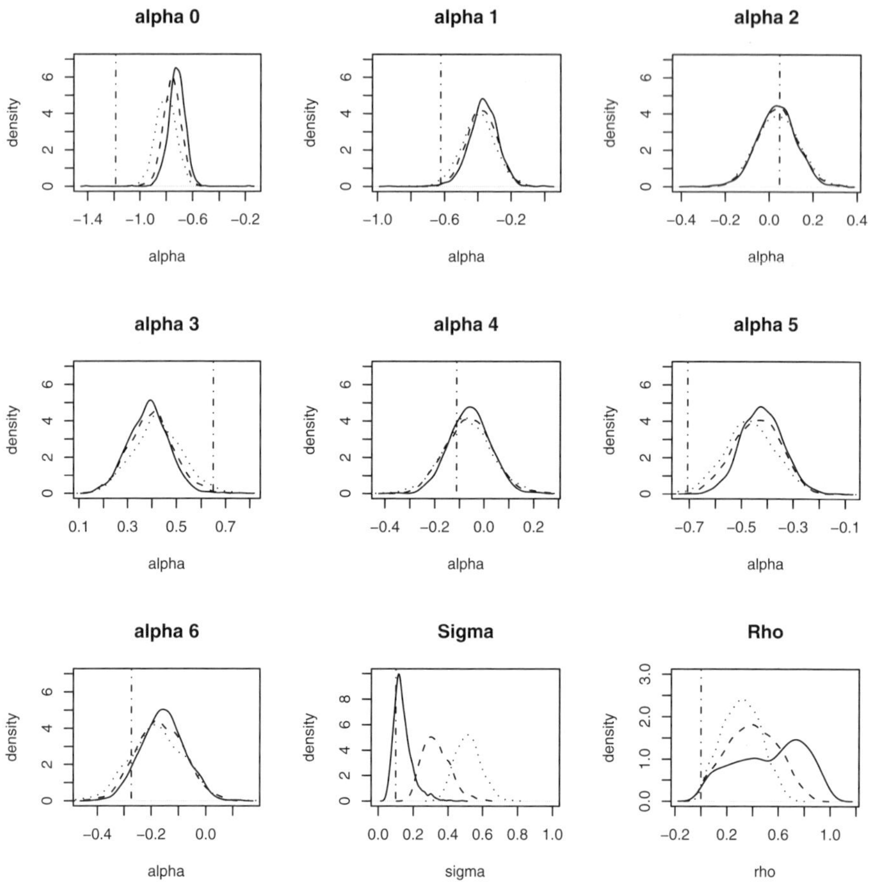

Fig. 11.4. Posterior density estimates for the rainfall data with $c = (5, 2, 1)$, corresponding to the solid, the broken, and the dotted lines.

model with $\rho = 1$. In conclusion, the state space mixed model has identified several important seasonal covariates to explain the variability in the rainfall probability over one year period, which are useful for forecasting. The average estimate of σ is around .32 over the three c levels.

In addition, the pointwise estimation of the rainfall probability π_t at day t, $t = 1, \ldots, 366$ is computed with $c = 5$, because this case corresponds to the smallest DIC. The solid line in Figure 11.5 shows the posterior mean estimates $\bar{\pi}_t$, $t = 1, \ldots, 366$, which was obtained by only using the deterministic component of the full model, together with its 90% credible bounds indicated by the dotted lines. The broken line, which is tightly tangled with the solid line, represents the posterior mean estimates of the probabilities computed by

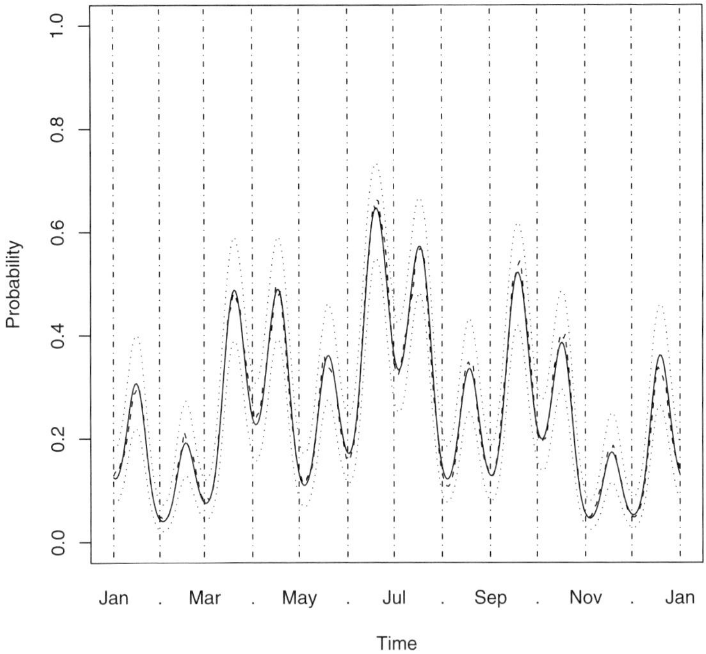

Fig. 11.5. Pointwise posterior mean estimates of the probabilities of rainfall with pointwise 90 % credible intervals.

using both deterministic component and random component θ_t. Again, this analysis suggest that the period of July–August is the most wet season of a year in Tokyo, and the periods of March–May and September also have high amounts of precipitation.

Example 11.4 (Infant Sleep Data).

The state space mixed model is now applied to analyze the infant sleep data. This new analysis includes two time-varying covariates collected together with the response of sleep status. They are, the number of body movements due not to sucking during minute t, x_{t1}, and the number of body movements during minute t, x_{t2}. The objective of this analysis is to investigate whether the probability of being in REM sleep status is significantly related to these two types of body movements x_{t1} and x_{t2}. If so, using the deterministic predictor, $\alpha_0 + \alpha_1 x_{t1} + \alpha_0 x_{t2}$, together with the random component, θ_t, would better interpret and predict the probability of REM sleep status. This leads to the observation equation of the following form:

$$\pi_t = \Phi(\alpha_0 + \alpha_1 x_{t1} + \alpha_2 x_{t2} + \theta_t),$$

where the state equation θ_t is the stationary AR(1) as before.

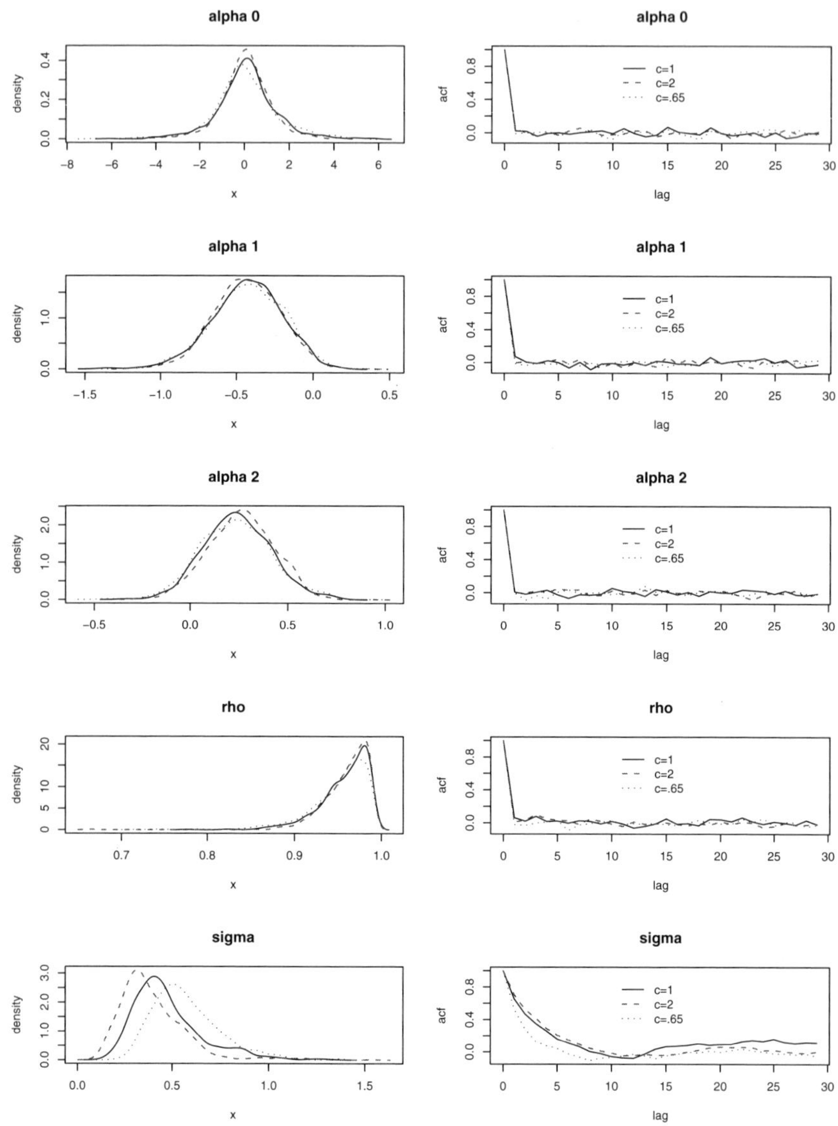

Fig. 11.6. Posterior density estimates and estimated ACF among the recorded MCMC iterates for the infant sleep data at different levels of ratio $c = (0.65, 1, 2)$.

The MCMC algorithm is applied with the conditional joint prior for (σ, ρ) at different values of c. A total of 20,000 iterations with every 20th iteration recorded were run. Various graphical examinations indicated that a burn-in of recorded 10 iterations was sufficient for this case. Figure 11.6 displays the estimated marginal posterior densities for the regression parameters $(\alpha_0, \alpha_1, \alpha_2)$, the autocorrelation (ρ), and the variance parameter (σ) together with the autocorrelation functions of the MCMC updates for the different c values. These autocorrelation functions suggested that the mixing of the MCMC algorithm was fast for the regression parameters and the correlation parameter, but slow for σ. Also, the posterior density estimates of the regression and correlation parameter are less sensitive to the ratio c than that of σ. The posterior mode estimate of σ increases as c decreases. It is further evident that a very high autocorrelation is present in the AR(1) process for state variables and thus in this binary time series.

Table 11.2. Posterior mean and quantiles for the infant sleep data.

Posterior	c	α_0	α_1	α_2	γ	σ
Mean	2.00	0.0624	−0.4315	0.2625	0.960	0.393
	1.00	0.1140	−0.432	0.2419	0.958	0.473
	.65	0.0540	−0.4225	0.2341	0.952	0.571
5% Quantile	2.00	−1.7610	−0.7765	−0.0137	0.915	0.181
	1.00	−2.1040	−0.830	−0.0281	0.904	0.252
	.65	−2.1023	−0.8122	−0.0594	0.887	0.339
10% Quantile	2.00	−1.2050	−0.7070	0.0462	0.927	0.213
	1.00	−1.3708	−0.717	0.0213	0.919	0.282
	.65	−1.5310	−0.7333	0.0163	0.909	0.373
50% Quantile	2.00	0.0755	−0.4291	0.2665	0.966	0.355
	1.00	0.0879	−0.427	0.2399	0.965	0.434
	.65	−0.0419	−0.4146	0.2333	0.961	0.543
90% Quantile	2.00	1.3204	−0.1632	0.4861	0.986	0.590
	1.00	1.6967	−0.144	0.4536	0.986	0.727
	.65	1.8113	−0.1389	0.4692	0.985	0.804
95% Quantile	2.00	1.8083	−0.0822	0.5303	0.988	0.667
	1.00	2.2723	−0.064	0.5352	0.988	0.849
	.65	2.6322	−0.0555	0.5523	0.987	0.925

Posterior means and quantiles are listed in Table 11.2. It is clear that the influence of the number of body movements (x_2) is marginal, since the corresponding 90% credible interval for α_2 contains the zero value. In contrast, the influence of the number of body movements not due to sucking (x_1) is detected. The negative value of the posterior mean for α_1 shows that a higher number of body movements not due to sucking will reduce the probability of the infant being in REM sleep. This conclusion is intuitively meaningful.

The top panel of Figure 11.7 shows the posterior mean estimates for the state variables $\{\theta_t\}$ with 90% pointwise credible intervals for $c = .65$ chosen according to the smallest DIC. This shows that the state process θ_t behaves as an underlying continuous "sleep state." The posterior mean estimates of the REM sleep state probabilities $p_t = \Phi(\alpha_0 + \alpha_1 x_{t1} + \alpha_2 x_{t2} + \theta_t)$ are shown in the bottom panel of Figure 11.7 for $c = .65$, together with the 90% credible bounds indicated by the dotted lines.

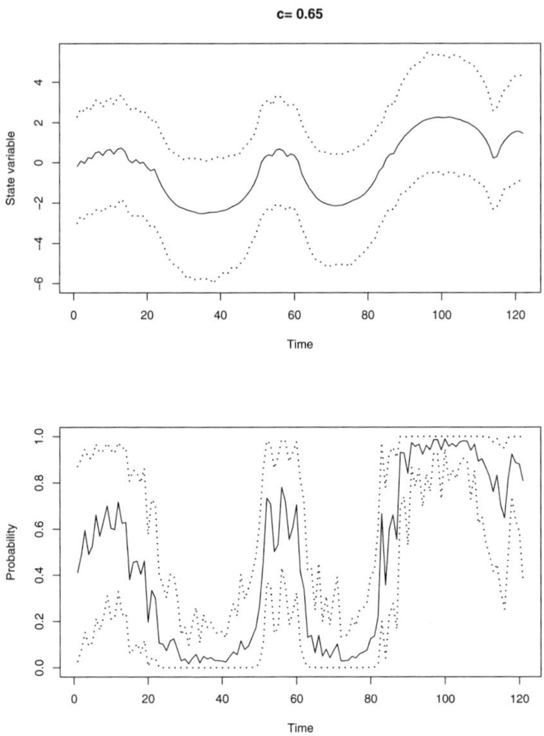

Fig. 11.7. Posterior mean estimates of the state variables (top panel) and posterior mean estimates of the REM sleep state probabilities (bottom panel), where dotted lines indicate 90% credible intervals, and $c = 0.65$.

To further demonstrate the usefulness of the state space mixed models, it is worth investigating their predictive ability, in comparison to the regular state space models. From the evidence presented below, it is clear that a state space model with no deterministic predictor η_t will have poor performance in prediction, caused simply by the fact that the state variables have zero expectation. On the other hand, the inclusion of covariates will in general improve the predictive power. To elucidate, three observation equations are considered: (i) the inclusion of both covariates x_{t1} and x_{t2}, (ii) the inclusion of only x_{t1}, and (iii) the exclusion of both covariates. The same MCMC algorithm was run as before in each of the three cases based only on the first 80 observations used. The out-of-sample predicted probabilities of REM sleep status were computed by

$$\hat{\pi}_t = \Phi(\mathbf{x}_t^T \bar{\boldsymbol{\alpha}} + \hat{\theta}_t), \ t > 80$$

where $\hat{\theta}_t = \bar{\rho}\hat{\theta}_{t-1}$. Here $\bar{\boldsymbol{\alpha}}$ and $\bar{\rho}$ denote the corresponding posterior mean estimates.

Figure 11.8 shows that the fitted probabilities for $t \leq 80$ and the predicted probabilities for $t > 80$. It is clear that for all three cases a reasonable fit of the probabilities ($t \leq 80$) is indicated. Moreover, compared to the fitted probabilities with $t < 80$ given in the bottom panel of Figure 11.8, the pure state space model has little predictive ability, while the model with both covariates shows better predictive power by utilizing the information from the both covariates over the period $t > 80$.

It is interesting to note that the DIC does not reflect the prediction capability. For example, at $c = .65$, the DIC values are computed for respective models, (i) 100.1, (ii) 100.3, and (iii) 99.57. These values are in fact very close to each other, which implies that the models are quite similar in terms of quality of fit, but these three models have indicated very different prediction power.

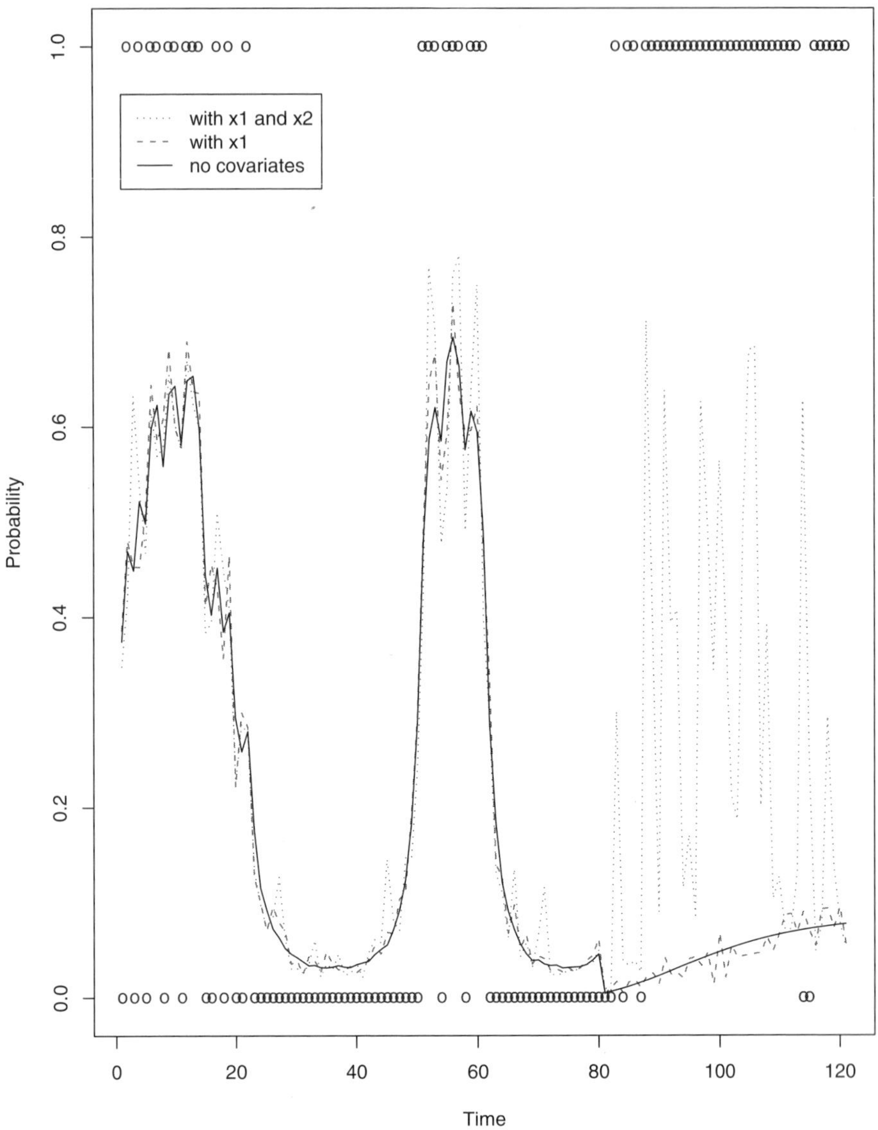

Fig. 11.8. Pointwise posterior mean estimates ($t \leq 80$) and predictions ($t > 80$) of REM sleep state probabilities for the infant sleep data.

12

Generalized State Space Models for Longitudinal Count Data

12.1 Introduction

This chapter discusses three inference approaches in the framework of generalized state space models (GSSM) for time series of counts. They are, the generalized estimating equation (GEE) method, Kalman estimating equation (KEE) method, and Monte Carlo EM (MCEM) algorithm. Because MCMC based inference has been studied in detail in Chapter 11 in the context of GSMMs for time series of binomial data, a similar inference procedure can derived in the setting of GSSMs for count data. Therefore, MCMC is not covered in this chapter.

Consider a longitudinal data $\{(\mathbf{Y}_t, \mathbf{x}_t), t = 1, \ldots, n\}$, where at time t a d-dimensional response vector of counts $\mathbf{Y}_t = (Y_{1t}, \ldots, Y_{dt})^T$ is observed. To set up a GSSM, model $\mathbf{M}_2$ (10.2) specifies the observation process as follows:

$$Y_{it}|\theta_t \sim \mathrm{Po}\left(a_{it}\theta_t\right), \; i = 1, \ldots, d; t = 1, \ldots, n \tag{12.1}$$

with the deterministic component

$$a_{it} = \exp(\mathbf{x}_{it}^T \boldsymbol{\alpha}_i), \; i = 1, \ldots, d; t = 1, \ldots, n$$

where $\boldsymbol{\alpha}_i$ is a p-element vector of regression parameters, and the initial state θ_0 is assumed to be degenerated at its mean just for convenience.

Also, model $\mathbf{M}_1$ (10.1) specifies the state process θ_t as a Markov process. As part of the mean in a Poisson distribution, the θ_t is constrained with being positive. In the literature, there are several versions of the $\mathbf{M}_1$ formulation, each being imposed for the suitability of a particular statistical inference approach. Some examples are given as follows.

Example 12.1 (Poisson Parameter-Driven Model).

The GEE method introduced in Chapter 5 is a quasi-likelihood inference based only on the first two moments of the underlying probability distribution. It is known that one advantage of this inference is the robustness against the misspecification of the parametric model in the part of $\mathbf{M}_1$. On this line, Zeger (1988) proposed a GSSM for time series of counts, termed *parameter-driven model* according to Cox (1981), in which the state process θ_t is assumed to be simply a weakly stationary process with $\mathrm{E}(\theta_t) = 1$ and the autocovariance function (ACVF) $\mathrm{cov}(\theta_t, \theta_{t+h}) = \sigma^2 \gamma_\theta(h), h = 0, 1, \cdots$. Here $\gamma_\theta(h)$ is the ACF of the stationary process θ_t, with $\gamma_\theta(0) = 1$ and $\sigma^2 = \mathrm{Var}(\theta_t)$.

It follows immediately that the marginal first two moments of $\mathbf{Y}_t$ are

$$\mu_{it} = \mathrm{E}(Y_{it}) = a_{it}, \; v_{it} = \mathrm{Var}(Y_{it}) = \mu_{it}(1 + \sigma^2 \mu_{it}),$$
$$\gamma_{y,ij}(t, h) = \mathrm{corr}(y_{i,t}, y_{j,t+h}) = \frac{\gamma_\theta(h)}{[\{1 + (\sigma^2 \mu_{it})^{-1}\}\{1 + (\sigma^2 \mu_{j,t+h})^{-1}\}]^{1/2}},$$

for $i, j = 1, \ldots, d$. It is interesting to note that the state process, θ_t, introduces autocorrelation, cross-component correlation, and overdispersion into the process $\{\mathbf{Y}_t\}$.

Example 12.2 (Poisson-Stationary Lognormal Model).

Chan and Ledolter (1995) considers a lognormal model for the state process, $\{\theta_t\}$, as follows. Let $\{W_t\}$ be a stationary Gaussian AR(1) process; that is, $W_t = \phi W_{t-1} + \epsilon_t$, where $\{\epsilon_t\}$ is *i.i.d.* $N(0, \sigma_\epsilon^2)$. Then, the state process θ_t is defined as $\theta_t = \exp(W_t)$, which is also stationary. Equivalently, $\log(\theta_t)$ follows a stationary Gaussian AR(1) process, or

$$\theta_t = \theta_{t-1}^\phi \varepsilon_t,$$

where $\epsilon_t = \log \varepsilon_t$ is a Gaussian white noise. That is, model $\mathbf{M}_1$ takes a multiplicative form of the first order stationary Markov process. Such a specification of model $\mathbf{M}_1$ gives rise to the ease of developing posterior densities at the E-step in the application of Monte Carlo EM algorithm.

One shortcoming of this multiplicative AR(1) process is the interpretation. For example, ϕ is no longer interpretable as the autocorrelation for the θ_t process, and the marginal mean of θ_t involves both ϕ and the variance of $\log \varepsilon_t$. These two parameters are in the second moments of the process, and from the modeling point of view, it seems unnatural to specify a model whose first moments are dependent on the second moments.

Example 12.3 (Poisson-Stationary Gamma Model).

When a parametric distribution is adopted in the specification of the θ_t process, a conjugate gamma distribution (*w.r.t.* Poisson) is commonly suggested in the literature. That is, assume marginally at time t, $\theta_t \sim \mathrm{Ga}^*(\mu = 1, \lambda)$, where Ga^* denotes an ED model of the additive form (see (2.13)) with the mean parameter equal to 1 and index parameter λ. This implies that the dispersion parameter is $1/\lambda$. Here the requirement of $\mu = 1$ is necessary for the sake of parameter identifiability *w.r.t.* the intercept term in the deterministic component a_{it} (Jørgensen and Song, 1998b). A stationary AR(1) gamma process (Lewis et al., 1989) for θ_t takes the following form:

$$\theta_t = B_t \theta_{t-1} + \varepsilon_t, \quad t = 1, \ldots, n, \tag{12.2}$$

where B_t is a beta random variable according to Beta$(\rho\lambda, (1-\rho)\lambda)$ with $|\rho| < 1$ and independent of θ_t process. The noise $\{\varepsilon_t\}$ is a sequence of *i.i.d.* innovations distributed as Ga$^*(1, (1-\rho)\lambda)$. Therefore, the property of convolution in Proposition 2.10 ensures the marginal stationarity, namely $\theta_t \sim \text{Ga}^*(1, \lambda)$. It can be shown that parameter ρ is the autocorrelation coefficient because the ACF of this gamma process defined in (12.2) is $\rho^{|h|}$ for lag h. Interested readers may refer to, for example, Jørgensen and Song (1998a) for more details.

This GSSM formulation is useful to model dynamics of infectious diseases. For example, θ_t may represent the volume of contagious material present in the environment, variable B_t is the proportion of contagious material that survives from time $t-1$ to time t, and ε_t is the amount of new contagious material introduced at time t. At a given time t, the amount of contagious material θ_t will trigger the number of affected subjects Y_t in the population via a Poisson log-linear model. Because of this interpretation, the Poisson-stationary gamma model will be applied to analyze the monthly polio incidences in USA introduced in Section 1.3.8.

Example 12.4 (Poisson-Nonstationary Gamma Model).

In many studies, effects of covariates may be lagged in time; some covariates may have an immediate effect on the response variable, and some may have a carry-over or lagged effect on the response variable. For example, in the analysis of Prince George air pollution data introduced in Section 1.3.10, meteorological covariates of temperature and humidity are more likely to have an acute (or short-term) effect on the daily counts of emergency room visits, while air pollution covariates of sulphur and particulates influence disease symptoms in a lagged (or long-term) fashion.

To reflect such a difference, it is appealing to distinguish and divide, if possible based on subject-matter knowledge, the covariates into different groups in the formulation of a state space model. Jørgensen et al. (1996b) suggested entering the meteorological covariates $\mathbf{x}_t$ into the deterministic component a_{it} of the Poisson mean in model $\mathbf{M}_2$ (12.1), while entering the air pollution covariates $\mathbf{z}_t$ into the state process θ_t Therefore, the $\mathbf{M}_1$ is specified as a gamma Markov process,

$$\theta_t|\theta_{t-1} \sim \text{Ga}\left(b_t\theta_{t-1}, \frac{\sigma^2}{\theta_{t-1}}\right),$$

where Ga(μ, δ^2) denotes the gamma distribution with mean μ and coefficient of variation δ. The parameter σ^2 is a dispersion parameter expressing the smoothness of the process, and $\log(b_t)$ is a linear predictor depending on the long-term covariates via their increments. That is,

$$b_t = \exp\left\{(\Delta\mathbf{z}_t)^T\boldsymbol{\beta}\right\}, \quad \text{with } \Delta\mathbf{z}_t = \mathbf{z}_t - \mathbf{z}_{t-1},$$

where $\mathbf{z}_t$ is the vector-valued process of multiple air pollution measurements and $\boldsymbol{\beta}$ is the parameter vector representing the long-term effects of air pollutants on morbidity rates of respiratory diseases.

12.2 Generalized Estimating Equation

Let us consider a simple case of a single time series of counts, $d = 1$, which is modeled by the Poisson parameter-driven model discussed in Example 12.1. The approach of GEE can be applied to estimate the vector of regression coefficients, $\boldsymbol{\alpha}$ in the log-linear Poisson model $\mathbf{M}_2$ (12.1). Let $\boldsymbol{\mu} = \mathrm{E}(\mathbf{Y}^n) = (\mu_1(\boldsymbol{\alpha}), \ldots, \mu_n(\boldsymbol{\alpha}))^T$, and let $\mathbf{V} = \mathrm{Var}(\mathbf{Y}^n) = \mathbf{G} + \sigma^2 \mathbf{G} \Gamma_\theta(\boldsymbol{\phi}) \mathbf{G}$, where $\mathbf{G} = \mathrm{diag}(\mu_1(\boldsymbol{\alpha}), \ldots, \mu_n(\boldsymbol{\alpha}))^T$ and $\Gamma_\theta(\phi)$ is an $n \times n$ matrix with (t, s)-element equal to $\gamma_\theta(|t-s|; \boldsymbol{\phi})$. Based on the first two moments specified above, the GEE takes the form:

$$\left(\frac{\partial \boldsymbol{\mu}^T}{\partial \boldsymbol{\alpha}}\right) \mathbf{V}^{-1}(\boldsymbol{\alpha}, \boldsymbol{\phi})(\mathbf{Y}^n - \boldsymbol{\mu}) = \mathbf{0}.$$

The nuisance parameter $\boldsymbol{\phi}$ is to be estimated separately. Let $\hat{\boldsymbol{\phi}}$ be a $\sqrt{n}$-consistent estimator of $\boldsymbol{\phi}$ depending possibly on $\boldsymbol{\alpha}$. Then the estimate of $\boldsymbol{\alpha}$, $\hat{\boldsymbol{\alpha}}$, is actually the solution to the following equation:

$$\left(\frac{\partial \boldsymbol{\mu}^T}{\partial \boldsymbol{\alpha}}\right) \mathbf{V}^{-1}(\boldsymbol{\alpha}, \hat{\boldsymbol{\phi}}(\boldsymbol{\alpha}))(\mathbf{Y}^n - \boldsymbol{\mu}) = \mathbf{0}. \tag{12.3}$$

However, the inversion of matrix $\mathbf{V}$ is usually difficult because the parameter-driven model does not have a stationary ACF. To overcome this, Zeger (1988) suggested using an approximation of the actual ACF Γ_θ by an ACF matrix of a stationary autoregressive process. That is, approximate $\mathbf{V}$ by a *working* ACVF,

$$\mathbf{V}_w = \mathbf{D}^{\frac{1}{2}} \Gamma_w(\boldsymbol{\psi}) \mathbf{D}^{\frac{1}{2}}$$

where $\mathbf{D} = \mathrm{diag}(\mu_t + \sigma^2 \mu_t^2)$ is a diagonal matrix of variances $\mathrm{Var}(Y_t)$, and Γ_w is the ACF matrix for a pre-specified working stationary AR process. Let $(\hat{\sigma}^2, \hat{\boldsymbol{\psi}})$ be $\sqrt{n}$-consistent estimators of σ^2 and ψ, respectively. Now define $\hat{\boldsymbol{\alpha}}_w$ as the solution to the estimating equation:

$$U(\boldsymbol{\alpha}) = \left(\frac{\partial \boldsymbol{\mu}^T}{\partial \boldsymbol{\alpha}}\right) \mathbf{V}_w^{-1}(\boldsymbol{\alpha}, \hat{\sigma}^2, \hat{\boldsymbol{\psi}}(\boldsymbol{\alpha}))(\mathbf{Y}^n - \boldsymbol{\mu}) = \mathbf{0}. \tag{12.4}$$

Inference function U in (12.4) is (asymptotically) unbiased. Under some mild regularity conditions, the estimator, $\hat{\boldsymbol{\alpha}}$, of $\boldsymbol{\alpha}$ produced by this GEE (12.4) is consistent, and $\sqrt{n}(\hat{\boldsymbol{\alpha}} - \boldsymbol{\alpha})$ is asymptotically normal with mean 0 and covariance matrix $\lim_n n \mathbf{j}_U^{-1}$, where $\mathbf{j}_U$ is the Godambe information matrix derived from the inference function $U(\boldsymbol{\alpha})$.

To estimate the nuisance parameters, Zeger (1988) suggested the method of moments. Utilizing the property $\text{Var}(Y_t) = \mu_t + \sigma^2\mu_t$, one can form a consistent estimator of σ^2 as follows:

$$\hat{\sigma}^2 = \sum_{t=1}^{n}\{(Y_t - \hat{\mu}_t)^2 - \hat{\mu}_t\}/\sum_{t=1}^{n}\hat{\mu}_t^2. \tag{12.5}$$

Using the residuals, one may estimate the ACF of the θ_t process by

$$\hat{\gamma}_\theta(h) = \hat{\sigma}^{-2}\sum_{t=h+1}^{n}\{(Y_t - \hat{\mu}_t)(Y_{t+h} - \hat{\mu}_{t+h})\}/\sum_{t=h+1}^{n}\hat{\mu}_t\hat{\mu}_{t+h}. \tag{12.6}$$

Then based on the estimated ACF, the Yule-Walker equations can be set up to estimate the AR parameter $\boldsymbol{\psi}$ in a given working AR process. For example, consider an AR(r) process of the form,

$$\theta_t = \psi_1\theta_{t-1} + \cdots + \psi_p\theta_{t-r} + \epsilon_t,$$

where ϵ_t's are white noise with variance σ_ϵ^2. The Yule-Walker equations estimate of $\boldsymbol{\psi} = (\psi_1, \ldots, \psi_r)^T$ is given by

$$\hat{\boldsymbol{\psi}} = \hat{\Gamma}_w^{-1}\hat{\boldsymbol{\gamma}},$$

where $\hat{\Gamma}_w$ is an $r \times r$ matrix with the (t,s)-element equal to $\hat{\gamma}(|t-s|), |t-s| = 0, 1, \ldots, r-1$, and $\hat{\boldsymbol{\gamma}}$ is an r-element vector with the t-th element being $\hat{\gamma}(t), t = 1, \ldots, r$. When $r = 1$ corresponding to the AR(1), the estimate of single ψ is simply $\hat{\gamma}(1)$.

12.3 Monte Carlo EM Algorithm

Consider again the simple case of $d = 1$, corresponding to single time series of counts, which is now modeled by Poisson-stationary lognormal model discussed in Example 12.2. Readers who are not familiar with EM algorithm may refer to Section 13.4.3. Treating all state variables θ_t or W_t as missing values, one can apply the EM algorithm to estimate parameter $\boldsymbol{\alpha}$. The augmented data consist of $(Y_t, W_t), t = 1, \ldots, n$. The augmented likelihood function is, subject to those terms independent of the model parameters,

$$\begin{aligned}\ell(\boldsymbol{\alpha}, \phi, \sigma_\epsilon^2|\mathbf{Y}^n, \mathbf{W}^n) &= \sum_{t=1}^{n}\{-\exp(W_t + \mathbf{x}_t^T\boldsymbol{\alpha}) + Y_tW_t + Y_t\mathbf{x}_t^T\boldsymbol{\alpha}\} \\ &\quad -\frac{n-1}{2}\log(\sigma_\epsilon^2) - \frac{1}{2\sigma_\epsilon^2}\sum_{t=1}^{n-1}(W_{t+1} - \phi W_t)^2 \\ &= \sum_{t=1}^{n}\{-\theta_t a_t + Y_t\log(a_t)\} + E \\ &\quad -\frac{n-1}{2}\log(\sigma_\epsilon^2) - \frac{1}{2\sigma_\epsilon^2}(B - 2\phi D + C\phi^2), \end{aligned} \tag{12.7}$$

where

$$B = \sum_{t=1}^{n-1} W_{t+1}^2, \; C = \sum_{t=1}^{n-1} W_t^2, \; D = \sum_{t=1}^{n-1} W_t W_{t+1}, \; E = \sum_{t=1}^{n} W_t Y_t.$$

Here the initial state W_0 is assumed to be known as 0 or θ_0 known as the marginal mean of 1. The utility of the lognormal distribution produces a nice property; that is, the augmented likelihood is linear in θ_t, as well as in B, C, D and E. All these terms are certain functions of missing data W_t and the observed data Y_t.

Given a sequence of Monte Carlo samples $\mathbf{W}^{n(1)}, \ldots, \mathbf{W}^{n(M)}$ drawn from the conditional distribution, $f(\mathbf{W}^n | \boldsymbol{\alpha}, \phi, \sigma_\epsilon^2, \mathbf{Y}^n)$, the Monte Carlo E-step is computed as follows:

$$\begin{aligned} Q^{(M)}(\cdot|\boldsymbol{\alpha}, \phi, \sigma_\epsilon^2) &= \sum_{t=1}^{n} \{-\bar{\theta}_t a_t + Y_t \log(a_t)\} + \bar{E} \\ &\quad - \frac{n-1}{2} \log(\sigma_\epsilon^2) - \frac{1}{2\sigma_\epsilon^2}(\bar{B} - 2\phi\bar{D} + \bar{C}\phi^2), \qquad (12.8) \end{aligned}$$

where

$$\bar{\theta}_t = \frac{1}{M}\sum_{m=1}^{M} \exp(W_t^{(m)}), \; \bar{B} = \frac{1}{M}\sum_{m=1}^{M}\sum_{t=1}^{n-1} W_{t+1}^{2(m)}, \; \bar{C} = \frac{1}{M}\sum_{m=1}^{M}\sum_{t=1}^{n-1} W_t^{2(m)},$$

$$\bar{D} = \frac{1}{M}\sum_{m=1}^{M}\sum_{t=1}^{n-1} W_t^{(m)} W_{t+1}^{(m)}, \; \bar{E} = \frac{1}{M}\sum_{m=1}^{M}\sum_{t=1}^{n} W_t^{(m)} Y_t.$$

The M-step updates the parameter values. Since parameters ϕ and σ_ϵ^2 are only involved in the second part of the augmented likelihood in the absence of θ_t, immediately their updates are given by

$$\begin{aligned} \hat{\phi} &= \frac{\bar{D}}{\bar{C}}, \\ \hat{\sigma}_\epsilon^2 &= \frac{1}{n-1}\left(\bar{B} - \frac{\bar{D}^2}{\bar{C}}\right). \end{aligned}$$

Likewise, parameter $\boldsymbol{\alpha}$ does not appear in the second part of the augmented likelihood, so updating $\boldsymbol{\alpha}$ is equivalent to running a cross-sectional Poisson log-linear regression with the linear predictor $\mathbf{x}_t^T \boldsymbol{\alpha}$ and offset $\log(\bar{\theta}_t)$. That is,

$$\log\{\mathrm{E}(Y_t)\} = \log(\bar{\theta}_t) + \mathbf{x}_t^T \boldsymbol{\alpha}, \; t = 1, \ldots, n.$$

Iterating the E-step and the M-step will produce the maximum likelihood estimates of $\boldsymbol{\alpha}$, ϕ, and σ_ϵ^2 when the algorithm achieves convergence.

Clearly, in this setting the MCEM algorithm is computationally simple, although the convergence rate of the EM algorithm is known slow. The technical challenge in this MCEM approach is really the procedure of sampling

$\mathbf{W}^n$ from its conditional distribution. Chan and Ledolter (1995) suggested a Gibbs sampler to generate the required samples from a Markov chain. Let $\mathbf{W}^n_{-t}$ be a subvector of the $\mathbf{W}^n$ with the t-th component W_t being omitted. It is easy to see that

$$f(w_t|\mathbf{W}^n_{-t}, \mathbf{Y}^n) \propto f_{\boldsymbol{\alpha}}(y_t|w_t) f_{\phi,\sigma_\epsilon}(w_t|\mathbf{W}^n_{-t}). \tag{12.9}$$

First, note that the conditional distribution of W_t, given $\mathbf{W}^n_{-t}$, must be Gaussian, simply because $\mathbf{W}^n$ is a joint Gaussian due to the fact that W_t follows a Gaussian AR(1) process. It can be shown that

$$W_t|\mathbf{W}^n_{-t} \sim \begin{cases} N(\phi W_1, \sigma_\epsilon^2), & \text{if } t = 1 \\ N(\phi(W_{t-1} + W_{t+1})/(1+\phi^2), \sigma_\epsilon^2/(1-\phi^2)), & \text{if } t = 2, \ldots, n-1 \\ N(\phi W_{n-1}, \sigma_\epsilon^2), & \text{if } t = n. \end{cases}$$

Denote the mean and variance parameters of the conditional distribution $f(w_t|\mathbf{W}^n_{-t})$ by u_t and v_t^2, respectively. Then, it is interesting to notice that it is more convenient to sample $Z_t = W_t + \mathbf{x}_t^T \boldsymbol{\alpha}$, based on the form given in (12.9). Subject to a constant,

$$\log f(z_t|\mathbf{W}^n_{-t}, \mathbf{Y}^n) = -\exp(z_t) + z_t Y_t - \frac{(z_t - \mu_t)^2}{2v_t^2}, \tag{12.10}$$

where $\mu_t = u_t + \mathbf{x}_t^T \boldsymbol{\alpha}$. This is a log-concave density, and sampling of Z_t can be easily implemented by a universal rejection algorithm (see, for example, Devroye, 1984).

12.4 KEE in Stationary State Processes

12.4.1 Setup

For convenience, consider again the case of $d = 1$. Kalman estimating equation (KEE) provides another way of implementing the EM algorithm, in which rather than applying Monte Carlo in the E-step, KEE invokes BLUP (see Chapter 9) based on Kalman filter and smoother. In other words, the E-step performs Kalman filter and smoother to evaluate the involved integrals, and the M-step solves an estimating equation derived from the augmented likelihood. Again, readers who are not familiar with EM algorithm may refer to Dempster et al. (1977) or first study Section 13.4.3. To elucidate, similar to the MCEM algorithm in Section 12.3, first treat the state variables as missing data. Let $(\boldsymbol{\alpha}, \boldsymbol{\zeta})$ denote the vector of all model parameters, where $\boldsymbol{\alpha}$ is the vector of regression coefficients in the observation process and $\boldsymbol{\zeta}$ is the vector of nuisance parameters in the gamma state process. The augmented log-likelihood takes the form

$$\ell(\boldsymbol{\alpha}, \boldsymbol{\zeta}|\mathbf{Y}^n, \boldsymbol{\theta}^n) = \sum_{t=1}^{n} \log f(Y_t|\theta_t, \boldsymbol{\alpha}) + \sum_{t=1}^{n-1} \log g(\theta_{t+1}|\theta_t, \boldsymbol{\zeta}) + \log g(\theta_0). \quad (12.11)$$

In order to update the parameter values via EM algorithm, maximizing the following objective function

$$Q(\boldsymbol{\alpha}, \boldsymbol{\zeta}|\boldsymbol{\alpha}', \boldsymbol{\zeta}') = \mathrm{E}\left\{\ell(\boldsymbol{\alpha}, \boldsymbol{\zeta}|\mathbf{Y}^n, \boldsymbol{\theta}^n)\right\}$$

is required, where the expectation is taken under the conditional distribution of $f(\boldsymbol{\theta}^n|\mathbf{Y}^n, \boldsymbol{\alpha}', \boldsymbol{\zeta}')$. Here $\boldsymbol{\alpha}'$ and $\boldsymbol{\zeta}'$ are given by the previous iteration. This maximization can be carried out by solving the score equations derived from this augmented likelihood, given respectively as follow:

$$\mathbf{s}_1(\boldsymbol{\alpha}, \boldsymbol{\zeta}) = \sum_{t=1}^{n} \mathbf{x}_t^T \{Y_t - a_t \mathrm{E}(\theta_t \mid \mathbf{Y}^n, \boldsymbol{\alpha}', \boldsymbol{\zeta}')\} = \mathbf{0}, \quad (12.12)$$

$$\mathbf{s}_2(\boldsymbol{\alpha}, \boldsymbol{\zeta}) = \sum_{t=1}^{n-1} \mathrm{E}\left\{\nabla_{\boldsymbol{\zeta}} g(\theta_{t+1}|\theta_t, \boldsymbol{\zeta})/g(\theta_{t+1}|\theta_t, \boldsymbol{\zeta}) \mid \mathbf{Y}^n, \boldsymbol{\alpha}', \boldsymbol{\zeta}'\right\} = \mathbf{0}, \quad (12.13)$$

where ∇ denotes the operation of gradient and the second score vector $\mathbf{s}_2$ can be simplified when a specific model of the state process is given. Here, instead of using Monte Carlo technique to calculate the conditional mean $\mathrm{E}(\theta_t \mid \mathbf{Y}^n, \boldsymbol{\beta}', \boldsymbol{\zeta}')$, an unbiased estimate based on BLUP would be an alternative. An obvious advantage for the utility of BLUP is the computational simplicity, and the related calculation can be carried out easily via the extended Kalman smoother developed in Chapter 9. The resulting estimating equations remain unbiased, so the standard theory of asymptotics for unbiased inference functions given in Chapter 3 is available to make needed statistical inference.

Let us first apply the KEE to the case of Poisson-Stationary gamma model described in Example 12.3. In this case, the conditional distribution of $\theta_t \mid \theta_{t-1}$ is a convolution of gamma $\mathrm{Ga}^*(\mu = 1, \rho\lambda)$ random variable and a $\mathrm{Beta}(\rho\lambda, (1-\rho)\lambda)$ random variable times θ_{t-1}. Because the expression of conditional density function $g(\theta_t \mid \theta_{t-1})$ is tedious, so the score $\mathbf{s}_2$ is analytically cumbersome. In contrast, the expressions of both conditional and marginal moments of this gamma process θ_t are neat and easily derived. Similar to the GEE in Section 12.2, Jørgensen and Song (1998b) suggested the following estimation procedure:

(a) estimate the nuisance parameter vector $\boldsymbol{\zeta} = (\rho, \lambda)$ of the state process by simply the method of moments; and
(b) given $\sqrt{n}$-consistent estimators, $\hat{\rho}$ and $\hat{\lambda}$, update $\boldsymbol{\beta}$ by the solution to the following estimating equation:

$$U(\boldsymbol{\alpha}) = \sum_{t=1}^{n} \mathbf{x}_t^T \{Y_t - a_t m_t^*(\boldsymbol{\alpha}', \hat{\boldsymbol{\zeta}})\} = \mathbf{0}, \quad (12.14)$$

where $m_t^*(\boldsymbol{\alpha}', \hat{\boldsymbol{\zeta}})$ is the Kalman smoother conditional on the observations $\mathbf{Y}^n$, the previous update $\boldsymbol{\alpha}'$, and the available estimate $\hat{\boldsymbol{\zeta}}$.

Note that estimating $\boldsymbol{\zeta}$ involves parameter $\boldsymbol{\alpha}$, so the implementation needs to iteratively update $\boldsymbol{\beta}$ and $\boldsymbol{\zeta}$ along iterations until convergence.

It is worth pointing out that the inference function $\mathbf{s}_1$ in (12.12) is $\boldsymbol{\zeta}$-insensitive, namely,

$$\mathrm{E}\left\{\frac{\partial \mathbf{s}_1(\boldsymbol{\alpha}, \zeta)}{\partial \zeta}\right\} = \mathbf{0}.$$

This holds because for each $m_t^*, t = 1, \ldots, n$,

$$\mathrm{E}\left\{\frac{\partial m_t^*}{\partial \zeta}\right\} = \mathbf{0}.$$

In effect, by the definition of BLUP, the Kalman smoother vector $\mathbf{m}^*$ is $\mathbf{m}^* = \mathrm{E}(\boldsymbol{\theta}) + \mathrm{cov}(\boldsymbol{\theta}, \mathbf{Y})\Sigma_{\mathbf{Y}}^{-1}(\mathbf{Y} - \mathrm{E}(\mathbf{Y}))$, and all the nuisance parameters in $\boldsymbol{\zeta}$ are involved only in the two terms, i.e., $\mathrm{cov}(\boldsymbol{\theta}, \mathbf{Y})$ and $\Sigma_{\mathbf{Y}}^{-1}$. Therefore, the expectation of the first order derivative of the $\mathbf{m}^*$ *w.r.t.* the parameter vector $\boldsymbol{\zeta}$ is zero because the expectation of the residual term $\mathbf{Y} - \mathrm{E}(Y)$ is always zero. This property of insensitivity ensures that the efficiency of the nuisance parameter estimators would have a marginal effect on that of the estimator of $\boldsymbol{\alpha}$.

12.4.2 Kalman Filter and Smoother

To derive Kalman filter and smoother, first acquire the conditional and marginal moments of the Poisson stationary gamma model.

Proposition 12.5. *(1) The conditional moments of the Poisson-stationary gamma model given in Example 12.3 are*

$$E(Y_t|\theta_t) = a_t\theta_t, \;\; Var(Y_t|\theta_t) = a_t\theta_t E(\theta_t|\theta_{t-1}) = \rho\theta_{t-1} + \bar{\rho},$$
$$Var(\theta_t|\theta_{t-1}) = F(\theta_{t-1}) + \bar{\rho}/\lambda,$$

where $\bar{\rho} = 1-\rho$ *and* $F(\theta_{t-1}) = Var\{G_t\theta_{t-1}\}$, *with mean* $E\{F(\theta_t)\} = \rho\bar{\rho}/\lambda$.
(2) The marginal moments of the Y_t *are*

$$E(Y_t) = a_t, \;\; Var(Y_t) = a_t + a_t^2/\lambda.$$

(3) The covariances are

$$cov(Y_t, \theta_t) = a_t/\lambda, \;\; cov(\theta_t, \theta_{t+h}) = \rho^h/\lambda, \;\; cov(Y_t, Y_{t+h}) = a_t a_{t+h}\rho^h/\lambda,$$

and

$$cov(Y_t, \theta_{t+h}) = a_t\rho^h/\lambda, \;\; cov(Y_{t+h}, \theta_t) = a_{t+h}\rho^h/\lambda.$$

(4) The ACF of the state process $\{\theta_t\}$ *is*

$$corr(\theta_t, \theta_{t+h}) = \rho^h,$$

and the ACF of the observation process $\{Y_t\}$ *is*

$$corr(Y_t,\, Y_{t+h}) = \frac{\rho^h}{\sqrt{(\lambda a_t^{-1} + 1)(\lambda a_{t+h}^{-1} + 1)}}.$$

Based on these moment properties, the Kalman filter and smoother presented in Section 9.3 are readily applied, which are needed to establish the KEE. Proposition 12.6 below is yielded immediately by the application of Theorem 9.4 in Section 9.3.

Proposition 12.6. *Suppose the filter at time* $t-1$ *is complete, namely*

$$\theta_{t-1} \left| \mathbf{Y}^{t-1} \sim [m_{t-1},\, c_{t-1}] \right. .$$

(1) The prediction is $Y_t|\mathbf{Y}^{t-1} \sim [f_t,\, Q_t]$ *with*

$$\begin{aligned} f_t &= \rho a_t m_{t-1} + \bar{\rho} a_t, \\ Q_t &= \rho^2 a_t^2 c_{t-1} + (1-\rho^2) a_t^2/\lambda + a_t = a_t^2 u_{t-1} + a_t, \end{aligned}$$

where $u_{t-1} = \rho^2 c_{t-1} + (1-\rho^2)/\lambda$.

(2) The forward Kalman filter recursion is

$$m_t = \rho m_{t-1} + \bar{\rho} + c_t\left(Y_t - f_t\right), \tag{12.15}$$

$$c_t = \frac{u_{t-1}}{1 + a_t u_{t-1}}. \tag{12.16}$$

The application of Theorem 9.5 in Section 9.3 leads to Proposition 12.7 for the Kalman smoother.

Proposition 12.7. *Suppose that all Kalman filters are available, namely*

$$\theta_t \left| \mathbf{Y}^t \sim [m_t, c_t] \right., \quad t = 1, \ldots, n,$$

and that the previous smoother is done,

$$\theta_t \left| \mathbf{Y}^n \sim [m_t^*, c_t^*] \right. .$$

The backward smoothing recursion is given by

$$m_t^* = m_t + \frac{\rho c_t}{u_t}\left(m_{t+1}^* - \rho m_t - \bar{\rho}\right) \tag{12.17}$$

and the prediction error is

$$c_t^* = \frac{1-\rho^2}{\lambda u_t} c_t + \left(\frac{\rho c_t}{u_t}\right)^2 c_{t+1}^*. \tag{12.18}$$

The recursion is started at $t = n$ *by* $m_n^* = m_n$ *and* $c_n^* = c_n$.

12.4.3 Godambe Information Matrix

Let $\hat{\boldsymbol{\alpha}}$ be the estimate obtained from the unbiased KEE (12.14) in that ρ and λ are replaced by their $\sqrt{n}$-consistent estimates. Under some mild regularity conditions, Jørgensen and Song (1998b) showed that $\hat{\boldsymbol{\alpha}}$ is consistent and $\sqrt{n}(\hat{\boldsymbol{\alpha}} - \boldsymbol{\alpha})$ is asymptotically Gaussian with mean zero and covariance matrix $\lim_{n\to\infty} n\,\mathbf{j}_U^{-1}$, where $\mathbf{j}_U = \mathbf{S}^T\mathbf{V}^{-1}\mathbf{S}$ is the Godambe information of inference function given in (12.14), with the sensitivity matrix $\mathbf{S}$ and the variability matrix $\mathbf{V}$ being detailed as follows.

The calculation of the Godambe information is done recursively. Let

$$\mathbf{L}_{\boldsymbol{\alpha}}(t) = \mathrm{E}\left(\nabla_{\boldsymbol{\alpha}} m_t\right), \text{ and } \mathbf{L}^*_{\boldsymbol{\alpha}}(t) = \mathrm{E}\left(\nabla_{\boldsymbol{\alpha}} m^*_t\right).$$

Then,

$$\mathbf{L}_{\boldsymbol{\alpha}}(t) = \rho\left(1 - \frac{a_t u_{t-1}}{a_t u_{t-1}+1}\right)\mathbf{L}_{\boldsymbol{\alpha}}(t-1) - \frac{a_t u_{t-1}}{a_t u_{t-1}+1}\mathbf{x}_t^T, \text{ with } \mathbf{L}_{\boldsymbol{\alpha}}(0) = 0,$$
$$\mathbf{L}^*_{\boldsymbol{\alpha}}(t) = \mathbf{L}_{\boldsymbol{\alpha}}(t) + \rho\frac{c_t}{u_t}\left\{\mathbf{L}^*_{\boldsymbol{\alpha}}(t+1) - \rho\mathbf{L}_{\boldsymbol{\alpha}}(t)\right\}, \quad \text{with } \mathbf{L}^*_{\boldsymbol{\alpha}}(n) = \mathbf{L}_{\boldsymbol{\alpha}}(n).$$

Thus, the sensitivity matrix is given by

$$\mathbf{S} = -\sum_{t=1}^{n} a_t^{-1}\mathbf{x}_t\left\{\mathbf{x}_t^T + \mathbf{L}^*_{\boldsymbol{\alpha}}(t)\right\}. \tag{12.19}$$

And, the variability matrix is $\mathbf{V} = \mathbf{X}\Sigma\mathbf{X}^T$, where $\Sigma = \mathrm{Var}(\mathbf{Y}^n - \mathbf{A}\mathbf{m}^*)$ is an $n \times n$ matrix whose (i,j)-th element takes the form

$$\mathrm{cov}(Y_i - a_i m^*_i, Y_j - a_j m^*_j) = \begin{cases} a_i - a_i^2 c^*_i, & i = j \\ -a_i a_j c^*_{i,\,j}, & i \neq j \end{cases}$$

where $c^*_{i,\,j}$ is the (i,j)-th element of the mean square error (MSE) matrix of the smoother, $\mathbf{C}^* = \mathrm{E}\left\{(\boldsymbol{\theta}^n - \mathbf{m}^*)(\boldsymbol{\theta}^n - \mathbf{m}^*)^T\right\}$. Jørgensen and Song (1998b) found that

$$c^*_{t,\,t+h} = \rho^h c^*_{t+h}\prod_{i=0}^{h-1}\frac{c_{t+i}}{u_{t+i}}.$$

The above recursions should be carried after the KEE algorithm has converged and produced the final updates, $\hat{\boldsymbol{\alpha}}$ and $\hat{\boldsymbol{\zeta}}$.

A method of moment estimate for the index parameter λ is given as follows. First, estimate the dispersion parameter $\sigma^2 = 1/\lambda$ by

$$\widehat{\sigma}^2 = \sum_{t=1}^{n}\left\{(Y_t - \hat{a}_t)^2 - \hat{a}_t\right\}/\sum_{t=1}^{n}\hat{a}_t^2, \tag{12.20}$$

then set $\hat{\lambda} = 1/\hat{\sigma}^2$.

In the meanwhile, a method of moment estimate for the autocorrelation parameter ρ is given by the property

$$\begin{aligned}\mathrm{cov}(m_t, m_{t-1}) &= \rho \mathrm{Var}(m_{t-1}) + c_t \mathrm{cov}(Y_t - f_t, m_{t-1}) \\ &= \rho\, c_{t-1},\end{aligned}$$

where the term second $\mathrm{cov}(Y_t - f_t, m_{t-1}) = 0$ because

$$\begin{aligned}\mathrm{E}\left\{\mathrm{cov}(Y_t - f_t, m_{t-1}|Y^{t-1})\right\} &= 0 \\ \mathrm{cov}\left(\mathrm{E}(Y_t - f_t|Y^{t-1}), \mathrm{E}(m_{t-1}|Y^{t-1})\right) &= 0.\end{aligned}$$

This leads to $\rho = \mathrm{cov}(m_t, m_{t-1})/\mathrm{Var}(m_{t-1})$, suggesting that the lag-1 autocorrelation of the Kalman filtering m_t or the lag-1 autocorrelation of the standardized filtering $m_t/\sqrt{c_t}$ may serve as an estimator of ρ.

A concern associated with these moment estimates is the possibility of the updated values of σ^2 and/or ρ falling outside of their admissible values. This is more likely to happen when the sample size is small. Whenever this happens, the KEE estimation algorithm stops. To overcome this issue, transformations on the parameters are helpful. For example, the log transformation may be invoked for the dispersion parameter σ^2 and Fisher's Z-transformation is common for the autocorrelation ρ, i.e., $\frac{1}{2}\log\frac{1+\rho}{1-\rho}$.

12.4.4 Analysis of Polio Incidences Data

To illustrate the three inference approaches discussed above, this example applies generalized state space models to analyze the polio incidence data introduced in Section 1.3.8. Figure 1.5 displays time series of monthly numbers of poliomyelitis cases in the USA from 1970 to 1983. The central question of the analysis is to investigate if the data provide evidence of a long-term decrease in the rate of polio infections in the USA; that is, whether the data indicate a significant decreasing time trend.

All three models (the Poisson parameter-driven model, the Poisson-Lognormal model, and the the Poisson-stationary gamma model) are applied to fit the data, respectively. The observation process takes the form

$$Y_t|\theta_t \sim \mathrm{Po}(a_t\theta_t), \text{ with } a_t = \exp(\mathbf{x}_t^T \boldsymbol{\alpha}),\ t = 1, \ldots, 168,$$

where $\mathbf{x}_t = (1, t, \cos(2\pi t/12), \sin(2\pi t/12), \cos(2\pi t/6), \sin(2\pi t/6))^T$. The inclusion of the intercept term in a_t implies that the mean μ of the state process is e^{α_0}.

Among the three model formulations, the stationary gamma AR(1) model has a nice interpretation for the beta-distributed thinning operator B_t; that is, it represents the volume of polio contagious material available in the environment. The solid line in Figure 12.1 shows the estimated state process from the Kalman smoother obtained by the KEE approach. It is clear that

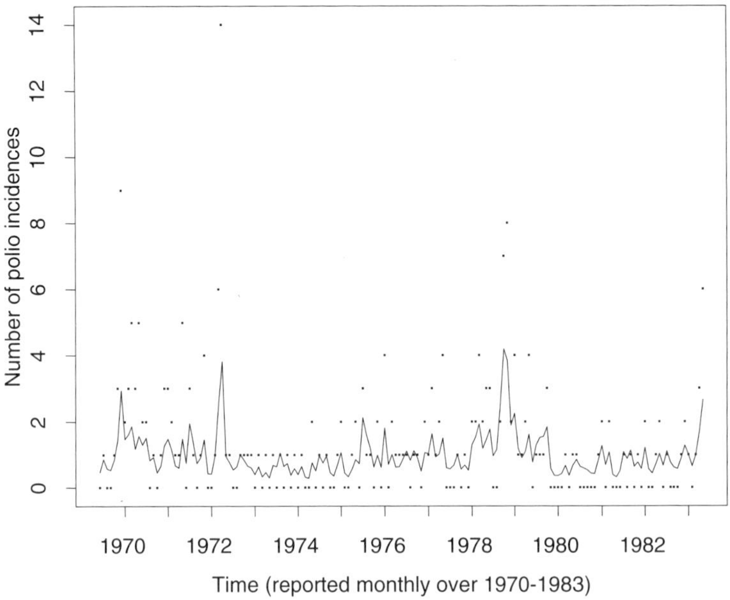

Fig. 12.1. Monthly counts of polio incidences. Dots show the observed counts and the solid line represents the estimated latent process by the Kalman smother.

various peaks in the latent state process nicely explain the episodes in the observed counts of polio incidences (denoted by dots in the figure) caused by the environmental risk. Note that this estimated state process is not available in the GEE approach.

Table 12.1 reports the estimation results from the GEE, the MCEM, and the KEE. Overall, the estimates from these three approaches are similar. Some differences are pointed out as follows.

The KEE estimates are closer to the MCEM estimates than the GEE estimates; for example, the estimates corresponding to the four seasonal effects are almost the same in (b) and (c). Note that the MCEM algorithm is fully likelihood-based and hence the resulting estimates should be most efficient. In this case, the KEE does not lose much estimation efficiency but it gains much computational efficiency. Because of invoking Gibbs sampler as well as the rejection sampling scheme, the MCEM is computationally intensive, besides the issue of convergence diagnosis for the Gibbs sampler.

With regard to the time trend, both the KEE method and the MCEM algorithm found that the slope is significantly different from zero, implying a significant decreasing time trend for the rate of polio incidences. The GEE reported no significance for the slope of the time trend at the 0.05 level.

Table 12.1. Estimates of coefficients and standard errors: (a) the Poisson parameter-driven model using the GEE; (b) the Poisson-lognormal model using the MCEM algorithm; (c) the Poisson-gamma model using the KEE.

	(a) GEE		(b) MCEM		(c) KEE	
	Estimate	Std Err	Estimate	Std Err	Estimate	Std Err
Intercept	.17	.13	.42		.49	.16
Trend $\times 10^{-3}$	−4.35	2.68	−4.62	1.38	−3.87	1.64
$\cos(2\pi t/12)$	−.11	.16	.15	.09	.13	.10
$\sin(2\pi t/12)$	−.48	.17	−.50	.12	−.50	.12
$\cos(2\pi t/6)$	.20	.14	.44	.10	.45	.10
$\sin(2\pi t/6)$	−.41	.14	−.04	.10	−.06	.10
$\hat{\sigma}^2$	.77		.54		.81	
$\hat{\rho}$	.77		.88		.36	

Standard errors for the KEE estimates corresponding to non-constant covariates are uniformly smaller than those given by the KEE, being nearly the same as the those obtained by the MCEM. Such differences are largely caused by the use of fully parametric modeling for the state process, which returns a desirable gain in efficiency if the parametric model is appropriately specified. Jørgensen and Song (1998b) performed a comprehensive model checking on the Poisson-gamma specification and validated all key assumptions imposed on the model. Interested readers can refer to Section 12.5.4 for model diagnostics developed for the nonstationary gamma process, where a similar residual analysis is conducted.

The estimate for the lag 1 autocorrelation, $\hat{\rho}$, is less than half the value reported by the GEE, and only 40% of the value obtained by the MCEM. The lag 1 autocorrelation for the raw data of polio counts is close to .3, so the KEE estimate .36 seems reasonable for the lag 1 autocorrelation of the state process, under the assumption that the observations are conditionally independent given the state process. One way to see the connection is to invoke property (4) of Proposition 12.5, with $\hat{\sigma}^2 = 0.81$,

$$\begin{aligned}\operatorname{corr}(Y_t, Y_{t+1}) &= \rho \times \frac{1}{\sqrt{\{1 + (\sigma^2 a_t)^{-1}\}\{1 + (\sigma^2 a_{t+1})^{-1}\}}} \\ &\approx \rho/1.39\end{aligned}$$

where the approximation is given by replacing the a_t's with the mean $\bar{a} = 1.32$ based on the $\hat{\alpha}_j$ in Table 12.1. This implies that the ρ parameter would be approximately equal to $1.39 * 0.3 = 0.417$.

12.5 KEE in Non-Stationary State Processes

This section focuses on the KEE estimation approach for generalized state space models in that the latent state process is nonstationary, proposed in Jørgensen et al. (1999). As discussed in Example 12.4, in many practical studies covariates are possibly divided into two types, long-term and short-term covariates, according to subject-matter knowledge. The distinction lies in the fact that effects of covariates may be different in lagged times. For the long-term covariates, the Markov structure of the state process creates a carry-over effect, which to some extent obviates the need for lagged covariates. On the other hand, the short-term covariates pertain to some immediate or acute effects on the mean of the observation process. When such a distinction is observed, the long-term covariates will enter the state process, while the short-term covariates will enter the observation process. The resulting state process is effectively nonstationary, requiring a further generalization of the classical state space models.

The Poisson-gamma model presented in this section is mainly inspired by the analysis of the Prince George data, where the response constitutes the number of emergency room visits for four categories of respiratory diseases (see Section 1.3.10). However, the KEE can be established in a more general GSSM framework than the Poisson-gamma model. This section follows the framework of Jørgensen et al. (1996a) to present a class of generalized state space models with nonstationary state processes, with the Poisson-gamma model as a special case. Statistical inference is based on the Kalman estimating equation approach, which is extended from that given in the previous Section 12.3 for a stationary state process. In addition, this section also gives a detailed discussion about the analysis of residuals for model diagnosis from both the observation and the state process, which is not available for either the GEE method or the MCMC method. The discussion of model diagnosis was ignored in Section 12.3, simply because of great similarity between the residual analysis in the GSSM with stationary state process to that with the nonstationary state process developed in this section.

Let $\mathbf{Y}_t$ be a d-dimensional response vector such that the components of $\mathbf{Y}_t$ all reflect the same underlying tendency θ_t, say, but where the individual components may not all have the same distribution. Assume that the state process θ_t is a Markov process including time-varying covariates and that the observations are conditionally independent given the state process, both across components and over time. A log-linear regression model for the marginal means $\mathrm{E}(\mathbf{Y}_t)$ is considered.

12.5.1 Model Formulation

Suppose a d-dimensional $\mathbf{Y}_t$ is recorded at equally spaced times $t = 1, \ldots, n$, and denote the full data vector by

$$\mathbf{Y}^n = (\mathbf{Y}_1^T, \ldots, \mathbf{Y}_n^T)^T.$$

Let θ_t be a univariate state process and denote the vector of θ_t by

$$\boldsymbol{\theta}^n = (\theta_0, \theta_1, \ldots, \theta_n)^T,$$

where θ_0 is an initializing variable.

In model $\mathbf{M}_2$ of equation (10.2), the conditional distribution of the ith component of $\mathbf{Y}_t$ given θ_t is assumed to follow a Tweedie exponential dispersion model (see Section 2.5),

$$Y_{it}|\theta_t \sim \mathrm{Tw}_{r_i}\left(a_{it}\theta_t, \frac{\nu_i^2}{\theta_t^{r_i-1}}\right), \tag{12.21}$$

for $i = 1, \ldots, d$, $t = 1, \ldots, n$, where $a_{it} = \exp\left(\mathbf{x}_t^T \boldsymbol{\alpha}_i\right)$ and r_i is the shape parameter. Here covariates $\mathbf{x}_t$ are assumed to be the same across all the d components, but the following development of statistical inference does not depend on such a specification of covariates, and hence can be extended to deal with the case with component-dependent covariates $\mathbf{x}_{it}$.

The short-term covariates represent modulating factors that have an immediate effect on Y_{it} relative to the value of θ_t. Note that the corresponding regression parameters $\boldsymbol{\alpha}_i \in \mathcal{R}^p$ may vary among the d components.

It is known that the shape parameters r_i of the Tweedie model must satisfy $r_i \geq 1$ or $r_i \leq 0$, but may vary from component to component, allowing each component to have a different distribution. The parameters ν_i^2 in (12.21) are dispersion parameters, so that even in the case where all r_i are equal, the d components may have different dispersions.

Note that $r_i \geq 1$ corresponds to non-negative distributions while $r_i \leq 0$ corresponds to distributions with support $\mathcal{R}$ (see Section 2.5). In particular, $r_i = 0$ gives the normal distribution, $r_i = 1$ gives the Poisson distribution, $1 < r_i < 2$ are compound Poisson distributions with a positive probability in zero, $r_i = 2$ is the gamma distribution, and $r_i = 3$ is the inverse Gaussian distribution. The Tweedie class hence accommodates a considerable range of distributions, including continuous, discrete, and mixed ones.

In the Poisson case ($r_i = 1$), (12.21) reduces to

$$Y_{it}|\theta_t \sim \mathrm{Tw}_1\left(a_{it}\theta_t, \nu_i^2\right) = \nu_i^2 \mathrm{Po}(a_{it}\theta_t/\nu_i^2),$$

so in this case the dispersion parameter $\nu_i^2 = 1$ in all formulas given in the rest of this section when applicable.

In (12.21) the conditional expectation and variance are respectively,

$$\mathrm{E}(Y_{it}|\theta_t) = a_{it}\theta_t, \quad \mathrm{Var}(Y_{it}|\theta_t) = \nu_i^2 a_{it}^{r_i}\theta_t.$$

Hence θ_t may be interpreted as the "intensity" of the observation $\mathbf{Y}_t$, being modified for the i-th component by the covariates $\mathbf{x}_t$ via a_{it}.

The state process of the state space model is assumed to be an exponential dispersion model Markov process, defined by the transition distribution

$$\theta_t|\theta_{t-1} \sim \mathrm{Tw}_l\left(b_t\theta_{t-1}, \frac{\sigma^2}{\theta_{t-1}^{l-1}}\right) \tag{12.22}$$

for $t = 1, \ldots, n$, where the shape parameter l is known and satisfies $l \geq 2$, making the state process positive. In particular, this excludes the possibility of a normal distribution for the state process. The conditional expectation and variance are respectively

$$\mathrm{E}(\theta_t|\theta_{t-1}) = b_t\theta_{t-1}, \quad \mathrm{Var}(\theta_t|\theta_{t-1}) = \sigma^2 b_t^l \theta_{t-1}.$$

The state process may be interpreted as a random walk with drift, and is hence non-stationary.

The parameter σ^2 in (12.22) is a dispersion parameter, and b_t depends on the long-term covariates $\mathbf{z}_t$ via their increments $\Delta\mathbf{z}_t = \mathbf{z}_t - \mathbf{z}_{t-1}$,

$$\mathbf{b}_t = \exp\left\{\left(\Delta\mathbf{z}_t^T\right)\boldsymbol{\beta}\right\}, \tag{12.23}$$

$\boldsymbol{\beta} \in \mathcal{R}^q$ being another regression parameter. For convenience, we assume from now on that $\mathbf{z}_0 = 0$, which may be obtained by subtracting $\mathbf{z}_0$ from all $\mathbf{z}_t$.

Note that the Markov structure of the state process produces a carry-over effect, such that an increase in the level of $\mathbf{z}$ at a given time t may have an effect on subsequent time-points. An important characteristic of long-term covariates is that they have the same effect on all d components of $\mathbf{Y}_t$, in contrast to short-term covariates whose effects may vary across categories.

For a single series of observations $\mathbf{Y}^n$, assume $\theta_0 \equiv g_0$ to be degenerate where g_0 is a certain positive constant. The marginal means of the state and observed processes are log-linear in the covariates, as shown by

$$\mathrm{E}(\theta_t) = \tau_t = gb_1 \cdots b_t = \exp\left\{\mathbf{z}_t^T\boldsymbol{\beta} + \log(g_0)\right\}$$

and

$$\mathrm{E}(Y_{it}) = \exp\left\{\mathbf{x}_t^T\boldsymbol{\alpha}_i + \mathbf{z}_t^T\boldsymbol{\beta} + \log(g_0)\right) .$$

The marginal variance of the observed vector $\mathbf{Y}_t$ consists of two terms,

$$\mathrm{Var}(\mathbf{Y}_t) = \Lambda_t\tau_t + \mathbf{a}_t\mathbf{a}_t^T\mathrm{Var}(\theta_t), \tag{12.24}$$

where $\Lambda_t = \mathrm{diag}(\nu_1^2 a_{1t}^{r_1}, \ldots, \nu_d^2 a_{dt}^{r_d})$, $\mathbf{a}_t = (a_{1t}, \ldots, a_{dt})^T$ and $\mathrm{Var}(\theta_t)$ is given by

$$\mathrm{Var}(\theta_t) = \sigma^2\phi_t\tau_t$$

where

$$\phi_t = b_t^{l-1} + b_t b_{t-1}^{l-1} + \cdots + b_t b_{t-1} \cdots b_1^{l-1}.$$

The second term in (12.24) shows overdispersion relative to the model (12.21) and shows that the correlation between components is positive. The covariance between two observation vectors separated by a lag of $s > 0$ is

$$\mathrm{cov}(\mathbf{Y}_t, \mathbf{Y}_{t+s}) = \mathbf{a}_t\mathbf{a}_{t+s}^T b_{t+1} \cdots b_{t+s}\mathrm{Var}(\theta_t).$$

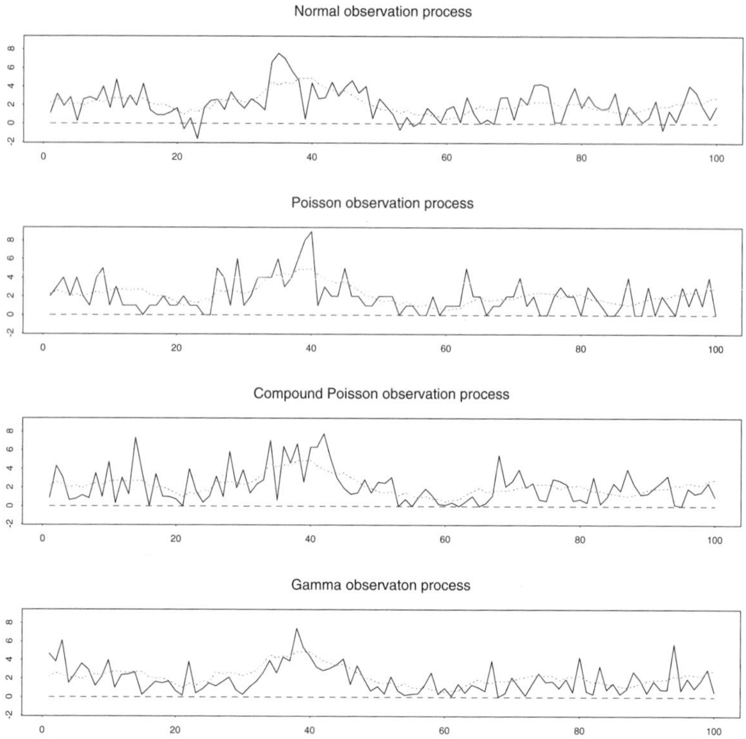

Fig. 12.2. Simulation of 100 observations from a process with four categories with $l = 2$, $r_i = 0, 1, 1.5, 2$, $\nu_i^2 = 1$, $\sigma^2 = 0.05$, and $\theta_0 = 2$ without any covariate effects. Observations are shown with solid lines, and the state processes with dashed lines.

Jørgensen et al. (1996a) considers a state space for multi-dimensional time series of mixed types. Figure 12.2 shows simulations of the model for $l = 2$ (a gamma state process) and four categories with $r_1 = 0$, $r_2 = 1$, $r_3 = 1.5$, and $r_4 = 2$. The four observation processes represent the following characteristics, respectively: symmetric continuous (normal observation process), discrete (Poisson observation process), positive skew with positive probability at zero (compound Poisson observation process), and positive skew (gamma observation process).

12.5.2 Kalman Filter and Smoother

Recall that the key in the Kalman estimating equation is the Kalman filter and smoother, which will be applied to approximate the E-step in the EM algorithm. The *innovations* for the state process are given by

$$\xi_t = \theta_t - \mathrm{E}\left(\theta_t | \boldsymbol{\theta}^{t-1}\right) = \theta_t - b_t\theta_{t-1}, \; t = 1, \ldots, n, \tag{12.25}$$

and for the observed process

$$\varphi_t = \mathbf{Y}_t - \mathrm{E}\left(\mathbf{Y}_t|\boldsymbol{\theta}^t\right) = \mathbf{Y}_t - \mathbf{a}_t\theta_t, \ t = 1, \ldots, n. \tag{12.26}$$

Using the innovations, the model may be rewritten as an additive form as follows, for $t = 1, \ldots, n$,

$$\begin{aligned}\mathbf{Y}_t &= \mathbf{a}_t\theta_t + \boldsymbol{\varphi}_t,\\ \theta_t &= b_t\theta_{t-1} + \xi_t,\end{aligned}$$

where $\boldsymbol{\varphi}_t$ and θ_t are uncorrelated, and so are ξ_t and θ_{t-1}. In addition, it is easy to show that all these $2n$ innovations, $\boldsymbol{\varphi}_t$'s and ξ_t's, are also uncorrelated. Hence the model is structured as that of an ordinary linear state space model given in Section 10.2. The marginal variances of the innovations are

$$\begin{aligned}\mathrm{Var}(\xi_t) &= \mathrm{E}\left(\xi_t^2\right) = \mathrm{E}\left\{\mathrm{Var}\left(\theta_t|\boldsymbol{\theta}^{t-1}\right)\right\} = \sigma^2 b_t^{l-1}\tau_t,\\ \mathrm{Var}(\boldsymbol{\varphi}_t) &= \mathrm{E}\left(\boldsymbol{\varphi}_t\boldsymbol{\varphi}_t^T\right) = \Lambda_t\tau_t.\end{aligned}$$

A direct application of results given by Theorem 9.4 and Theorem 9.5 in Section 9.3 yields Kalman recursions as follows. Let

$$\theta_{t-1}|\mathbf{Y}^{t-1} \sim [m_{t-1}; C_{t-1}],$$

and

$$D_t = b_t C_{t-1} + \sigma^2 b_t^{l-1}\tau_{t-1}.$$

The prediction is then given by

$$\mathbf{Y}_t|\mathbf{Y}^{t-1} \sim [\mathbf{f}_t; \mathbf{Q}_t], \tag{12.27}$$

where

$$\mathbf{f}_t = \mathbf{a}_t b_t m_{t-1}; \qquad \mathbf{Q}_t = b_t D_t \mathbf{a}_t \mathbf{a}_t^T + \Lambda_t\tau_t.$$

The filtered values of the state process are, for $t = 1, \ldots, n$

$$\theta_t|\mathbf{Y}^t \sim [m_t; C_t], \tag{12.28}$$

starting with $m_0 = g_0$ and $C_0 = 0$, where

$$\begin{aligned}m_t &= b_t\left\{m_{t-1} + D_t\mathbf{a}_t^T\mathbf{Q}_t^{-1}(\mathbf{Y}_t - \mathbf{f}_t)\right\},\\ C_t &= b_t D_t\left(1 - b_t D_t \mathbf{a}_t^T\mathbf{Q}_t^{-1}\mathbf{a}_t\right).\end{aligned} \tag{12.29}$$

Given all n observations, the smoothed version of the state process is given by the following backward recursion for $t = n-1, \ldots, 0$,

$$\theta_t|\mathbf{Y}^n \sim [m_t^*; C_t^*], \tag{12.30}$$

starting with $m_n^* = m_n$ and $C_n^* = C_n$, where

$$m_t^* = m_t + \frac{C_t}{D_{t+1}} \left(m_{t+1}^* - b_{t+1} m_t \right),$$

and

$$C_t^* = \sigma^2 \frac{C_t}{D_{t+1}} b_{t+1}^{l-2} \tau_{t+1} + \frac{C_t^2}{D_{t+1}^2} C_{t+1}^*.$$

The mean squared error matrix $\mathbf{C}^* = \mathrm{E}\left\{ (\boldsymbol{\theta} - \mathbf{m}^*) (\boldsymbol{\theta} - \mathbf{m}^*)^T \right\}$. Jørgensen et al. (1999) showed that the diagonal elements C_t^* and off-diagonal elements of the $\mathbf{C}^*$ are given by

$$C_{t\ t+s}^* = C_{t+s}^* \prod_{i=1}^{s} \frac{C_{t+i-1}}{D_{t+i}}.$$

12.5.3 Parameter Estimation

As before, the KEE method is built on a set of unbiased inference functions for regression coefficients, $(\boldsymbol{\alpha}, \boldsymbol{\beta})$, while the dispersion parameters ν_i^2 and σ^2, denoted collectively by $\boldsymbol{\zeta}$, are treated as nuisance parameters that are estimated separately by the method of moments. To make the KEE work, the estimates of these nuisance parameters have to be $\sqrt{n}$-consistent.

As discussed in Section 12.4, the KEE is essentially an approximate EM algorithm, in which the E-step is approximate by the Kalman smoother via BLUP and the M-step is equivalent to conducting a cross-sectional Poisson regression. It is easy to see that for the nonstationary state space models, the observation process leads to a similar score equation to that given in (12.12), and the state process results in a score equation, given by respectively

$$\mathbf{s}_1(\boldsymbol{\alpha}, \boldsymbol{\beta}, \boldsymbol{\zeta}) = \sum_{t=1}^{n} \sum_{i=1}^{d} \mathbf{x}_t a_{it}^{1-r_i} \left(Y_{it} - a_{it} \theta_t \right) = \mathbf{0}, \tag{12.31}$$

$$\mathbf{s}_2(\boldsymbol{\alpha}, \boldsymbol{\beta}, \boldsymbol{\zeta}) = \sum_{t=1}^{n} b_t^{-1} (\theta_t - b_t \theta_{t-1}) \Delta \mathbf{z}_t = \mathbf{0}. \tag{12.32}$$

It is interesting to note that both score equations are linear in the state variables, which makes the application of the Kalman filter and smoother really straightforward. That is, the respective unbiased estimating equations are obtained by replacing the state variables by their Kalman smoothers $m_t^*(\boldsymbol{\zeta})$, leading to

$$U_1(\boldsymbol{\alpha}, \boldsymbol{\beta}, \boldsymbol{\zeta}) = \sum_{t=1}^{n} \sum_{i=1}^{d} \mathbf{x}_t a_{it}^{1-r_i} \left\{ Y_{it} - a_{it} m_t^*(\boldsymbol{\zeta}) \right\} = \mathbf{0}, \tag{12.33}$$

$$U_2(\boldsymbol{\alpha}, \boldsymbol{\beta}, \boldsymbol{\zeta}) = \sum_{t=1}^{n} b_t^{-1} \{ m_t^*(\boldsymbol{\zeta}) - b_t m_{t-1}^*(\boldsymbol{\zeta}) \} \Delta \mathbf{z}_t = 0. \tag{12.34}$$

Clearly, $\mathbf{U} = (U_1^T, U_2^T)^T$ is an unbiased inference function, and the estimates $(\hat{\boldsymbol{\alpha}}, \hat{\boldsymbol{\beta}})$ are the solutions of the equation $\mathbf{U} = 0$. The standard theory of inference functions applies to the KEE estimating function $\mathbf{U}$, and the asymptotic standard errors of the estimator $(\hat{\boldsymbol{\alpha}}, \hat{\boldsymbol{\beta}})$ can be calculated from the inverse of the Godambe information matrix $\mathbf{j} = \mathbf{S}^T\mathbf{V}^{-1}\mathbf{S}$, where $\mathbf{S} = \mathrm{E}(\nabla\mathbf{U})$ is the sensitivity matrix and $\mathbf{V} = \mathrm{E}(\mathbf{U}\mathbf{U}^T)$ is the variability matrix.

Again, by a similar argument given at the end of Section 12.4.1, inference function U_1 is $\boldsymbol{\zeta}$-insensitive. This means that the efficiency of the estimator of $\boldsymbol{\zeta}$ will not affect the efficiency of the estimators of $\boldsymbol{\alpha}$ and $\boldsymbol{\beta}$ much.

The Newton scoring algorithm, a parallel to Fisher's scoring method, is defined as the Newton algorithm applied to solve the equation $\mathbf{U} = 0$, in which iterations proceed as follows:

$$\begin{pmatrix} \boldsymbol{\alpha}' \\ \boldsymbol{\beta}' \end{pmatrix} = \begin{pmatrix} \boldsymbol{\alpha} \\ \boldsymbol{\beta} \end{pmatrix} - \mathbf{S}(\boldsymbol{\alpha}, \boldsymbol{\beta})^{-1}\mathbf{U}(\boldsymbol{\alpha}, \boldsymbol{\beta}),$$

with the sensitivity matrix S. An advantage of this algorithm is that the calculation of $\mathbf{S}$ can be done recursively in parallel with the calculation of the Kalman smoother, in a similar fashion as in equation (12.19). The initial values of the parameters may be obtained by fitting cross-sectional GLMs under the independence correlation, and the initial smoother may be simply the average of the moving averages of individual observed series.

The estimation of the nuisance parameters $\boldsymbol{\zeta}$ may be obtained by the method of moments. Jørgensen et al. (1999) suggested that an unbiased estimate for σ^2 is

$$\hat{\sigma}^2 = \frac{1}{n}\sum_{t=1}^{n}\frac{(m_t^* - b_t m_{t-1}^*)^2}{b_t^{l-1}\tau_t} + \frac{1}{n}\sum_{t=1}^{n}\frac{C_t^* + b_t^2 C_{t-1}^* - 2b_t C_{t-1} C_t^*/D_t}{b_t^{l-1}\tau_t}, \quad (12.35)$$

where the last term corrects for the bias introduced by using the smoother for the state process.

Similarly, estimates for ν_i^2 are given by

$$\hat{\nu}_i^2 = \frac{1}{n}\sum_{t=1}^{n}\frac{(Y_{it} - a_{it}m_t^*)^2}{\tau_t a_{it}^{r_i}} + \frac{1}{n}\sum_{t=1}^{n}\frac{a_{it}^2 C_t^*}{\tau_t a_{it}^{r_i}}. \quad (12.36)$$

12.5.4 Model Diagnosis

Residual analysis for both observed and unobserved parts of the model is crucial for any parametric modeling, such as the generalized state space models presented in this chapter. Although the development below is primarily for the GSSM with a nonstationary state process, a similar diagnosis is available for the model with stationary state process (Jørgensen and Song, 1998b). The basic idea is to use plots of standardized residuals (i.e., residuals divided by their standard errors) in much the same way as in the theory of generalized linear models, in order to check the distributional assumptions and the regression part of the model. Also, it is possible to check the correlation structure of the model by various methods from the theory of time-series analysis. Model diagnosis may proceed with each series separately, or with a combination of residuals from several series.

The main type of residuals are the *conditional residuals*, defined as the predicted values of the innovations φ_t and ξ_t, based on either the Kalman filter or smoother. All residuals have means 0. The properties derived in the following do not take into account the effect of substituting parameter estimates.

The predicted values of the innovations based on the Kalman filter are

$$\widehat{\varphi}_t = \mathbf{Y}_t - \mathbf{f}_t \qquad \text{and} \qquad \widehat{\xi}_t = m_t - b_t m_{t-1}.$$

The prediction errors $\widehat{\varphi}_t$ are mutually uncorrelated over time, and since $\widehat{\xi}_t = b_t D_t \mathbf{a}_t^T \mathbf{Q}_t^{-1} \widehat{\varphi}_t$, then so are the predicted innovations of the state process. This property of the filter residuals $\widehat{\varphi}_t$ and $\hat{\xi}_t$ makes them especially useful for residual plots. Moreover, $\widehat{\xi}_t$ is uncorrelated with f_{it}, and $\widehat{\xi}_t$ is uncorrelated with $b_t m_{t-1}$. The variances of the two sets of residuals are, respectively,

$$\mathrm{Var}\left(\widehat{\varphi}_t\right) = \mathbf{Q}_t \qquad \text{and} \qquad \mathrm{Var}\left(\widehat{\xi}_t\right) = b_t D_t - C_t.$$

A disadvantage of the filter residuals is that their variances may be large for the first few observations of each series. This disadvantage is not shared by residuals based on the Kalman smoother, which in fact have smaller variances than the corresponding filter residuals. These residuals are

$$\widehat{\varphi}_t^* = \mathbf{Y}_t - \mathbf{a}_t m_t^* \qquad \text{and} \qquad \widehat{\xi}_t^* = m_t^* - b_t m_{t-1}^*.$$

However, the smoother residuals do not have the same uncorrelatedness properties as the filter residuals, making the smoother residuals less useful for some purposes. Their variances are

$$\mathrm{Var}\left(\widehat{\varphi}_t^*\right) = \Lambda_t \tau_t - \mathbf{a}_t \mathbf{a}_t^T C_t^*$$

and

$$\mathrm{Var}\left(\widehat{\xi}_t^*\right) = \sigma^2 b_t^{l-1} \tau_t - C_t^* - b_t^2 C_{t-1}^* + 2b_t \frac{C_{t-1}}{D_t} C_t^*.$$

Standardized residuals (having unit variance) are denoted $\hat{\hat{\xi}}_t$ and so on. The residuals $\hat{\varphi}_t$ are standardized componentwise. The correlation structure of the model is easy to check by means of the autocorrelation functions of the standardized filter residuals (the $\hat{\hat{\varphi}}_{it}$ are considered separately for each i). One may also plot $\hat{\hat{\xi}}_t$ against $\hat{\hat{\xi}}_{t-1}$, to check the Markov assumption for the state process, and $\hat{\hat{\varphi}}_{it}$ against $\hat{\hat{\varphi}}_{i,t-1}$, to check the conditional independence of the counts over time.

To check the assumption that the components are conditionally independent given the state process, one may consider the empirical variance-covariance matrices of the vector of standardized prediction errors,

$$\frac{1}{n}\sum_{t=1}^{n}\mathbf{Q}_t^{-1/2}\widehat{\varphi}_t\widehat{\varphi}_t^T\mathbf{Q}_t^{-T/2}, \tag{12.37}$$

whose expectations are all the $d \times d$ identity matrix. The off-diagonal elements have asymptotic standard errors of $n^{-1/2}$. Note that the standardization depends on the version of the square-root matrix $\mathbf{Q}_t^{1/2}$ chosen.

As in generalized linear models, the form of the variance functions (and hence distributional forms) may be checked by plotting the standardized residuals against the corresponding log fitted values. A "megaphone" shape would indicate that the chosen shape parameter in the corresponding variance function is incorrect.

Plots of standardized residuals against each covariate are useful for detecting nonlinearity of the model. The smoother residuals are a better choice of residuals for this purpose. To check the log link assumption, one may plot $\log Y_{it}$ against the log fitted values $\log(\mathbf{a}_t m_t^*)$. This plot should show a horizontal linear relationship; a curvature or other unusual shape would indicate inadequacy of the log link assumption.

Residual analysis may also help to determine whether a covariate is long-term or short-term. The basic idea is that a short-term covariate would show an association with the observation residuals ($\widehat{\varphi}_t$ or $\widehat{\varphi}_t^*$) if it had been incorrectly fitted as a long-term covariate, and similarly, a long-term covariate would show association with the state process residuals ($\widehat{\xi}_t$ or $\widehat{\xi}_t^*$) if it had been incorrectly fitted as a short-term covariate.

12.5.5 Analysis of Prince George Data

This section presents an application of the nonstationary KEE in the regression analysis of Prince George data introduced in Section 1.3.10. This is a simpler version of the analysis than that given by Jørgensen et al. (1996b). The response variable $\mathbf{Y}_t$ is a four-dimensional vector, consisting of daily counts of emergency room visits for the four respiratory disease categories. The main idea in the model is to consider the number of daily emergency room visits to be a Poisson process driven by a latent Markov process, denoted $\{\theta_t\}$. The components of $\mathbf{Y}_t$ are assumed to be conditionally independent given θ_t, and the conditional distribution for the i-th component of $\mathbf{Y}_t$ follows a Poisson distribution,

$$Y_{it}|\theta_t \sim \text{Po}\left(a_{it}\theta_t\right), \ i = 1, 2, 3, 4,$$

where $\log(a_{it})$ is a linear predictor depending on a common set of short-term covariates $\mathbf{x}_t$ via $a_{it} = \exp\left(\mathbf{x}_t^T\boldsymbol{\alpha}_i\right)$, where $\boldsymbol{\alpha}_i$ denotes the parameter vector for the i-th disease category.

The latent morbidity process θ_t represents the "overall potential" for air pollution to cause emergency room visits with respiratory symptoms. It is assumed to follow a gamma Markov process, defined by

$$\theta_t|\theta_{t-1} \sim \text{Ga}\left(b_t\theta_{t-1}, \frac{\sigma^2}{\theta_{t-1}}\right),$$

where $\text{Ga}(\mu, \delta^2)$ denotes the gamma distribution with mean μ and coefficient of variation δ. The parameter σ^2 is a dispersion parameter, and $\log(b_t)$ is a linear predictor depending on the long-term covariates *via* their increments. That is,

$$b_t = \exp\left\{(\Delta\mathbf{z}_t^T)\boldsymbol{\beta}\right\}, \quad \text{with } \Delta\mathbf{z}_t = \mathbf{z}_t - \mathbf{z}_{t-1},$$

where $\mathbf{z}_t$ is the vector process of air pollution measurements and $\boldsymbol{\beta}$ is the parameter vector. The state process is standardized by taking $\theta_0 = 1$, and the long-terms covariates are centralized by subtracting $\mathbf{z}_1$ from all values and taking $\mathbf{z}_0 = \mathbf{0}$.

Knight et al. (1989) transformed Sulphur and Particulates to respectively $\log(0.5+\text{Sulphur})$ and log(Particulates), and both transformed covariates will be used in the present analysis. For convenience, refer to $\log(0.5 + \text{Sulphur})$ as simply log-Sulphur. A logarithmic transformation of the minimum and maximum humidity is made, and the difference and the sum of high and low log-humidities are used as short-term covariates.

The three meteorological covariates are centralized by subtracting their respective means. In this way, the intercept parameter represents the log expected number of emergency room visits for a day with average temperature and humidity, and with air pollution as on April 1, 1984. The intercept is allowed to depend on the day of the week.

To investigate the need for lagging the log-Sulphur covariate, according to Knight et al. (1989), the proposed model includes up to lag 2 for log-Sulphur. It does not seem reasonable to include lags for log-Particulates, because of the linear interpolation used for this covariate (Jørgensen et al., 1996b).

The preliminary model for the data includes the covariates listed below.

Short-term covariates $\mathbf{x}_t$:

log-temperature,
difference of log-humidities,
sum of log-humidities,
7 day-of-week factors.

Long-term covariates $\mathbf{z}_t$:

lag 0, 1 and 2 of log-Sulphur,
log-Particulates.

The KEE estimates of the coefficients for the air pollution covariates are reported in Table 12.2. To verify the need for lag 1 and lag 2 for log-Sulphur, the Wald test given in Section 5.2.3 is used. The result was $W = 10.251$, which, compared with a $\chi^2(2)$-distribution, gave a p-value of 0.006. This confirms the need for the inclusion of the two lagged log-Sulphur in the model.

The Kalman filter and smoother allow to define residuals for both the observed and latent processes, to be used for model checking. There are two

Table 12.2. Estimates and standard errors for air pollution effects.

	log-Sulphur	log-Particulates
lag 0	0.029 (0.016)	−0.120(0.058)
lag 1	−0.023 (0.017)	—
lag 2	0.048 (0.016)	—

types of conditional residuals for the i-th observed process, defined in terms of the Kalman filter and smoother by

$$\hat{\varphi}_{it} = Y_{it} - f_{it}, \text{ and } \hat{\varphi}^*_{it} = Y_{it} - a_{it}m^*_t,$$

respectively. For the latent process, the conditional residuals are

$$\hat{\xi}_t = m_t - b_t m_{t-1},$$

based on the Kalman filter. The residuals are then standardized to have zero mean and unit variance. All references to residuals from now on are to these standardized residuals with parameter estimates inserted.

The standardized residuals may be used for checking the correlation structure, the distributional assumptions, and the covariate structure of the model, as shown below. The residuals $\hat{\xi}_t$ and $\hat{\varphi}_{it}$ are particularly useful for checking the correlation structure of the data, because they are uncorrelated over time.

To investigate the possibility of a seasonality effect, the plots of the second year's residuals against the first year's are shown, respectively, for both $\hat{\xi}_t$ and $\hat{\varphi}_{it}$ in Figure 12.3 and Figure 12.4.

In the presence of seasonality not accounted for by the model, one would expect to see a correlation between the first and second year's residuals. None of these five plots indicated any such correlation, but it should be kept in mind that a two-year period may be too short to detect seasonality. However, it seems that the meteorological variables have accounted for all seasonality variation in the present data.

In order to verify the main model assumptions, now the residual analysis is performed. Among many checks, here only a couple of key diagnoses are reported.

To check the correlation structure for the state process, the autocorrelation function (ACF) for $\hat{\xi}_t$ is displayed in Figure 12.5. The plot shows that the autocorrelations for lags 1, 16, and 27 fall slightly outside the asymptotic 95% confidence bands. However, the confidence bands are approximate, even for Gaussian time-series, and it seems that the significance would be slight in any case, confirming the Markov assumption for the state process.

Similar ACF plots for $\hat{\varphi}_{it}$ for each of the four categories shown in Figure 12.6, did not show anything unexpected. This confirms that the $\hat{\varphi}_{it}$ are uncorrelated over time for each of the four categories and moreover confirms the

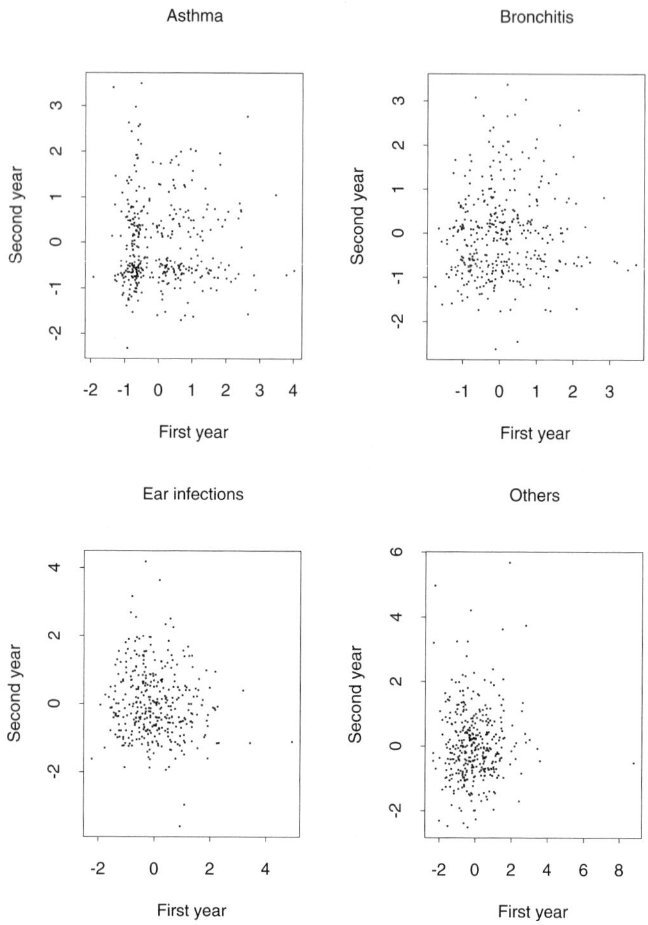

Fig. 12.3. First-year residuals against second-year residuals for the four categories.

adequacy of assuming conditional independence of the Poisson counts over time, given the state process.

The KEE estimates of the short-term effects are listed in Table 12.3, with asymptotic standard errors in brackets. The interpretation of the model is summarized as follows.

(a) The Wald test for the joint significance lag 0, 1 and 2 of log-Sulphur had the value $W = 11.706$, which gave a p-value of 0.0085 compared with a $\chi^2(3)$-distribution. Hence, the effect of log-Sulphur on emergency room visits is highly significant.

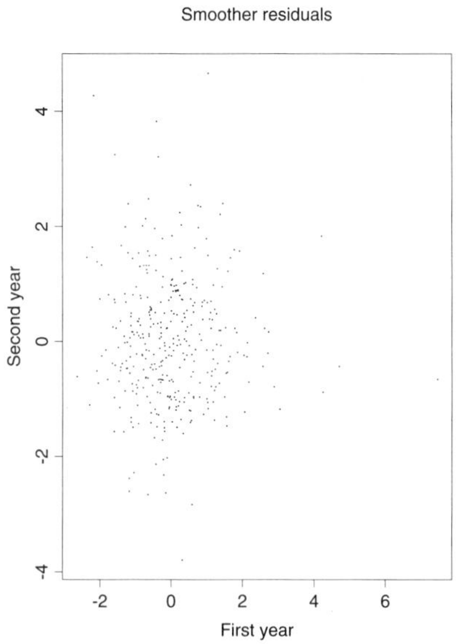

Fig. 12.4. First-year residuals against second-year residuals for latent process.

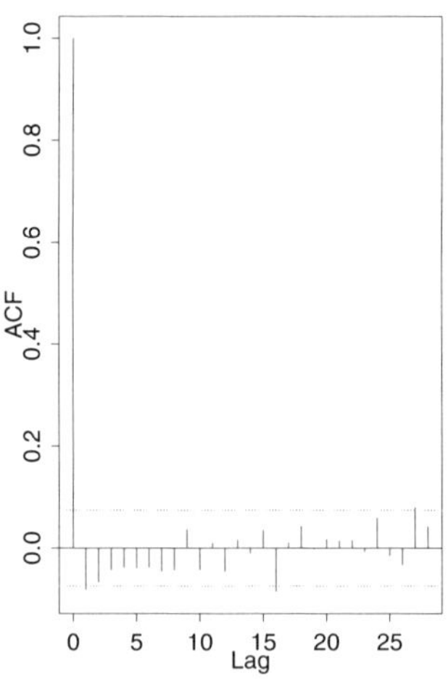

Fig. 12.5. Autocorrelation function of residuals from state process.

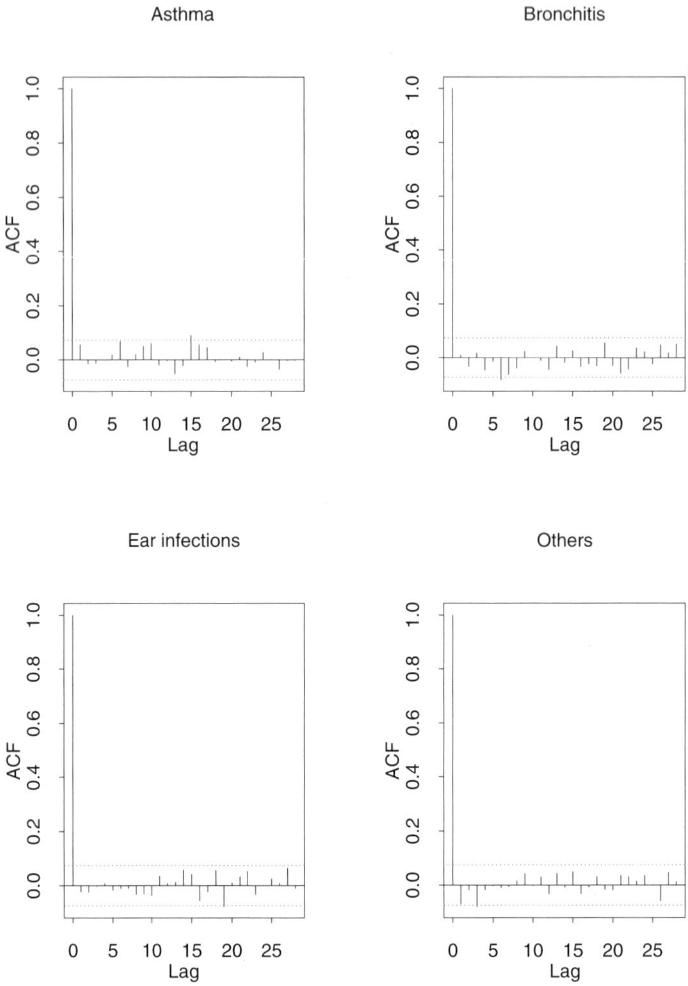

Fig. 12.6. Autocorrelation functions of residuals for the four categories.

(b) The coefficient for log-Particulates is negative, but only slightly significant. Moreover, due to the use of linear interpolation, this result should be interpreted with care. When the above model is fitted without covariate log-Particulates, in order to see if correlation between log-Particulates and log-Sulphur could have influenced the coefficient of log-Sulphur, the resulting estimates for the effect of log-Sulphur are close to those of Table 12.2. The actual correlation between log-Particulates and log-Sulphur was

Table 12.3. Estimates and standard errors for short-term effects.

	Asthma	Bronchitis	Ear	Others
Temperature	0.0106 (0.0060)	−0.0147 (0.0045)	0.0002 (0.0039)	– 0.0018 (0.0035)
Diff. humid	0.0682 (0.2336)	0.2964 (0.1763)	0.0903 (0.1168)	– 0.0569 (0.0906)
Sum humid	−0.0388 (0.1590)	0.1780 (0.1207)	−0.0099 (0.0816)	−0.1638 (0.0646)
Sunday	0.0796 (0.1760)	0.9099 (0.1620)	1.8208 (0.1552)	2.4185 (0.1531)
Monday	−0.1050 (0.1807)	0.3878 (0.1695)	1.1237 (0.1601)	1.8075 (0.1554)
Tuesday	−0.3667 (0.1893)	0.1634 (0.1745)	1.0236 (0.1614)	1.5843 (0.1571)
Wednesday	−0.6121 (0.1986)	0.4878 (0.1685)	1.1242 (0.1606)	1.7100 (0.1564)
Thursday	−0.5163 (0.1951)	0.3026 (0.1722)	1.0648 (0.1613)	1.6419 (0.1570)
Friday	−0.1725 (0.1834)	0.1881 (0.1741)	1.1538 (0.1604)	1.8060 (0.1561)
Saturday	0.1798 (0.1747)	1.0689 (0.1612)	1.9462 (0.1554)	2.5418 (0.1535)
	$\hat{\sigma}^2 = 0.019$			

0.36, a modest value, confirming that the effect of log-Particulates on the coefficient for log-Sulphur is likely to be of minor importance.

(c) Figure 12.7 shows the Kalman smoother estimate of the state process, which estimates the potential morbidity due to air pollution without regard to meteorological conditions. The estimated process is highly variable, and the estimate of the dispersion parameter ($\hat{\sigma} = 0.13$, from Table 12.3) shows that the stochastic variation of the state morbidity process θ_t itself is high.

(d) There is evidence of major episodes in December, 1985 and late March, 1986 of elevated risk of respiratory morbidity due to air pollution, and minor episodes in November–December, 1984 and April, 1985. These episodes are to some extent evident in the counts for three of the categories (Bronchitis, Ear, and Others), seen in Figure 1.7 in Section 1.3.10.

(e) Figure 12.8 shows a plot of the day-of-the-week effects based on the estimated in Table 12.3 for the four categories, which are very similar to the plot of daily averages shown by Knight et al. (1989). This plot shows a fairly clear weekly cycle, with more emergency room visits during the weekends than on workdays.

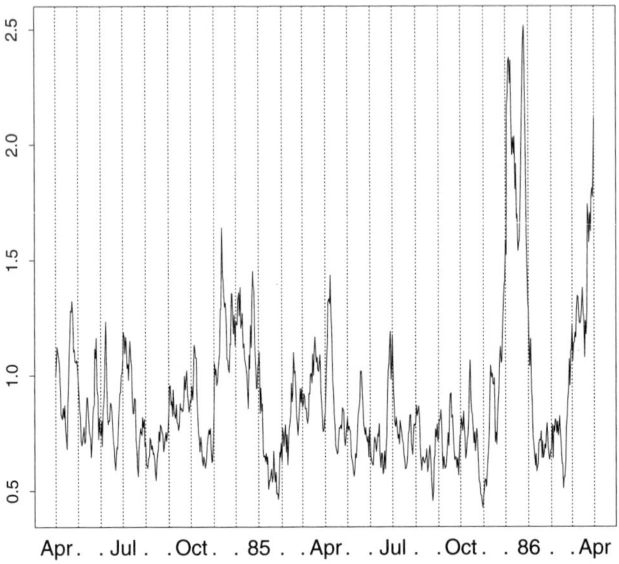

Fig. 12.7. The estimated state process from Kalman smoother.

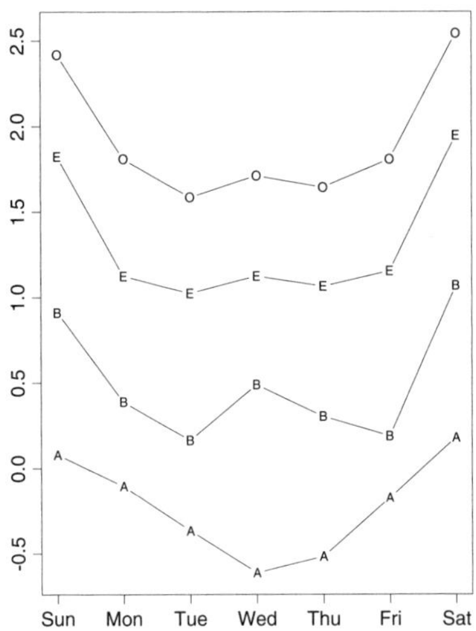

Fig. 12.8. Day-of-the-week effects.

13

Missing Data in Longitudinal Studies

13.1 Introduction

Missing values are omnipresent in longitudinal studies due to various reasons. In general, a missing value refers to such an observation that is intended to be recorded according a sampling or experimental protocol but failed to be observed. Missing data present a potential source of bias for data collection, especially when the number of missing values is substantial. It is ambiguity introduced by missing values that possibly violates randomness and representativeness of the collected sample from the study population, which, as a result, could lead to misleading conclusions in data analysis.

Handling missing values is not a trivial task, as related statistical procedures may be complicated by the following factors:

(a) The level of data incompleteness can vary from a study to another. Typically, observational studies are more likely to involve substantial amounts of missing data than clinical trials or experiment-based studies. This is simply because the former has essentially no control on the procedure of data collection, as opposed to the latter in which some efforts can be made to reduce the chance of getting void measurements on subjects. In addition, sometimes in clinical trials it is possible to track patients for reasons behind the missing values. This extra information regarding reasons of missingness is very useful for data analysts to determine a proper strategy of handling missingness. However, such information is unavailable in observational studies.

(b) The mechanism of missingness varies from case to case. In the literature, three types of missing data mechanisms have been widely adopted. They are, missing completely at random (MCAR), missing at random (MAR), and not missing at random (NMAR). Among the three types, NMAR is the most difficult case to deal with, because this is the situation where causes of missingness depend on missing values themselves. Unfortunately, NMAR does not occur rarely in practice. To properly handle NMAR, a certain probability model is inevitably required to describe the underlying mechanism of NMAR. In the building of this model, however, the lack of adequate information make this task almost impossible. Therefore, any model built for the mechanism of missingness needs to go over an exhaustive process of validation through a sensitivity analysis. More details will be supplied in the consequent sections of this chapter.

(c) Reasons leading to incomplete data are manifold and variable. One reason for missingness may be participants' relocation to another location. This is a reason that is unrelated to the study. More often reasons for missingness are indeed related to the study. For example, in a quality of life study, self-assessed questionnaires contain some questions that are personal and highly sensitive, and hence patients are unwilling or feel uncomfortable to answer them. This results in the so-called *nonresponse*. *Dropouts* occur in a longitudinal study when some subjects drop out from the study due to various reasons. For instance, when a patient experiences adverse treatment effects, this patient has to leave the study. It is not uncommon in practice to encounter intermittent missingness, which is a missing data pattern different from dropouts. *Intermittent missing* pattern refers to the scenario in that a subject completes the study but skips a few occasions in the middle of the study period. One example of this pattern is that a scheduled date of clinical visit happens to be a patient's birthday, so he/she decides not to visit the clinic. In general, a study should make every effort to collect and document reasons behind each missing value because such information is very valuable for data analysts to determine proper strategies of handling missing values.

Developing suitable strategies of handling missing values is challenged by many complicating factors, including those discussed above. There are no universal criteria or approaches available to dealing with missing data. Each case has to be dealt with under specific circumstances related to that case. Essentially, understanding the probability mechanism of how missing data are generated is the key to draw valid statistical inference in the analysis of incomplete data.

Data attrition can affect statistical analysis critically. For instance,

(1) Data attrition would reduce the power of statistical inference. This is particularly a concern in clinical trials where typically the sample size is determined beforehand, and increasing sample size in the middle of a clinical trial, in order to compensate the loss of data, is nontrivial and costly. Some statistical techniques, such as imputation method introduced in this chapter, may help to partially overcome this difficulty.

(2) When part of data information is lost, data variability would be very likely to be under-estimated on available data only. In some situations, this can severely affect the estimation of standard errors and hence result in a misleading conclusion about statistical significance.

(3) Data attrition may incur bias in representativeness of the study sample in relation to the study population. Therefore, estimation of treatment effects is biased and the comparability of treatment groups is based on unbalanced data information.

More detailed discussions on the effects of missing values can be found in many books concerning missing data problems, such as in Little and Rubin (2002), Schafer (1997), and Rubin (1987).

The present chapter will discuss two principled methods to handle missing values in the context of longitudinal data analysis: the multiple imputation based method and likelihood based method. These principled methods are closely relied on missing data mechanisms, and in contrast some simple methods, such as the method of *last observation carry over*, do not need to assume any missing data processes. Note that a modeling approach can only partially retrieve in a systematic way the uncertainty of missing information in the data. It is also hard to fully validate a model for a missing data mechanism. Therefore, one must be always cautious to take any model-based approaches in the analysis of incomplete longitudinal data. Sensitivity analysis is always recommended in a serious investigation; see, for example, Little (1994), Rubin et al. (1994), and Verbeke and Molenberghs (2000). Essentially, a sensitivity analysis aims to examine how robust the results would be when some of key model assumptions are intentionally perturbed in a set of "typical" scenarios. A criterion widely used to assess robustness is the so-called *local influence* proposed by Cook (1986). See Verbeke and Molenberghs (2000) for some developments of local influence in the modeling of missing data mechanisms.

Throughout this chapter, the schizophrenia trial data introduced in Section 1.3.5 will be used as a running example. Again, missing data behave very differently in different settings, and the procedures presented in the analysis of the data should not be assumed to be applicable everywhere.

In summary, there are three major steps required in analyzing incomplete longitudinal data (or in any incomplete data analysis):

(1) Understand and model the underlying mechanism of missingness.
(2) Incorporate the understood mechanism to build a valid statistical inference.
(3) Conduct a sensitivity analysis and assess influences of some key model assumptions on results of estimation and inference.

13.2 Missing Data Patterns

13.2.1 Patterns of Missingness

Let us begin by exploring missing data patterns, which is one of the most important initial steps for understanding/modeling missing data mechanisms. To illustrate some common missing data patterns encountered in practice, consider a hypothetical example given as follows. A longitudinal study involves six subjects, each having three visits. Half of them are randomized into the standard treatment and the other half into the new treatment. Blood pressure is the outcome variable of interest.

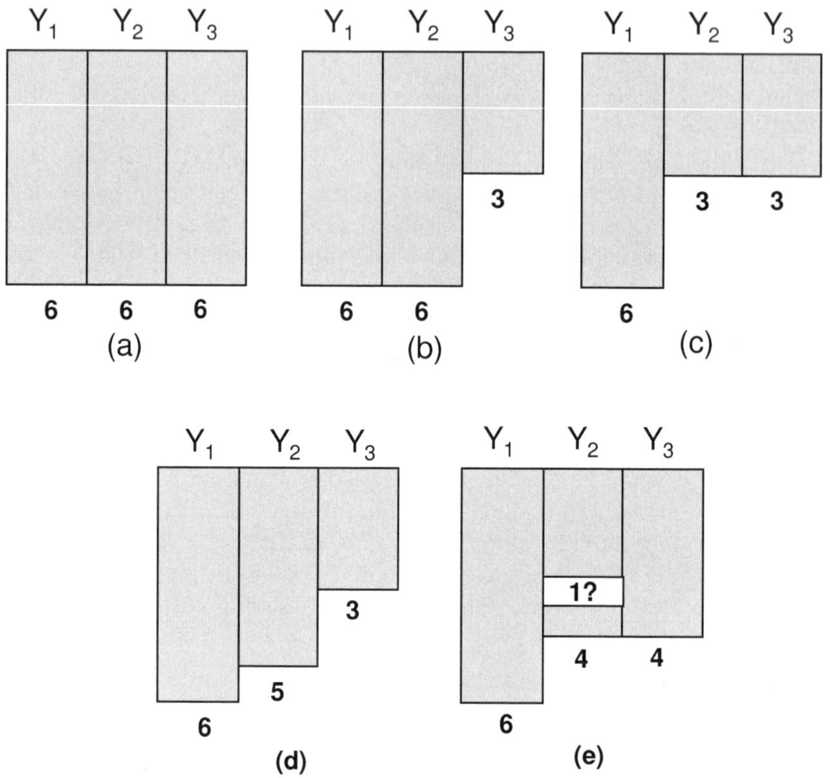

Fig. 13.1. Graphic representation of five missing data patterns.

A complete data pattern refers to the case with no missing values, as shown in Table 13.1 and in panel (a) of Figure 13.1.

A univariate (response) missing pattern refers to the situation where missing values only occur at the last visit, as shown in Table 13.2 and panel (b) of Figure 13.1. This is a special case of dropout pattern.

Table 13.3 and panel (c) of Figure 13.1 show *a uniform missing pattern*, in which missing values occur in a joint fashion. That is, the measurements at last two visits are either both observed or both missing simultaneously. This is a kind of dropout mechanism and the dropout time is uniform across all subjects.

Table 13.1. Complete data pattern.

Subject	Time	Treatment	Blood pressure	Subject	Time	Treatment	Blood pressure
1	1	New	155	4	1	Standard	123
1	2	New	191	4	2	Standard	201
1	3	New	192	4	3	Standard	188
2	1	New	104	5	1	Standard	115
2	2	New	131	5	2	Standard	107
2	3	New	178	5	3	Standard	110
3	1	New	98	6	1	Standard	118
3	2	New	166	6	2	Standard	158
3	3	New	134	6	3	Standard	131

Table 13.2. Univariate missing data pattern.

Subject	Time	Treatment	Blood pressure	Subject	Time	Treatment	Blood pressure
1	1	New	155	4	1	Standard	123
1	2	New	191	4	2	Standard	201
1	3	New	??	4	3	Standard	188
2	1	New	104	5	1	Standard	115
2	2	New	131	5	2	Standard	107
2	3	New	??	5	3	Standard	??
3	1	New	98	6	1	Standard	118
3	2	New	166	6	2	Standard	158
3	3	New	134	6	3	Standard	131

Table 13.4 and panel (d) of Figure 13.1 display *a monotonic missing pattern*, where if one observation is missing, then all the observation after it will be unobserved. This is a general and important kind of dropout mechanism that allows subjects to have different dropout times. As a matter of fact, all the above cases (b)-(d) are monotonic missing patterns.

An arbitrary missing pattern refers to the case in that missing values may occur in any fashion, an arbitrary combination of intermittent missing values and dropouts. Table 13.5 and panel (e) of Figure 13.1 demonstrate a possible scenario for a mixture of intermittent missing (on Subject 5) at the second visit and some dropouts (on both Subject 1 and Subject 2).

To fully describe the nature of data structure, it is convenient to introduce an indicator variable for the status of missingness. An intended complete data is denoted by $\mathbf{Y} = (\mathbf{Y}_{obs}, \mathbf{Y}_{mis})$, where $\mathbf{Y}_{obs}$ is the subset of all observed data

Table 13.3. Uniform missing data pattern.

Subject	Time	Treatment	Blood pressure	Subject	Time	Treatment	Blood pressure
1	1	New	155	4	1	Standard	123
1	2	New	??	4	2	Standard	201
1	3	New	??	4	3	Standard	188
2	1	New	104	5	1	Standard	115
2	2	New	??	5	2	Standard	??
2	3	New	??	5	3	Standard	??
3	1	New	98	6	1	Standard	118
3	2	New	166	6	2	Standard	158
3	3	New	134	6	3	Standard	131

Table 13.4. Monotonic missing data pattern.

Subject	Time	Treatment	Blood pressure	Subject	Time	Treatment	Blood pressure
1	1	New	155	4	1	Standard	123
1	2	New	191	4	2	Standard	201
1	3	New	??	4	3	Standard	188
2	1	New	104	5	1	Standard	115
2	2	New	??	5	2	Standard	107
2	3	New	??	5	3	Standard	??
3	1	New	98	6	1	Standard	118
3	2	New	166	6	2	Standard	158
3	3	New	134	6	3	Standard	131

and $\mathbf{Y}_{mis}$ is the subset of all missing values. Here missing observations are assumed to have their respective true underlying values and do not include missing potential outcomes. Let R be the indicator of observedness. So, the full set of observed data are

$$\{Y_{ij,obs}, R_{ij}\}, j = 1, \ldots, n_i; i = 1, \ldots, K.$$

So far missing values are simply assumed to occur in response variables. For example, the vector of (intended) observations for Subject 1 given in Table 13.5 is $\mathbf{Y}_1 = (155, \text{NA}, \text{NA})$, and the corresponding missing indicator vector is $\mathbf{R}_1 = (1, 0, 0)$. Similarly, the above discussion and notation can be extended to cases of missing covariates.

Table 13.5. Arbitrary missing data pattern.

Subject	Time	Treatment	Blood Pressure	Subject	Time	Treatment	Blood Pressure
1	1	New	155	4	1	Standard	123
1	2	New	??	4	2	Standard	201
1	3	New	??	4	3	Standard	188
2	1	New	104	5	1	Standard	115
2	2	New	131	5	2	Standard	??
2	3	New	??	5	3	Standard	110
3	1	New	98	6	1	Standard	118
3	2	New	166	6	2	Standard	158
3	3	New	134	6	3	Standard	131

13.2.2 Types of Missingness and Effects

According to Rubin (1976), in the context of likelihood inference there are three types of missingness, defined as follows. These definitions are based really on the conditional probability model of R given the data, namely $f(R \mid \mathbf{Y}_{obs}, \mathbf{Y}_{mis}, \phi)$, where ϕ is a generic parameter involved in the modeling of missing data process.

(1) *Missing completely at random* (MCAR) refers to the missing data process that does not depend on either observed or missing values, namely

$$f(R \mid \mathbf{Y}_{obs}, \mathbf{Y}_{mis}, \phi) = f(R \mid \phi).$$

(2) *Missing at random* (MAR) refers to the missing data process that does not depend on missing values, namely

$$f(R \mid \mathbf{Y}_{obs}, \mathbf{Y}_{mis}, \phi) = f(R \mid \mathbf{Y}_{obs}, \phi).$$

(3) *Not missing at random* (NMAR) refers to the missing data process that can depend on observed and missing values, namely

$$f(R \mid \mathbf{Y}_{obs}, \mathbf{Y}_{mis}, \phi) = f(R \mid \mathbf{Y}_{obs}, \mathbf{Y}_{mis}, \phi).$$

These definitions can be better understood by examining the effects of such mechanisms of missingness on likelihood inference. The following analytical derivations provide a direct assessment.

First, the full likelihood function of model parameters $\eta = (\theta, \phi)$ is

$$L_{full}(\eta) = \int f(\mathbf{Y}_{obs}, \mathbf{Y}_{mis} \mid \theta) f(R \mid \mathbf{Y}_{obs}, \mathbf{Y}_{mis}, \phi) d\mathbf{Y}_{mis}, \tag{13.1}$$

where θ is related to the measurement process and ϕ is associated with the missing data process.

In the case of MCAR, the full likelihood function can be simplified as follows:

$$\begin{aligned}
L_{full}(\eta) &= \int f(\mathbf{Y}_{obs}, \mathbf{Y}_{mis} \mid \theta) f(R \mid \mathbf{Y}_{obs}, \mathbf{Y}_{mis}, \phi) d\mathbf{Y}_{mis} \\
&= \int f(\mathbf{Y}_{obs}, \mathbf{Y}_{mis} \mid \theta) f(R \mid \phi) d\mathbf{Y}_{mis} \\
&= \int f(\mathbf{Y}_{obs}, \mathbf{Y}_{mis} \mid \theta) d\mathbf{Y}_{mis} \, f(R \mid \phi) \\
&= f(\mathbf{Y}_{obs} \mid \theta) f(R \mid \phi).
\end{aligned}$$

This result implies that the *complete-case analysis* is valid if two parameters θ and ϕ are distinct. Here the validity means that the large-sample performance of the MLE obtained from the $f(\mathbf{Y}_{obs} \mid \theta)$ is equivalent to that obtained from the full dataset (if missing data were observed). However, the small sample performance might be different, to some extent, because of the attrition of observations.

The complete-case dataset is generated by the so-called method of *listwise deletion*, which deletes all subjects involving missing values from the data collection. In parallel, the *available-case dataset* is obtained by deleting all of those individual visits (not subjects if partially observed) at which missing values occur. For example, in Table 13.4, the complete-case dataset would contain the observations collected only from the three completers, namely Subjects 3, 4, and 6; all observations of the other three Subjects 1, 2, and 5 are deleted because of the missing values. The resulting complete-case dataset is then given in Table 13.6.

Table 13.6. Complete-case dataset.

Subject	Time	Treatment	Blood pressure	Subject	Time	Treatment	Blood pressure
				4	1	Standard	123
				4	2	Standard	201
				4	3	Standard	188
3	1	New	98	6	1	Standard	118
3	2	New	166	6	2	Standard	158
3	3	New	134	6	3	Standard	131

Most statistical softwares such as `SAS` set the method of listwise deletion as a default procedure of dealing with (in fact, cleaning up) missing values. Obviously, this default analysis is valid only when R follows MACR.

On the other hand, the available-case dataset comprises of all subjects, either completely or partially observed, and only those visits involving missing values will be deleted. For the example of Table 13.4, the resulting available-case dataset is given in Table 13.7.

Table 13.7. Monotonic missing data pattern.

Subject	Time	Treatment	Blood pressure	Subject	Time	Treatment	Blood pressure
1	1	New	155	4	1	Standard	123
1	2	New	191	4	2	Standard	201
				4	3	Standard	188
2	1	New	104	5	1	Standard	115
				5	2	Standard	107
3	1	New	98	6	1	Standard	118
3	2	New	166	6	2	Standard	158
3	3	New	134	6	3	Standard	131

An issue in the analysis of available-case data is that the within-subject correlation might be distorted in the connection to intermittent missingness. For example, when the data of three repeated measurements were observed, the AR-1 correlation structure is

$$\mathrm{corr}(Y_t, Y_{t-h}) = \rho^h, h = 0, 1, 2,$$

where $\rho \in (-1, 1)$ is the correlation coefficient. If Y_2 is missing and deleted from the dataset to create an available-case dataset, the AR-1 correlation between Y_1 and Y_3 would be inflated to be ρ, rather than the originally ρ^2. However, the interchangeable correlation structure will not be affected by the intermittent missingness. In contrast, for the case of monotonic dropouts, both AR-1 and exchangeable correlation structure will be preserved. In general, within-subject correlation structure is more easily made adaptive for the pattern of monotonic dropouts to some of the existing models and methods in the longitudinal data analysis than for other missing data patterns. This *available-case analysis* allows more observations to remain in model fit, so it is a better choice than the complete-case analysis, if the correlation structure is not damaged or repairable. Nevertheless, this analysis is again valid only when R follows MARC.

Second, when R follows MAR, the likelihood function can also be simplified as follows.

$$
\begin{aligned}
L_{full}(\eta) &= \int f(\mathbf{Y}_{obs}, \mathbf{Y}_{mis} \mid \theta) f(R \mid \mathbf{Y}_{obs}, \mathbf{Y}_{mis}, \phi) d\mathbf{Y}_{mis} \\
&= \int f(\mathbf{Y}_{obs}, \mathbf{Y}_{mis} \mid \theta) f(R \mid \mathbf{Y}_{obs}, \phi) d\mathbf{Y}_{mis} \\
&= \int f(\mathbf{Y}_{obs}, \mathbf{Y}_{mis} \mid \theta) d\mathbf{Y}_{mis} \, f(R \mid \mathbf{Y}_{obs}, \phi) \\
&= f(\mathbf{Y}_{obs} \mid \theta) f(R \mid \mathbf{Y}_{obs}, \phi).
\end{aligned}
$$

This result suggests that the complete-case analysis is valid if θ and ϕ are distinct. That is, the large-sample performance of the MLE obtained from $f(\mathbf{Y}_{obs} \mid \theta)$ is equivalent to that obtained from the full data likelihood. The small-sample performance can be affected by the shrinkage of sample size.

However, it is important to point out that when a quasi-likelihood inference (such as the GEE) is considered, the above arguments and conclusions concerning MCAR and MAR are questionable. In effect, it is known that a quasi-likelihood inference is still valid if R follows MCAR, but it is invalid if R follows MAR.

Third, NMAR is often the case in practice. When R follows the NMAR mechanism, the likelihood function cannot be further simplified,

$$
L_{full}(\eta) = \int f(\mathbf{Y}_{obs}, \mathbf{Y}_{mis} \mid \theta) f(R \mid \mathbf{Y}_{obs}, \mathbf{Y}_{mis}, \phi) d\mathbf{Y}_{mis}.
$$

This means that neither the complete-case analysis nor the available-case analysis is valid. As a default the complete-case analysis, implemented in most statistical softwares, will unfortunately produce misleading results. Although NMAR seems to be more plausible in many practical studies, it is very difficult to justify specific assumptions made in the modeling of NMAR process since there is often no strong evidence in the data against MAR. Sensitivity analysis is sometimes useful to provide certain clues for making a choice between NMAR and MAR. Interested readers may refer to Verbeke and Molenberghs (2000) for details regarding sensitivity analysis.

13.3 Diagnosis of Missing Data Types

Although it is difficult, understanding missing data patterns is the first step in incomplete data analysis. In other words, for given a dataset with missing values, one has to examine which and how the missing data mechanism, MCAR, MAR or NMAR, is plausible. In the current literature, little investigation has been made on this issue. Most methodological developments or data analyses start with an assumption on missing data mechanism, either MAR or NMAR, with no data-driven justification for the validity of the assumption. Also, most of existing works in the literature are based only on the mechanism of dropouts, a monotonic missing data pattern. Statistical procedures inferring mechanisms of general missing data processes are much needed in order to handle missing data properly.

This section introduces a couple of methods that may assist in understanding missing data patterns.

13.3.1 Graphic Approach

A simple way to inspect missing data patterns is plotting summary trajectories across time or response variables. Take the schizophrenia trial data given in Section 1.3.5 as an example. Figure 13.2 demonstrates average profiles of patients' dropout of the trial, where only the data collected at the medium dose level are used in the summary. In this plot, the mean BPRS scores are calculated within each of the dropout cohorts determined in terms of their dropout times. This groping criterion is chosen according to the belief that patients who withdrew at the same time from the study tended to have similar medical conditions, and hence they can be considered as a clinically relevant cohort. Note that the solid line (representing the completion cohort) in the figure indicates a decreasing trend, reflecting on average the completers had gained a certain improvement on their symptom over the course of the trial. In other words, those patients, either in ST cohort and NT treatment, who have benefited from the treatment are more likely to stay in the trial till the end. In contrast, the mean BPRS scores of the dropout groups rise up right prior to each dropout time, which seems to suggest the patients who did not benefit from the trial tended to withdraw from the trial.

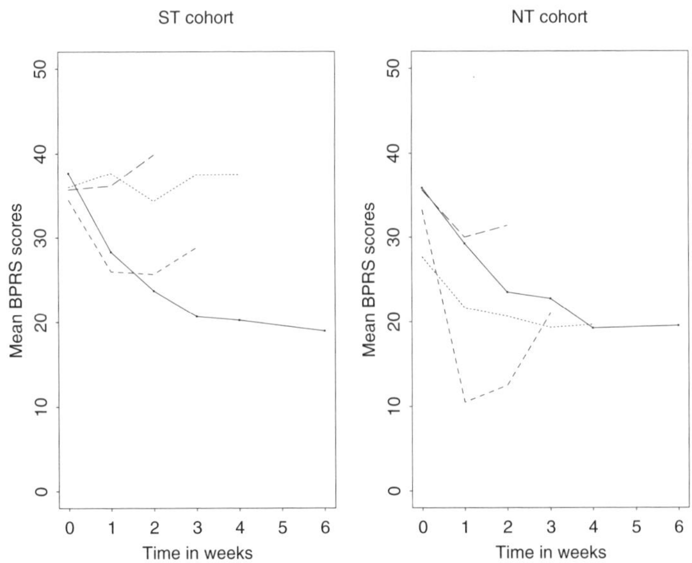

Fig. 13.2. Summary plot of average response profiles across different times of dropout.

This study documented three major reasons for patients' dropout: dropout due to lack of treatment effect, dropout due to adverse experience, and dropout due to other reasons. Table 1.4 summarizes the withdrawal patterns for the two treatment groups. Based on the empirical evidence, it appears reasonable to suspect that BPRS and patients' intent of withdrawal are related. This means that the dropout mechanism is NMAR and thus non-ignorable. To conduct a proper comparability study between the two treatments, the informative dropout must be taken into account in statistical inference. To further confirm the dependence of the dropout mechanism on the BPRS score, one may run a logistic regression of R_{ij} on Y_{ij}, by treating missing data (NA) as a category. A mixture model may be needed to modify the standard logistic regression in this setting.

13.3.2 Testing for MCAR

The graphic approach only provides an informal check on the missing data mechanism, most useful for monotonic missing data patterns. It is appealing to conduct a formal statistical test for whether a given missing data type is MCAR, MAR, or NMAR.

Diggle (1989) developed a method to test the null hypothesis that the dropouts are completely random (MCAR); that is, the probability that a subject drops out at time t_j is independent of the observed sequence of outcomes on that subject at time $t_1, \ldots, t_{j-1}$. Let $\pi_{ij} = P(R_{ij} = 0)$ be the probability that subject i drops out at time t_j. Under the null hypothesis, π_{ij} cannot depend on the observed measurements of subject i, but may be dependent on some covariates such as treatment, time and so on. To screen if there is any evidence from the data against the null hypothesis, Diggle (1989) suggested first conducting tests separately at each time within each treatment group and then investigating the resulting sample of p-values for any departure from the uniform distribution on $(0, 1)$.

The construction of individual tests at each time within each treatment group requires a critical choice of a statistic such that it can properly reflect the actual dependence of the dropout probabilities on the observed measurement history. For example, a partial sum of Y_{ij}'s may be used. With no doubt, the conclusion drawn from this approach may vary over different choices of test statistic. Moreover, the power of this approach is usually low because the individual tests are typically based on very few cases.

Qu and Song (2002) proposed a method that seems appealing to test for MCAR with the utility of quadratic inference functions (QIF). Unlike Diggle's approach that avoids any parameter assumptions about the process generating the measurement data, Qu and Song consider the testing procedure more specifically in the context of estimating equations based inferences. Their idea essentially involves testing for whether the zero-mean (or unbiasedness) property of estimating equations holds true. It is well known from the standard theory of inference functions in Chapter 3 that the zero-mean assumption is a crucial condition that ensures consistency for estimators obtained from estimating equations. If the unbiasedness holds in the presence of missing values, modeling or not modeling the missing data mechanism will not matter, so a complete-case analysis is proper. This case is generally referred to as the *ignorable* missingness. Otherwise, modeling the missing data mechanism is necessary, and in such a situation missing data mechanism is *nonignorable*.

In the context of quasi-likelihood or estimating equations based inferences, because likelihood function is unavailable, deriving the effects of missing data patterns discussed in Section 13.2.2 is no longer applicable. Thus in such a quasi-likelihood setting, it seems reasonable to simply classify the missing data types only into two categories: ignorable or nonignorable, corresponding to no-need or need of modeling the missing data mechanism. A similar idea has been also discussed in Chen and Little (1995) who developed a Wald-type test in the GEE setting for testing for MCAR. Qu and Song's test is a generalized score-type test based on QIF, which essentially examines whether there exists a common parameter under which the mean of estimating equations for different missing patterns is zero.

Suppose first that there are two disjoint groups of subjects, one with complete data and the other with incomplete data. Two vectors of estimating equations, denoted by $\boldsymbol{\psi}_1$ and $\boldsymbol{\psi}_2$, respectively, can be constructed for each group. Let $\boldsymbol{\psi} = (\boldsymbol{\psi}_1, \boldsymbol{\psi}_2)^T$. The ignorable missingness refers to the situation where there exists a common parameter $\boldsymbol{\eta}$ such that

$$\mathrm{E}_{\boldsymbol{\eta}}(\boldsymbol{\psi}) = 0, \text{ i.e., } \mathrm{E}_{\boldsymbol{\eta}}(\boldsymbol{\psi}_1) = \mathrm{E}_{\boldsymbol{\eta}}(\boldsymbol{\psi}_2) = \mathbf{0}. \tag{13.2}$$

The compatibility of two sets of estimating equations under two missing patterns ensures that a consistent estimator can be obtained by solving $\boldsymbol{\psi}_1 = 0$ alone, with the incomplete observations being ignored.

To check the compatibility, a test for the zero-mean assumption in (13.2) is utilized. Consider a quadratic inference function of the form

$$Q = \begin{pmatrix} \boldsymbol{\psi}_1 \\ \boldsymbol{\psi}_2 \end{pmatrix}^T \begin{pmatrix} C_1 & 0 \\ 0 & C_2 \end{pmatrix}^{-1} \begin{pmatrix} \boldsymbol{\psi}_1 \\ \boldsymbol{\psi}_2 \end{pmatrix} = \boldsymbol{\psi}_1{}^T C_1^{-1} \boldsymbol{\psi}_1 + \boldsymbol{\psi}_2{}^T C_2^{-1} \mathbf{s}_2, \tag{13.3}$$

where C_1 and C_2 are the estimated variances of $\boldsymbol{\psi}_1$ and $\boldsymbol{\psi}_2$, respectively. Note that the covariance between $\boldsymbol{\psi}_1$ and $\boldsymbol{\psi}_2$ is 0, because they are dependent on two distinct groups of subjects. A test statistic is naturally $Q(\hat{\boldsymbol{\eta}})$, which was previously proposed to evaluate goodness-of-fit in Chapter 5, where $\hat{\boldsymbol{\eta}}$ is the minimizer of the Q function. Under the null hypothesis of the existence of a common parameter satisfying (13.2), $Q(\hat{\boldsymbol{\beta}})$ asymptotically follows χ^2 with $\{\dim(\boldsymbol{\psi}) - \dim(\boldsymbol{\eta})\}$ degrees of freedom, according to Theorem 3.14. Using this chi-squared test, one can diagnose whether ignorable missingness is present in the data.

Now extend this strategy to more general cases where there are m distinct missing patterns instead of two. Note that subjects i and j are in the same missing pattern if $\mathbf{R}_i = \mathbf{R}_j$. The vectors of missing status, $\mathbf{R}_i, i = 1, \ldots, K$, contain not only dropout but also intermittent missingness. If there are m missing-data patterns, then there are m unique missing indicator vectors, denoted by $\mathbf{e}_l$, $l = 1, \ldots, m$. Let $\mathbf{e}_l(\mathbf{X}_i)$ and $\mathbf{e}_k(\mathbf{Y}_i)$ denote observed covariates and observed responses with respect to the l-th missing pattern.

According to Qu and Song (2002), the estimating function for the l-th missing pattern is then given by

$$\boldsymbol{\psi}^l(\boldsymbol{\eta}) = \sum_{i=1}^{K} \boldsymbol{\psi}_i^l\{\mathbf{e}_l(\mathbf{X}_i), \mathbf{e}_k(\mathbf{Y}_i), \boldsymbol{\eta}\} I(\mathbf{R}_i = \mathbf{e}_l),$$

where I is an indicator function. The dimensions of the estimating functions $\boldsymbol{\psi}^l$ do not have to be the same for different l. Under the null hypothesis $H_0 : \mathrm{E}_{\boldsymbol{\eta}}(\boldsymbol{\psi}^l) = 0$, $l = 1, \ldots, m$, the generalized score-type test statistic, defined in the form of the QIF by

$$Q(\hat{\boldsymbol{\eta}}) = \sum_{l=1}^{m} \{\boldsymbol{\psi}^l(\hat{\boldsymbol{\eta}})\}^T C_l^{-1}(\hat{\boldsymbol{\eta}}) \boldsymbol{\psi}^l(\hat{\boldsymbol{\eta}}), \text{where } C_l = \hat{\mathrm{Var}}(\boldsymbol{\psi}^l), \tag{13.4}$$

follows the chi-squared distribution with $\{\sum_{l=1}^{m} \dim(\boldsymbol{\psi}^l)\} - \dim(\boldsymbol{\eta})$ degrees of freedom asymptotically.

The flexibility of permitting $\boldsymbol{\psi}^l$ to have different dimensions for different missing patterns provides advantages in estimation and testing with no involvement of the so-called maximum identifiable parameter transformations as required in Chen and Little's (1995) Wald-type test. See more detailed discussions in Qu and Song (2002).

Example 13.1 (Schizophrenia Trial Data Analysis).

Now let us apply Qu and Song's generalized score-type test and Chen and Little's Wald-type test to a schizophrenia data example. By treating the BPRS at week 0 as the baseline score, the marginal model with the identity link function is

$$\mu = \mathrm{E}(Y) = \beta_0 + \beta_1 \,\mathrm{trt} + \beta_2 \,\mathrm{base} + \beta_3 \,\mathrm{week},$$

where `trt` is 1 or 0 for the new or standard drug, `base` is the baseline score, and the possible value of `week` is 1, 2, 3, 4, or 6.

As shown in Table 1.4, nearly half of the patients did not complete the trial for various reasons. Obviously, only the dropout type of missingness occurs in this study, possibly happening at follow-up time 2, 3, 4 or 6. Therefore $m = 5$, including a pattern with complete observations.

Figure 13.2 displays the mean summary profiles for different missing patterns including the complete one. As concluded in the preliminary graphic screening, it appears there are very different trends for patients who completed the trial and for those who withdrew from the study. This suggests that the mechanism of patients' dropouts might not be ignorable.

To provide rigorous statistical evidence as to whether ignorable missing occurs for this trial, the generalized score-type test is applied, in which generalized estimating equations with independence working correlation for $m = 5$. The use of independence working correlation is tenable as far as consistency concerns. The quadratic inference function is

$$Q(\boldsymbol{\beta}) = \sum_{l=1}^{5} (\boldsymbol{\psi}^l)^T C_l^{-1} \boldsymbol{\psi}^l,$$

where

$$\boldsymbol{\psi}^l = \sum_{i=1}^{124} \mathbf{X}_i^T (\mathbf{Y}_i - \boldsymbol{\mu}_i(\boldsymbol{\beta})) I(\mathbf{R}_i = \mathbf{e}_l)$$

and $\mathbf{X}_i$ is a 4×5 covariate matrix whose t-th column is $(1, \text{trt}_{it}, \text{base}_{it}, \text{week}_{it})^T$. Table 13.8 lists estimates of regression coefficients $\boldsymbol{\beta} = (\beta_0, \beta_1, \beta_2, \beta_3)^T$, the corresponding standard errors, and the Z-statistics.

Table 13.8. Estimates and test statistics for schizophrenia trial data.

	Intercept	trt	base	week
Estimates	5.49	0.89	0.62	−1.76
Std Err	3.45	1.91	0.09	0.25
Z-statistic	1.59	0.47	7.17	−7.04

The score-type test is found to be

$$Q(\hat{\boldsymbol{\beta}}) = 0.41 + 5.41 + 8.58 + 10.36 + 6.73 = 31.49,$$

with $(5 - 1) \times 4 = 16$ degrees of freedom, and p-value$= 0.012$ under the χ^2_{16} distribution. Thus there exists strong evidence that the missing data are nonignorable. In particular, for missing patterns $\mathbf{e}_3 = (1, 1, 1, 0, 0)^T$ and $\mathbf{e}_4 = (1, 1, 0, 0, 0)^T$, missing data contribute more than 18 out of the total test statistic of 31.49, or 57%. This implies that the patients who drop out of the study at the mid-periods of the trial (i.e., $\mathbf{e}_3$ and $\mathbf{e}_4$) experience very different treatment effectiveness in comparison to the other patients. Hence a separate medical investigation for these patients may be required.

In the presence of nonignorable missing data, the estimators obtained from minimizing the quadratic inference function are biased. Statistically this means that there does not exist a common set of regression coefficients that explain mechanisms of the complete and incomplete data generations.

A useful property of the score-type test using the QIF is its flexibility in pooling or collapsing data in the light of missing patterns. Since the Q-statistic has an additive form and each component in the form uses different patients or clusters, the chi-squared distribution is valid for different ways of splitting the data. For instance, one could combine all patients who drop out at different follow-up times together into one set, and those who complete the trial to another set. Then two sets of estimating equations can be constructed, one for each missing pattern. Under such a pooling, the minimum Q is found as

$$Q(\hat{\boldsymbol{\beta}}) = 0.89 + 10.50 = 11.39,$$

where 0.89 and 10.50 are the test statistics from the completion group and the drop-out group, respectively. The number of degrees of freedom is reduced from 16 to 4, and the corresponding p-value is 0.023 under χ^2_4-distribution. The conclusion is drawn here again that missing is nonignorable.

13.4 Handling MAR Mechanism

There are a few approaches proposed in the literature to handling missing data process that follows MAR mechanism. If a likelihood-based inference is invoked, the complete-case analysis produces estimators that hold valid large-sample properties. The weakness of the complete-case analysis is really the shrinkage of sample size, which can sometimes seriously affect small-sample performance.

In the context of correlated data analysis, a likelihood-based inference requires the availability of proper parametric families of multivariate distributions for the data. As far as handling missing values concerns, vector generalized linear models in Chapter 6 and mixed-effects models in Chapter 7 furnish two needed venues in that maximum likelihood inference is available. In contrast, a quasi-likelihood inference based on estimating functions, such as GEE and QIF in Chapter 5, is more troublesome in handling missing values of MAR type because in this case MAR is nonignorable missingness. Although MAR mechanism favors likelihood-based inference, in reality it is difficult to determine if, for a given dataset, a missing data process follows MAR or NMAR. A practical recommendation is to start with MAR mechanism, an much easier mechanism to handle than NMAR, and then conduct a sensitivity analysis to confirm the appropriateness of this MAR assumption, by perturbing the dependence between the measurement process and missing data process. In practice, there are no general rules as to how the perturbation is undertaken; but the current consensus in the literature is to perturb some components in an assumed missing data model, such as selection model and pattern-mixture model introduced later in Section 13.5, in a rather restrictive manner. See Minini and Chavance (2004). More substantial investigations are needed on the research of sensitivity analysis.

13.4.1 Simple Solutions and Limitations

Complete-case analysis is the simplest strategy to handle MAR, in which only those subjects with complete records will be used in likelihood-based inference. As a rule of thumb, in practice when the resulting sample size of the complete dataset is 10% of or less than that of the original sample size, the complete-case analysis may be acceptable and the related small sample performance may be little affected. Obviously, the complete-case analysis can waste substantial amount of information, as those subjects that contain partially observed measurements are removed from the analysis. Again, this method is vulnerable to deviations from the assumption that the missing data process is not related to the measurement process. For example, in the above schizophrenia trial data analysis, patients who completed the trial cannot be regarded as a random sample from the study population; as a result, the complete-case analysis will produce a biased inference for the comparability of two treatments.

An available-case analysis retains subjects who are partially observed, which is otherwise deleted in a complete-case analysis. As pointed out before, a caveat for the available-case analysis is that the correlation structure may be distorted by the resulting available data. For example, in longitudinal data analysis, a serial correlation structure will be damaged by intermittent missing types, but in clustered data analysis, an exchangeable correlation structure may remain valid by the data restructuring.

Another simple method is the so-called *last observation carried forward* (LOCF), proposed to handle monotonic missing data type of dropout. The method of LOCF creates a complete dataset by carrying the last observation forward. This is essentially a single-value imputation approach, so it suffers most of the drawbacks that have been understood in the imputation paradigm (see Section 13.4.2). For example, LOCF underestimates variation; that is, it treats imputed values like observed measurements, but actual measurements are apparently much less certain. In general, this method is not recommended without cautious scrutiny and serious sensitivity analysis. See Cook et al. (2004) for more discussions about LOCF.

13.4.2 Multiple Imputation

This section gives an introduction to the multiple imputation procedure, and interested readers can find more discussions from Schafer (1997) and Rubin (1987). This method works only for the case that the missing data process is MAR.

The primary objective of an imputation method is to fill out missing values with imputed data, so a complete dataset is created. Multiple Imputation (MI) is developed to overcome some of shortcomings of single-value imputation, including the key weakness–underestimation of variation. To some extent, MI can be regarded as a procedure that really "imputes variation." Sampling variation is a familiar concept in the setting of data analysis. That is, a different sample gives different parameter estimates. Standard error reflects such a variation. To estimate sampling variation, one needs to measure variation across multiple samples, which may be generated by the method of bootstrapping (Efron and Tibshirani, 1993).

Imputation variation is similar. Imputing different values leads to different parameter estimates, and standard error should reflect this variation as well. To estimate imputation variation, one needs to measure variation across multiple imputed datasets. Figure 13.3 gives a graphic presentation of MI procedure. First, M copies of the set of missing values, $\mathbf{Y}_{mis}$, is sampled from a conditional distribution $f(\mathbf{Y}_{mis}|\mathbf{Y}_{obs}, \boldsymbol{\theta})$. Then, each of M copy fills in the missing part of the dataset to produce an imputed dataset. In total, there are M imputed datasets of the original full size.

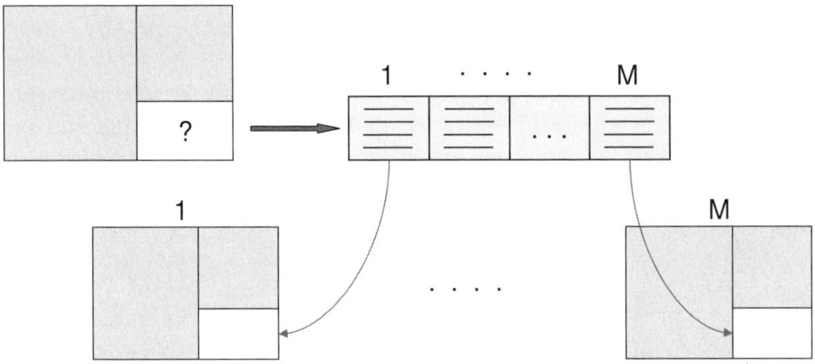

Fig. 13.3. Graphic representation of multiple imputation procedure.

In principle, MI procedure works for arbitrary missing data patterns and for missing responses and/or missing covariates. Markov chain Monte Carlo (MCMC) facilitates MI for arbitrary missing patters, in that both model parameters $\boldsymbol{\theta}$ and missing values $\mathbf{Y}_{mis}$ are treated as unknowns. At the l-th iteration, the algorithm proceeds by

drawing $\mathbf{Y}_{mis}^{(l)}$ from $f(\mathbf{Y}_{mis} \mid \mathbf{Y}_{obs}, \boldsymbol{\theta}^{(l-1)})$, and then
drawing $\boldsymbol{\theta}^{(l)}$ from $f(\boldsymbol{\theta} \mid \mathbf{Y}_{obs}, \mathbf{Y}_{mis}^{(l)})$.

After the burn-in is reached, the algorithm generates M samples from the stationary distribution $f(\boldsymbol{\theta}, \mathbf{Y}_{mis} \mid \mathbf{Y}_{obs})$, and the simulated copies of $\mathbf{Y}_{mis}$ can therefore be used to form M imputed datasets. To implement this MCMC sampling scheme, usually noninformative priors are assumed for $\boldsymbol{\theta}$. This MCMC-based MI procedure is now available in SAS PROC MI as a default imputation approach. PROC MI has the following features:

- It outputs multiple copies of imputed datasets with the default of 5 sets.
- It works only for the multivariate normal distribution of the measurement process. This imputing is undertaken directly from the distribution with no modeling (such as linear regression models) at the mean component or the variance component.
- It allows three imputation methods: regression, propensity, and MCMC, with MCMC being the default. The first two methods work only for monotone missing patterns, but the MCMC-based method works for arbitrary missing patterns, including the scenario of missing covariates. Since the multivariate normal distribution is assumed, all variables (either responses or covariates) involved in imputation are presumably continuous.

A simpler version of MI is to sample M copies of $\mathbf{Y}_{mis}$ directly from $f(\mathbf{Y}_{mis} \mid \mathbf{Y}_{obs}, \hat{\boldsymbol{\theta}}_{EM})$, where the parameter $\boldsymbol{\theta}$ is replaced its MLE, $\hat{\boldsymbol{\theta}}_{EM}$, obtained from the EM algorithm (see the next section 13.4.3), provided that this estimate $\hat{\boldsymbol{\theta}}_{EM}$ is available. Of course, when the $\hat{\boldsymbol{\theta}}_{EM}$ is already obtained, there is no need to carry out MI procedure as far as parameter estimation concerns. In fact, the MI-based estimation is a finite dimensional retrieval of missing information and a discrete version of the EM-algorithm based MLE. In other words, when imputed datasets covered all the sample space of $\mathbf{Y}_{mis}$ given $\mathbf{Y}_{obs}$, the MI estimation would be the same as the MLE obtained by the EM algorithm. However, in many cases, finding MLE via the EM algorithm is difficult, especially when missing values occur at both response variables and covariates, and when the measurement process is high dimensional and non-normal.

After the completion of imputation, for each imputed dataset, one simply conducts the complete-case analysis, which produces estimates, $\hat{\boldsymbol{\theta}}_l, l = 1, \ldots, M$, of the model parameter $\boldsymbol{\theta}$ and the corresponding sampling covariances (called *the within-imputation variance*), denoted by $\mathbf{W}_l = \text{Var}(\hat{\boldsymbol{\theta}}_l), l = 1, \ldots, M$. On the other hand, the *between-imputation variance* is given by

$$\mathbf{B} = \frac{1}{M-1} \sum_{l=1}^{M} (\hat{\boldsymbol{\theta}}_l - \bar{\boldsymbol{\theta}})(\hat{\boldsymbol{\theta}}_l - \bar{\boldsymbol{\theta}})^T,$$

where the average estimate is

$$\bar{\boldsymbol{\theta}} = \frac{1}{M} \sum_{l=1}^{M} \hat{\boldsymbol{\theta}}_l.$$

Clearly, covariance $\mathbf{B}$ measures the variation due to imputation. Likewise, the average of within-imputation variances is

$$\bar{\mathbf{W}} = \frac{1}{M} \sum_{l=1}^{M} W_l,$$

which reflects the variation due to sampling of the data. Combining the two types of variances in a componentwise fashion, the total variance of the average estimator, $\bar{\theta}_i$, of the i-th component θ_i of the $\boldsymbol{\theta}$, is given by

$$T_i = [\bar{\mathbf{W}}]_{ii} + \left(1 + \frac{1}{M}\right)[\mathbf{B}]_{ii},$$

where $[\mathbf{A}]_{ii}$ denotes the i-th diagonal element of matrix $\mathbf{A}$. Moreover, for the parameter θ_i, it is shown that

$$T_i^{-1/2}(\bar{\theta}_i - \theta_i) \overset{asy.}{\sim} t_{v_{i,M}}, \text{ as } M \to \infty$$

a t-distribution with the degrees of freedom equal to

$$v_{i,M} = (M-1)\left[1 + \frac{[\bar{W}]_{ii}}{(1+M^{-1})[\mathbf{B}]_{ii}}\right]^2.$$

This large-sample property allows us to easily establish related statistical inferences for parameters θ_i's.

In addition, to assess the efficiency of multiple imputation, first define the ratio r_i, called *the relative increase in variance due to nonresponse*,

$$r_i = \frac{(1+M^{-1})[\mathbf{B}]_{ii}}{[\bar{W}]_{ii}},$$

and λ_i, called *the fraction of missing information about* θ_i

$$\lambda_i = \frac{r_i + 2/(v_{i,M}+3)}{r_i + 1}.$$

Then, the *relative efficiency* of using the finite M-copies imputation estimator, rather than using an infinite copies of the fully efficient imputation, in units of variance, is approximately a function of M and λ_i, given by

$$RE_i = \left(1 + \frac{\lambda_i}{M}\right)^{-1}.$$

`SAS` PROC MIANALYZE combines results of individual complete-case analyses and produces valid statistics useful for inference. This `SAS` procedure is applicable to many different statistical analyses, including linear mixed-effects models (`PROC MIXED`) and generalized linear models (`PROC GENMOD`) with or without GEE option.

Example 13.2 (SAS Illustration). Consider the hypothetical example of blood pressure data discussed in Tables 13.1–13.5. Suppose that one applies a linear mixed-effects model to analyze the data via `SAS` procedures MI and MIANALYZE. To proceed, first invoke `SAS` PROC MI to create $M = 5$, *say*, imputed datasets:

```
proc mi data=bpress seed=12345 out=miout nimpute=5
     minimum = 0 0 0
     maximum = 300 300 300 .
     round = 0.1 0.1 0.1 1;
     var bp1 bp2 bp3 age;
     mcmc nbiter=200 niter=100 chain=single;
run;
```

Since covariate `treatment` is categorical, it is not used in the calculation. A continuous covariate `age` that was not shown in Tables 13.1–13.5 is added here just for the illustration. One way to add `treatment` in the method of MI is to separately run the above PROC MI for the two treatment cohorts, one for the subpopulation of new treatment and the other for the subpopulation of standard treatment.

After the imputed datasets are made available, run the complete-case analysis for individual dataset by PROC MIXED:

```
proc mixed data=miout method=ml;
     class id trt;
     model y=age trt /s;
     repeat / type = un subject=id;
     by _imputation_;
     ods output solutionF=fixparms CovB=fixbcov;
run;
```

To combine individual results, run PROC MIANALYZE as follows:

```
proc mianalyze parms=fixparms covb=fixbcov;
     var intercept age trt;
run;
```

13.4.3 EM Algorithm

In short, EM algorithm introduced by Dempster et al. (1977) is a general approach to iterative computation of maximum-likelihood estimate when the data are incomplete. Such an algorithm is comprised of an *expectation step* followed by a *maximization step* at each iteration, so it is referred to as *EM algorithm*. For the sake of ease for presentation, a Bayesian perspective is taken to introduce this algorithm.

Suppose data are observed from a parametric model $Y \sim f(y|\theta)$. In the context of Bayesian inference for θ, the central task is to find the posterior mode $\hat{\theta}$, namely, a statistic $\hat{\theta}(y_1, \ldots, y_K)$ that maximizes the posterior $f(\theta|Y)$.

The basic idea behind EM algorithm is to augment the observed data Y with latent (or missing) data Z so that the augmented posterior distribution $p(\theta|Y, Z)$ is "simple" in the sense that for instance, it is easy to carry out sampling, calculating, or maximizing the observed posterior $p(\theta|Y)$.

EM algorithm is detailed as follows. Let $\theta^{(l)}$ be the current estimate of the mode of posterior $p(\theta|Y)$. Then the next iteration requires:

- E-step: Compute an objective function of the form

$$\begin{aligned} Q(\theta, \theta^{(l)}) &= \mathrm{E}\{\log p(\theta|Z, Y)\}, \quad w.r.t.\ p(Z|\theta^{(l)}, Y) \\ &= \int_{\mathcal{Z}} \log\{p(\theta|Z, Y)\} p(Z|\theta^{(l)}, Y) dZ, \end{aligned}$$

  where $\mathcal{Z}$ is the sample space of the latent data Z.
- M-step: Maximize the Q function $w.r.t.$ parameter θ to obtain $\theta^{(l+1)}$. The related optimization may be carried out by algorithms such as Newton-Raphson, quasi-Newton, or downhill simplex.

The algorithm is iterated until a certain difference of, for example,

$$||\theta^{(l+1)} - \theta^{(l)}|| \text{ or } ||Q(\theta^{(l+1)}, \theta^{(l)}) - Q(\theta^{(l)}, \theta^{(l)})||$$

is sufficiently small.

Why does the algorithm work? It is because EM algorithm is in nature a fixed point algorithm with the *ascent property*; that is, each next iteration pushes the update value closer towards the true mode of posterior $p(\theta|Y)$ by steadily increasing the posterior relative to the current value. That is,

$$p\left(\theta^{(l+1)}|Y\right) \geq \left(\theta^{(l)}|Y\right).$$

The theoretical justification is given below. Note first that

$$1 = \frac{p(\theta, Z, Y)}{p(\theta, Z, Y)} = \frac{p(\theta|Z, Y)p(Z, Y)}{p(Z|\theta, Y)p(\theta|Y)p(Y)} = \frac{p(\theta|Z, Y)}{p(Z|\theta, Y)} \frac{1}{p(\theta|Y)} p(Z|Y).$$

Taking logarithm on both sides yields

$$0 = \log p(\theta|Z, Y) - \log p(Z|\theta, Y) - \log p(\theta|Y) + \underbrace{\log p(Z|Y).}_{\text{constant } w.r.t.\ \theta.}$$

Therefore,

$$\log p(\theta|Y) = \log p(\theta|Z, Y) - \log p(Z|\theta, Y) + \text{constant}.$$

Now integrating both sides $w.r.t.$ $p(Z|Y, \theta)$ gives

$$\begin{aligned} \log p(\theta|Y) = &\int_{\mathcal{Z}} \log p(\theta|Z, Y) p(Z|Y, \theta) dZ - \int_{\mathcal{Z}} \log p(Z|\theta, Y) p(Z|\theta, Y) dZ \\ &+ \int_{\mathcal{Z}} \log p(Z|Y) p(Z|\theta, Y) dZ, \end{aligned}$$

where the last term is always a constant when $\theta = \theta^*$ (given from the previous iteration) in $p(Z|\theta, Y)$.

Define Q function and H function, respectively, as follows:

$$Q(\theta, \theta^*) = \int_{\mathcal{Z}} \log p(\theta|Z, Y) p(Z|\theta^*, Y) dZ,$$

and

$$H(\theta, \theta^*) = \int_{\mathcal{Z}} \log p(Z|\theta, Y) p(Z|\theta^*, Y) dZ.$$

The likelihood gain relative to the previous iteration $\theta = \theta^{(l)}$ is given by

$$\begin{aligned} \log\{p(\theta^{(l+1)}|Y)\} - \log\{p(\theta^{(l)}|Y)\} &= Q(\theta^{(l+1)}, \theta^{(l)}) - Q(\theta^{(l)}, \theta^{(l)}) \\ &- \underbrace{(H(\theta^{(l+1)}, \theta^{(l)}) - H(\theta^{(l)}, \theta^{(l)}))}_{\text{always } \leq 0, \text{ due to Rao(1973, 1e6.6)}}. \end{aligned}$$

Therefore, if we select an update $\theta^{(l+1)}$ such that $Q(\theta^{(l+1)}, \theta^{(l)}) > Q(\theta^{(l)}, \theta^{(l)})$, which is what exactly the M-step does, then

$$p\left(\theta^{(l+1)}|Y\right) \geq p\left(\theta^{(l)}|Y\right),$$

unless

$$Q(\theta^{(l+1)}, \theta^{(l)}) = Q(\theta^{(l)}, \theta^{(l)}).$$

When the latter happens, the algorithm stops and the convergence is declared.

Example 13.3 (Genetic Linkage Model).

Genetic linkage model (Rao, 1973) is an example used in many books to illustrate the implementation of EM algorithm. Suppose 197 animals are distributed into four genetic categories as

$$Y = (y_1, y_2, y_3, y_4) = (125, 18, 20, 34), \text{ (e.g., genotypes AA, AB, BA, BB)}$$

with cell probabilities

$$\left(\frac{1}{2} + \frac{\theta}{4}, \frac{1}{4}(1-\theta), \frac{1}{4}(1-\theta), \frac{\theta}{4}\right), \text{ implicitly confined } \theta \in (0, 1).$$

The direct approach is to use a flat prior $\theta \sim U(0, 1)$, and for this case the posterior is, via multinomial distribution,

$$\begin{aligned} p(\theta|y_1, y_2, y_3, y_4) &= \frac{p(y_1, y_2, y_3, y_4|\theta)p(\theta)}{\int p(y_1, y_2, y_3, y_4|\theta)p(\theta)d\theta} \\ &\propto p(y_1, y_2, y_3, y_4|\theta)p(\theta) \\ &\propto (2+\theta)^{y_1}(1-\theta)^{y_2+y_3}\theta^{y_4}. \end{aligned}$$

Finding the posterior mode of $p(\theta|y_1, y_2, y_3, y_4)$ is equivalent to finding maximizer of the polynomial $(2+\theta)^{y_1}(1-\theta)^{y_2+y_3}\theta^{y_4}$.

The latent data approach is to augment the observed data by splitting the first cell into two cells with probabilities $\frac{1}{2}$ and $\frac{\theta}{4}$, respectively. The augmented data are then given by

$$X = (x_1, x_2, x_3, x_4, x_5)$$

such that

$$\begin{aligned} x_1 + x_2 = y_1 &= 125, \\ x_{i+1} &= y_i, \quad i = 2, 3, 4. \end{aligned}$$

Also using a flat prior $\theta \sim U(0,1)$, the posterior conditional on the augmented data is given by, through a similar augment based on a multinomial distribution as above,

$$\begin{aligned} p(\theta | x_1, x_2, x_3, x_4, x_5) &\propto (\frac{1}{2})^{x_1} \theta^{x_2} (1-\theta)^{x_3} (1-\theta)^{x_4} \theta^{x_5} \\ &\propto \theta^{x_2 + x_5} (1-\theta)^{x_3 + x_4}. \end{aligned}$$

By working with the augmented posterior, one can realize a simplification in functional form.

Now the EM algorithm is invoked for estimation in this model. The E-step computes

$$\begin{aligned} Q(\theta, \theta^{(l)}) &= \mathrm{E} \log p(\theta | Z, Y) \\ &= \mathrm{E}\{(x_2 + x_5) \log \theta + (x_3 + x_4) \log(1-\theta) | X_2, Y\} \end{aligned}$$

where

$$\begin{aligned} p(x_2 | \theta^{(l)}, Y) &= p(x_2 | \theta^{(l)}, x_1 + x_2) \\ &\sim \text{Binomial}\left(125, \tfrac{\theta^{(l)}}{\theta^{(l)}+2}\right). \end{aligned}$$

So, the objective function takes the form

$$Q(\theta, \theta^{(l)}) = \{\mathrm{E}(x_2 | \theta^{(l)}, Y) + x_5\} \log \theta + (x_3 + x_4) \log(1-\theta),$$

which is linear in all of the latent data, with

$$\mathrm{E}\left(x_2 | \theta^{(l)}, Y\right) = 125 \frac{\theta^{(l)}}{\theta^{(l)} + 2}. \tag{13.5}$$

The M-step needs to find an update $\theta^{(l+1)}$ as the solution to the following equation

$$\left. \frac{\partial Q(\theta, \theta^{(l)})}{\partial \theta} \right|_{\theta^{(l+1)}} = 0.$$

It follows that the equation has a closed form expression given by

$$\frac{\mathrm{E}(x_2|\theta^{(l)}, Y) + x_5}{\theta^(l+1)} - \frac{x_3 + x_4}{1 - \theta^{(l+1)}} = 0,$$

and the solution is

$$\theta^{(l+1)} = \frac{\mathrm{E}(x_2|\theta^{(l)}, Y) + x_5}{\mathrm{E}(x_2|\theta^{(l)}, Y) + x_3 + x_4 + x_5},$$

where $\mathrm{E}(x_2|\theta^{(i)}, Y)$ is given by (13.5).

Starting at $\theta^0 = 0.5$ (a naive guess as the middle point of the parameter space), the EM-algorithm converges to $\theta^* = 0.6268$ (the observed posterior mode) after 4 iterations.

Having arrived at the observed posterior mode, θ^*, one wants to evaluate the observed Fisher information given by

$$-\left.\frac{\partial^2 \log p(\theta|Y)}{\partial \theta^2}\right|_{\theta=\theta^*}.$$

Numerical differentiation may be used in this calculation. However, in practice the direct numerical calculation of the observed Fisher information can be numerically unstable.

The Louis' Formula (1982) is based on the following identity relation,

$$-\frac{\partial^2 \log p(\theta|Y)}{\partial \theta^2} = -\int_{\mathcal{Z}} \frac{\partial^2 \log p(\theta|Y, Z)}{\partial \theta^2} p(Z|Y, \theta) dZ - \mathrm{Var}\left\{\frac{\partial \log p(\theta|Y, Z)}{\partial \theta}\right\}$$

where the variance is taken under $p(Z|Y, \theta)$. In some situations, it may be difficult to analytically compute the first term on the right-hand side

$$\int_{\mathcal{Z}} \frac{\partial^2 \log p(\theta|Y, Z)}{\partial \theta^2} p(Z|Y, \theta) dZ,$$

which can be approximated by using the Monte Carlo simulation method,

$$\int_{\mathcal{Z}} \frac{\partial^2 \log p(\theta|Y, Z)}{\partial \theta^2} p(Z|Y, \theta) dZ \approx \frac{1}{M} \sum_{l=1}^{M} \frac{\partial^2 \log p(\theta|Y, z_l)}{\partial \theta^2}$$

where $z_1, \ldots, z_M \overset{iid}{\sim} p(Z|\theta^*, Y)$.

Similarly, one can approximate the variance term on the right-hand side by

$$\mathrm{Var}\left\{\frac{\partial \log p(\theta|Y, Z)}{\partial \theta}\right\} \approx \frac{1}{M} \sum_{l=1}^{M} \left\{\frac{\partial \log p(\theta|Y, z_l)}{\partial \theta}\right\}^2 - \left\{\frac{1}{M} \sum_{l=1}^{M} \frac{\partial \log p(\theta|Y, z_l)}{\partial \theta}\right\}^2.$$

Example 13.4 (Genetic Linkage Model Continued).

In the Example 13.3 of the genetic linkage model, set

$$\theta^* = 0.6268,\ M = 10,000,\ K = 125,\ p = \frac{\theta^*}{\theta^* + 2}.$$

The Monte Carlo simulation estimated the variance as

$$\widehat{\text{Var}}\left(\frac{\partial \log p(\theta|Y, Z)}{\partial \theta}\right) = 57.8.$$

Example 13.5 (Linear Regression Analysis).

Consider a linear regression model for n-dimensional response

$$\mathbf{Y} = X^T\boldsymbol{\beta} + \boldsymbol{\varepsilon}$$

where $\boldsymbol{\varepsilon} \sim \text{MVN}_n(0, \Sigma)$ and $\mathbf{X}^T = (\mathbf{x}_1^T, \ldots, \mathbf{x}_n^T)$. Data consist of independent pairs $(\mathbf{Y}_i, \mathbf{X}_i), i = 1, \ldots, K$.

Here consider a simple case that missing values occur only in the responses and no occurrence of missing values in covariates. For each data point, set $\mathbf{Y}_i = (\mathbf{Y}_{i,obs}, \mathbf{Y}_{i,mis}),\ i = 1, \ldots, K$. Let $\mathbf{Y} = (\mathbf{Y}_1, \ldots, \mathbf{Y}_K)$, $\mathbf{Y}_{obs} = (\mathbf{Y}_{1,obs}, \ldots, \mathbf{Y}_{K,obs})$, and $\mathbf{Y}_{mis} = (\mathbf{Y}_{1,mis}, \ldots, \mathbf{Y}_{K,mis})$. Thus, write $\mathbf{Y} = (\mathbf{Y}_{obs}, \mathbf{Y}_{mis})$. Also, let $\boldsymbol{\mu}_i = \text{E}(\mathbf{Y}_i) = (\mu_{i1}, \ldots, \mu_{in})^T$, and let $\boldsymbol{\theta}$ be the vector of parameters to be estimated, including the regression coefficients $\boldsymbol{\beta}$ and distinct variance-component elements in the covariance matrix Σ.

It is easy to show that for the multivariate normal distribution,

$$S(\mathbf{Y}) = \{\mathbf{Y}_i, i = 1, \ldots, K; \mathbf{Y}_i\mathbf{Y}_i^T, i = 1, \ldots, K\}$$

gives a set of sufficient statistics. So, to estimate $\boldsymbol{\theta}$, the EM algorithm is formulated as follows.

- E-Step: Calculate conditional expectations of the sufficient statistics:

$$\begin{aligned}\mathbf{Y}_i^{(l)} &= \text{E}(\mathbf{Y}_i|\mathbf{Y}_{obs}, \mathbf{X}, \boldsymbol{\beta}^{(l)}, \Sigma^{(l)}) \\ &= \begin{cases} Y_{ij}, & \text{if } Y_{ij} \text{ is observed} \\ \mathbf{x}_{ij}^T\boldsymbol{\beta}^{(l)}, & \text{if } Y_{ij} \text{ is missing.} \end{cases}\end{aligned}$$

And the (j,k)-th element of matrix $(\mathbf{Y}_i\mathbf{Y}_i^T)^{(l)} = \text{E}(\mathbf{Y}_i\mathbf{Y}_i^T|\mathbf{Y}_{obs}, \mathbf{X}, \boldsymbol{\beta}^{(l)}, \Sigma^{(l)})$ is given by

$$\left[(Y_iY_i^T)^{(l)}\right]_{j,k} = \begin{cases} Y_{ij}Y_{ik}, & Y_{ij} \text{ and } Y_{ik} \text{ observed} \\ \left[\Sigma^{(l)}\right]_{jk} + Y_{ij}^{(l)}Y_{ik}^{(l)}, & Y_{ij} \text{ and } Y_{ik} \text{ missing} \\ Y_{ij}Y_{ik}^{(l)}, & Y_{ij} \text{ observed but } Y_{ik} \text{ missing} \\ Y_{ij}^{(l)}Y_{ik}, & Y_{ij} \text{ missing but } Y_{ik} \text{ observed.} \end{cases}$$

- M-step: Find the MLE based on the sufficient statistics $S(\mathbf{Y}^{(l)})$ of the full data $\mathbf{Y}^{(l)}$ obtained from the E-step. For the regression coefficient $\boldsymbol{\beta}$,

$$\boldsymbol{\beta}^{(l+1)} = \left(\sum_{i=1}^{K} \mathbf{X}_i^T \Sigma^{(l)-1} \mathbf{X}_i \right)^{-1} \sum_{i=1}^{K} \mathbf{X}_i^T \Sigma^{(l)-1} \mathbf{Y}_i^{(l)}.$$

And, for the variance matrix Σ,

$$\begin{aligned} \Sigma^{(l+1)} = & \frac{1}{K} \sum_{i=1}^{K} (\mathbf{Y}_i \mathbf{Y}_i^T)^{(l)} - \frac{1}{K} \sum_{i=1}^{K} \mathbf{X}_i^T \boldsymbol{\beta}^{(l+1)} \mathbf{Y}_i^{(l)T} \\ & - \frac{1}{K} \sum_{i=1}^{K} \mathbf{Y}_i^{(l)} \boldsymbol{\beta}^{(l+1)T} \mathbf{X}_i + \frac{1}{K} \sum_{i=1}^{K} \mathbf{X}_i^T \boldsymbol{\beta}^{(l+1)}) \boldsymbol{\beta}^{(l+1)T} \mathbf{X}_i. \end{aligned} \tag{13.6}$$

Iterate the E and M steps until convergence. The initial values may be specified as follows. First conduct an available-data analysis under the independence correlation structure ($\Sigma = I$) and obtain $\boldsymbol{\beta}^{(0)}$. Then acquire $\Sigma^{(0)}$ by applying the formula (13.6).

13.4.4 Inverse Probability Weighting

Handling MAR missing mechanism becomes a lot harder when likelihood-based inference is not available. For example, a complete-case analysis based on GEE, when the missing data process is MAR, produces biased estimation for effects of covariates. Therefore, in order to conduct a valid inference using GEE approach or as such, it is necessary to adjust for sampling bias induced by missing observations. Typically, in the context of estimating (or inference) functions, adjusting MAR type of missingness is facilitated through weighting subjects' probabilities of being sampled from the study population. This section introduces the so-called inverse probability weighting procedure proposed by Robins et al. (1994,1995) that gives rise to unbiased estimating equations under MAR.

For convenience, the following presentation focus only on dropout missing data pattern. Let $\pi_{ij} = P(R_{ij} = 1 | Y_i, \mathbf{x}_i)$ be the probability of observing subject i at time point j, given the response vector Y_i and covariates $\mathbf{x}_i$. Here probability π_{ij} may involve the parameter vector $\boldsymbol{\alpha}$, of dimension q, *say*, which parametrizes the missing data process. Let $\Delta_i = \text{diag}\{R_{ij}/\pi_{ij}, j = 0, 1, ..., n_i\}$ be a matrix of weights accommodating missingness, and let $\mathbf{D}_i = \partial \boldsymbol{\mu}_i / \partial \boldsymbol{\beta}^T$ be an $n_i \times p$ matrix of the first-order derivatives of the mean vector $\boldsymbol{\mu}_i$ *w.r.t.* the regression parameter vector $\boldsymbol{\beta}$ of p dimension, $i = 1, 2, ..., K$. Denote by $\mathbf{V}_i = \text{Var}(Y_i)$ the covariance matrix of the response vector Y_i. Let $\gamma_{ij} = Y_{ij} - \mu_{ij}$, and $\boldsymbol{\gamma}_i = (\gamma_{i1}, ..., \gamma_{in_i})^T$. Set an estimating function of the form

$$U(\boldsymbol{\beta}, \boldsymbol{\alpha}) = \sum_{i=1}^{K} U_i(\boldsymbol{\beta}, \boldsymbol{\alpha}) = \sum_{i=1}^{K} \mathbf{D}_i^T \mathbf{V}_i^{-1} \Delta_i \boldsymbol{\gamma}_i, \tag{13.7}$$

with

$$U_i(\boldsymbol{\beta}, \boldsymbol{\alpha}) = \mathbf{D}_i^T \mathbf{V}_i^{-1} \Delta_i \boldsymbol{\gamma}_i, \; i = 1, 2, ..., K.$$

The resulting estimating equations $U(\boldsymbol{\beta}, \boldsymbol{\alpha}) = 0$ is called the inverse probability weighted GEE (IPWGEE). Note that the inclusion of R_{ij} in the expression helps to select all observed data in the formulation of IPWGEE. Therefore, this corresponds effectively to an available-case analysis. Under some regularity conditions, IPWGEE produces consistent estimation for the $\boldsymbol{\beta}$, because as shown in the following Proposition 13.6, the $U(\cdot)$ is unbiased.

Proposition 13.6. *Suppose that the weight matrices $\Delta_i(\boldsymbol{\alpha}_0)$ are correctly specified under the true value $\boldsymbol{\alpha}_0$ and that the first moment of the measurement process, $\boldsymbol{\mu}_{ij} = E(Y_{ij}|\mathbf{x}_i)$, is correctly modeled. Then the estimating function $U(\boldsymbol{\beta}, \boldsymbol{\alpha}_0)$ given in (13.7) is unbiased, namely $E(U) = 0$.*

Proof. First let

$$\begin{aligned} \mathbf{A}_i &= \mathrm{E}_{\mathbf{R}|(Y,\mathbf{x})}(\Delta_i) \\ &= \mathrm{diag}\{\mathrm{E}(R_{ij}|\mathbf{Y}, \mathbf{x})/\pi_{ij}, j = 0, 1, ..., n_i\}. \end{aligned}$$

It follows that

$$\begin{aligned} \mathrm{E}\{U_i(\boldsymbol{\beta}, \boldsymbol{\alpha}_0)\} &= \mathrm{E}_{Y|\mathbf{x}} \left\{ \mathrm{E}_{\mathbf{R}|(Y,\mathbf{x})}(\mathbf{D}_i^T \mathbf{V}_i^{-1} \Delta_i \boldsymbol{\gamma}_i) \right\} \\ &= \mathrm{E}_{Y|\mathbf{x}}(\mathbf{D}_i^T \mathbf{V}_i^{-1} \mathbf{A}_i \boldsymbol{\gamma}_i) \\ &= \mathrm{E}_{Y|\mathbf{x}}(\mathbf{D}_i^T \mathbf{V}_i^{-1} \boldsymbol{\gamma}_i) \\ &= 0, \end{aligned}$$

where the second last equality holds by the fact that $\mathbf{A}_i$ is the identity matrix if the probabilities π_{ij} are correctly specified; that is, $\pi_{ij}(\boldsymbol{\alpha}_0) = \mathrm{E}(R_{ij}|\mathbf{Y}, \mathbf{x})$ under the true value $\boldsymbol{\alpha}_0$. Moreover, the last equality is due to the assumption that the mean structure of the measurement process is correctly specified; that is, $\mathrm{E}_{Y|\mathbf{x}}(\boldsymbol{\gamma}_i) = 0$.

This weighting strategy was originally proposed in the literature of survey sampling (Horvitz and Thompson, 1952) to handle unequal probability sampling. The basic idea is that each subject's contribution to the weighted available-case analysis at time j is replicated in a scale of $\frac{1}{\pi_{ij}}$, where π_{ij} is the probability that subject i is sampled (or observed) at time t_j. The weight $\frac{1}{\pi_{ij}}$ counts once for himself/herself and $\left(\frac{1}{\pi_{ij}} - 1\right)$ for those subjects with the same history of responses and covariates, but who are not observed at time t_j.

One constraint in the utility of this IPWGEE is the proper modeling of probabilities π_{ij} of the dropout process, which is unknown in practice. The estimator $\hat{\boldsymbol{\beta}}$ obtained from IPWGEE (13.7) is consistent only if $\sqrt{K}$-consistent estimates $\hat{\pi}_{ij}$ are plugged in the equation. Another limitation for the IPWGEE is that the probabilities π_{ij} must be bounded from a certain lower limit. This lower bound cannot be too small; otherwise very few observations that have small (close to zero) probabilities of occurrence will have very large weights and hence dominate the performance of the IPWGEE. In practice, however, this condition is not always satisfied or controlled properly, especially when the modeling of the dropout probabilities is involved. One possible approach to dealing with this issue is to trim or tune the probabilities π_{ij} under a certain criterion, such as a trade-off between bias and efficiency of the resulting estimators. This is worth a further exploration.

There are two modeling strategies for missing data processes that are widely used in practice. The first one is a *transition model* in that the propensities for dropout are dependent on the history of dropout process. Let $\lambda_{ij} = P(R_{ij} = 1|R_{i,j-1} = 1, \mathbf{x}_i, Y_i)$ be the conditional probability that subject i is observed at time t_j, given covariates $\mathbf{x}_i$ (containing no missing values), the response vector Y_i, and that subject i stays at the study up to the previous time point $j-1$. Let $H_{ij}^y = \{Y_{i1}, ..., Y_{i,j-1}\}$ denote the history of observed responses up to (but not including) time point t_j.

A logistic regression model may be used for the dropout probabilities,

$$\text{logit}(\lambda_{ij}) = \mathbf{u}_{ij}^T \boldsymbol{\alpha}, \tag{13.8}$$

where $\mathbf{u}_{ij}$ is a vector consisting of both information concerning covariates $\mathbf{X}_i$ and observed responses H_{ij}^y. This model implies that the conditional probability λ_{ij} is determined by the covariates $\mathbf{X}_i$ and the historic responses H_{ij}^y, given subject i is in the study at the previous time point t_{j-1}, and hence this model (13.8) characterizes an MAR mechanism. This approach has been widely used in the literature for the modeling of dropout processes. See Diggle and Kenward (1994), Robins et al. (1995), Fitzmaurice et al. (1996), and Molenberghs et al. (1997).

Estimation of parameter $\boldsymbol{\alpha}$ can be proceeded by running a logistic regression analysis. Let D_i be the random dropout time for subject i, and d_i be a realization, $i = 1, 2, ..., K$. Denote

$$L_i(\boldsymbol{\alpha}) = (1 - \lambda_{id_i}) \prod_{j=2}^{d_i - 1} \lambda_{ij},$$

where λ_{ij} follows the model (13.8). Then the log-likelihood is given by

$$\ell(\boldsymbol{\alpha}) = \prod_{i=1}^{K} \log L_i(\boldsymbol{\alpha}).$$

Then, the score equation is $\mathbf{s}(\boldsymbol{\alpha}) = \sum_{i=1}^K \partial \log L_i(\boldsymbol{\alpha})/\partial \boldsymbol{\alpha}$, which will be unbiased if model (13.8) is correctly assumed. Moreover, the resulting marginal probabilities $\pi_{ij} = \pi_{ij}(\boldsymbol{\alpha}) = \prod_{t=2}^{j} \lambda_{it}(\boldsymbol{\alpha})$ can be consistently estimated, when the $\boldsymbol{\alpha}$ is replaced by a consistent estimator $\hat{\boldsymbol{\alpha}}$.

The second model is a *marginal model* in that the marginal probabilities of the dropout process are directly specified as a function of historic responses and covariates. A logistic model takes the form

$$\text{logit}(\pi_{ij}) = \mathbf{u}_{ij}^T \boldsymbol{\alpha}, \tag{13.9}$$

where $\mathbf{u}_{ij}$ is similar to that defined above in (13.8). Note that model (13.9) states that marginally the probability of observing subject i at time t_j depends on the observed responses and covariates $\mathbf{X}_i$. In contrast, jointly the probability of observing subject i over the entire time course could or could not depend on unobserved responses, depending on the imposed association structure for observed indices R_{ij}. If the association structure depends only on the observed responses and covariates $\mathbf{X}_i$, then (13.9) models MAR mechanism; otherwise, it models NMAR mechanism. Under this setup, estimation of $\boldsymbol{\alpha}$ can be proceeded by applying a usual GEE approach on observed indices R_{ij}, in which the covariance matrix of $\mathbf{R}_i = (R_{i1}, \ldots, R_{in_i})^T$ may be further modeled by an setup similar to GEE2 presented in Section 5.3 (e.g., Yi and Cook, 2002). Some alternatives to the IPWGEE are also proposed in the literature, such as Paik (1997) and Zorn (2001).

13.5 Handling NMAR Mechanism

This section gives a brief review on several parametric and semiparametric approaches to handling dropouts that are potentially NMAR. Follmann and Wu (1995) and Ibrahim et al. (2001) gave considerable attention to the modeling of longitudinal binary data with nonignorable missing values for GLMMs, while Paik (1997), Lipsitz and Ibrahim (2000), and Fitzmaurice and Laird (2000) analyzed nonignorable missing data using GEEs, among others. Interested readers can find more details regarding parametric modeling of the dropout process from Diggle et al. (2002) and Verbeke and Molenberghs (2000). This section focuses on a semiparametric pattern mixture model proposed by Sun and Song (2001).

13.5.1 Parametric Modeling

In the literature, there are three parametric approaches to modeling of longitudinal data with potentially informative dropouts. The first one is the so-called *selection model approach*, proposed by Diggle and Kenward (1994). This model corresponds to the following factorization of the joint distribution (or the likelihood):

$$\begin{aligned} f(\mathbf{Y}, \mathbf{R}) &= f_\theta(\mathbf{Y}) f_\phi(\mathbf{R}|\mathbf{Y}) \\ &= \int f_\theta(\mathbf{Y}_{obs}, \mathbf{Y}_{mis}) d\mathbf{Y}_{mis} \int f_\phi(\mathbf{R}|\mathbf{Y}_{obs}, \mathbf{Y}_{mis}) d\mathbf{Y}_{mis}, \end{aligned}$$

which essentially claims that the missing data mechanism is dependent on the measurement process. In a selection model, probabilities of dropout are specified as a function of the measurement process via a marginal model of the form, for example,

$$\text{logit}P(R_{ij} = 1|\mathbf{Y}) = \phi_0 + \phi_1(H_{i,j-1}) + \phi_2(Y_{ij}), \qquad (13.10)$$

where $\phi_l, l = 1, 2$ are certain suitable functions. Clearly, when $\phi_l \equiv 0, l = 1, 2$, (13.10) corresponds to MCAR; when $\phi_2 \equiv 0$, (13.10) corresponds to MAR; and $\phi_2 \neq 0$, (13.10) corresponds to NMAR.

The application of selection models should be cautious. When the dropout process is NMAR, the parameters θ and ϕ cannot be separately identifiable in the likelihood, and the resulting likelihood inference becomes nontrivial. More critically, validating the assumed model from the observed data is very difficult or even impossible due to poor identifiability of the model parameters. See, for example, Fitzmaurice et al. (1996) and Verbeke and Molenberghs (2000).

The second approach is based on a *shared random effects model* proposed by Wu and Carroll (1988). Intuition behind this modeling strategy is that both measurement process and dropout process are driven by a common set of unobserved characteristics, $\mathbf{U}$, of subjects, and assume that conditional on such a shared set of random effects (describing those unobserved characteristics) $\mathbf{U}$, $\mathbf{Y}, \mathbf{R}|\mathbf{U}$ are independent. The measurement process can be modeled by the familiar mixed-effects models in Chapter 7, and the dropout process may be specified as follows:

$$\text{logit P}(R_{ij} = 1|\mathbf{U}) = \phi_0 + \phi_1(H_{i,j-1}) + \phi_2(Y_{ij}) + u_i \qquad (13.11)$$

where $\phi_l, l = 1, 2$ are certain suitable functions and u_i is the random effect. Similar to the selection model, when $(\phi_1, \phi_2) \equiv (0, 0)$, (13.11) corresponds to MCAR; when $\phi_2 \equiv 0$, (13.11) corresponds to MAR; and $\phi_2 \neq 0$,(13.11) corresponds to NMAR.

Since both measurement and dropout processes depend on random effects, related inferential procedures are very challenging. Little progress has been seen in the literature regarding statistical inference in such a model setting.

Pattern mixture models is the third parametric modeling approach proposed by Little (1993). It takes a completely opposite view of factorizing the joint distribution (or the likelihood) to that of the selection model:

$$f(\mathbf{Y}, \mathbf{R}) = f_\theta(\mathbf{Y}|\mathbf{R}) f_\phi(\mathbf{R}) = f_\phi(\mathbf{R}) \int f(\mathbf{Y}_{obs}, \mathbf{Y}_{mis}|\mathbf{R}) d\mathbf{Y}_{mis},$$

which essentially claims that the measurement process is driven by the dropout process. In other words, measurements are generated under different missing data patterns that classify subjects into different subpopulations. On the basis of this point of view, one may conduct a stratified analysis conditional on dropout patterns if the dropout process is NMAR. In the situation of clinical trials, this parametric modeling approach is relevant to ask what is the difference between two treatments on patients who drop out from the trial with similar reasons. Interested readers may find more discussion about pattern mixtures models in the mixed-effects models in Verbeke and Molenberghs (2000).

13.5.2 A Semiparametric Pattern Mixture Model

This section presents a semiparametric version of pattern mixture model to handle informative dropouts, proposed by Sun and Song (2001). This example is motivated from the schizophrenia clinical trial described in Section 1.3.5. The main objective of the trial is to evaluate a new treatment against a standard treatment for the disease.

Suppose a longitudinal study prespecifies a fixed number of visits, say n, at times $t_1 < \ldots < t_n$ on K subjects. For subjet i, let $Y_i(t)$ denote the measurement process and R_i be the time of dropout due to lack of treatment effect, and C_i the time of withdrawal for reasons unrelated to the measurement process. In practice, $Y_i(t)$ is observed only if $t \leq \min\{R_i, C_i\}$ and for the withdrawal (censoring) time, one observes $\tilde{R}_i = \min\{R_i, C_i\}$ and an indicator $\xi_i = I\{R_i \leq C_i\}$.

The dropout time R_i can take possible values in $\{t_1, \ldots, t_n\}$. Let $\mathbf{x}_i = (1, 0, \ldots, 0),\ (0, 1, \ldots, 0),\ \ldots,\ (0, \ldots, 0)$ be $(p-1)$-dimensional vectors of treatment indicators for subjects in treatment groups $1, 2, \ldots, p$, respectively.

To compare treatment groups, Hogan and Laird (1997) suggested to stratify subjects according to their dropout patterns (or dropout times). Therefore, all subjects can be classified into $n+1$ groups according to their values of R_i. This gives groups of subjects with $R_i = t_l$, $l = 1, \ldots, n$, and the group of subjects who complete the study or withdraw for reasons unrelated to treatment effect ($\xi_i = 0$), respectively. For convenience, for subjects who finish the study, define $R_i = t_{n+1} = \infty$. For subjects with $R_i = t_l$ and $\xi_i = 1$, define vector $\{\delta_{i,k},\ k = 1, ..., n+1\}$ as its classification group indicators (0 or 1) such that $\delta_{il} = 1$ and $\sum_{k=1}^{n+1} \delta_{ik} = 1$. For subjects with $\xi_i = 0$ and who finish the study, define $\delta_{ik} = 0$, $k = 1, ..., n$, and $\delta_{i,n+1} = 1$.

Let $Y_i = \{Y_i(t_1), ..., Y_i(t_{k_i})\}$ denote the observed values of $Y_i(t)$, $1 \leq k_i \leq n$. Then the observed data consist of $\{(Y_i, \mathbf{x}_i, \tilde{R}_i, \xi_i, \{\delta_{ik}, k = 1, \ldots, n+1\})\,;\ i = 1, \ldots, K\}$. Although this idea is applicable to general missing data patterns, the presentation below focuses on the pattern of dropout only. Let $\omega_{il} = P\{R_i = t_l\}$, $l = 1, \ldots, n$, and $\omega_{i\,n+1} = 1 - \sum_{l=1}^{n} \omega_{il}$ (the probability that a subject completes the study), $i = 1, \ldots, K$. Here the ω_{il} are assumed to be identical for the subjects within the same treatment group, but could differ for the subjects in different treatment groups. In addition, assume that conditional on R_i, the missing mechanism of the response is MAR.

The first objective is to test for the hypothesis H_0: there is no treatment effect. A natural way to test this H_0 is to estimate the underlying treatment-specific process of the clinical outcomes and then to compare the estimated processes. Another method is to summarize each subject's outcomes with a single summary statistic and then to compare the summary statistics. Sun and Song (2001) proposed a Wilcoxon-type statistic:

$$U = \frac{1}{K^{3/2}} \sum_{l=1}^{n+1} \sum\sum_{i<j} \hat{\omega}_{il}\hat{\omega}_{jl}\, \delta_{il}\delta_{jl}\, (\mathbf{x}_i - \mathbf{x}_j) \sum_{u=1}^{k_i} \sum_{v=1}^{k_j} [\, Y_i(t_u) - Y_j(t_v)\,] \tag{13.12}$$

where the $\hat{\omega}_{il}$ are the treatment-specific product-limit estimators of the corresponding ω_{il} based on the time-to-event data $\{(\tilde{R}_i, \xi_i,);\ i = 1, \dots, K\}$, which is well known in the survival analysis (see for example Klein and Moeschberger, 1997, Section 4.2).

Let $S_l = \{i : \delta_{il} = 1\}$, $l = 1, \dots, n+1$, and $\bar{Y}_i = \sum_{u=1}^{k_i} Y_i(t_u)$, $i = 1, \dots, K$. Then the statistic in (13.12) can be rewritten as

$$U = \frac{1}{K^{3/2}} \sum_{l=1}^{n+1} \sum\sum_{i<j \in S_l} \hat{\omega}_{il}\hat{\omega}_{jl}(\mathbf{x}_i - \mathbf{x}_j)(k_j \bar{Y}_i - k_i \bar{Y}_j), \tag{13.13}$$

which is a weighted summation of treatment differences over different withdrawal groups. Note that the estimated weights $\hat{\omega}_{il}$ are used in U to take into account the possible differences of withdrawal distributions among different treatment groups.

It is easy to show that under the null hypothesis H_0, U has expectation zero. Also, under H_0, U is asymptotically equivalent to the U-statistic $U^* = K^{-1/2} \sum_{i=1}^{K} \mathbf{a}_i$, where

$$\mathbf{a}_i = \sum_{l=1}^{n+1} \omega_{il}\, \delta_{il}\, \bar{Y}_i\, (\, \bar{k}_l\, \mathbf{x}_i - \bar{\mathbf{x}}_l\,)\,, \tag{13.14}$$

where,

$$\bar{k}_l = \frac{1}{K} \sum_{j=1}^{K} \omega_{jl}\, \delta_{jl}\, k_j$$

and

$$\bar{\mathbf{x}}_l = \frac{1}{K} \sum_{j=1}^{K} \omega_{jl}\, \delta_{jl}\, k_j\, \mathbf{x}_j\ .$$

It follows that for large K, the distribution of U can be approximated by a multivariate normal distribution with mean zero and covariance matrix $\hat{\Sigma} = K^{-1} \sum_{i=1}^{K} \hat{\mathbf{a}}_i\, \hat{\mathbf{a}}_i^T$, where $\hat{\mathbf{a}}_i$ is the $\mathbf{a}_i$ in (13.14) with the replacement of the ω_{jl} by their estimates. Thus, the null H_0 is rejected at significance level α if the observed Wald statistic $W = U^T\, \hat{\Sigma}^{-1}\, U > \chi^2_{p-1}(\alpha)$, where $\hat{\Sigma}^{-1}$ denotes the inverse of $\hat{\Sigma}$ and $\chi^2_{p-1}(\alpha)$ the level α critical value of the χ^2 distribution with $p - 1$ degrees of freedom.

The second objective is, after the null H_0 is rejected, to estimate the magnitude of differenced treatment effects. To proceed, Sun and Song (2001) assumed that given $\mathbf{x}_i$ and R_i, the mean function of the underlying measurement process for subject i follows a semi-parametric model of the from

$$\mu_i(t) = \mathrm{E}[\,Y_i(t)\,|\,\mathbf{x}_i, R_i = t_l\,] = \mu_0^{(l)}(t)\,\exp\{\,\mathbf{x}_i^T\boldsymbol{\beta}\,\}\,, \tag{13.15}$$

$i = 1, \ldots, K$, $l = 1, \ldots, n+1$. Here $\mu_0^{(l)}(t)$ denotes the underlying baseline mean function for all individuals with $R_i = t_l$ and is assumed to be a continuous and completely unspecified function, and $\boldsymbol{\beta}$ is a $(p-1)$ dimensional vector of regression parameters representing treatment effects on the mean function of the $Y_i(t)$.

Model (13.15) can be thought of essentially as a semiparametric pattern mixture model or as a stratified marginal model. It implies that all subjects having the same dropout time (or the same dropout pattern) share the same underlying mean function and that treatments have multiplicative effects on the conditional mean functions of the underlying measurement process. Model (13.15) also states that the treatment effect does not differ across different dropout groups. It should be noted that here $\boldsymbol{\beta}$ represents conditional treatment effects; that is, it gives the effect of treatments on subjects who remain in the study for the same period of time. In other words, it models treatment effects separately according to dropout patterns. As pointed out by Shih and Quan (1997), in clinical trials, the conditional effect could be more relevant than the unconditional effect. For example, a medical doctor may be more interested in answers to the questions such that what would be the difference between two treatments if a patient took them for the same period of time.

If the baseline mean functions $\mu_0^{(l)}$ are identical for different dropout patterns or the dropout probabilities are the same for all subjects, then the regression parameters $\boldsymbol{\beta}$ also represent unconditional marginal treatment effects. This follows since

$$\mathrm{E}\,[\,Y_i(t)\,|\,\mathbf{x}_i\,] = \exp\{\,\mathbf{x}_i^T\boldsymbol{\beta}^T\,\}\sum_{l=1}^{n+1}\mu_0^{(l)}(t)\,\omega_{il}\,,$$

which can be used to estimate the unconditional marginal treatment effect.

An estimating function similar to GEE is proposed to estimate $\boldsymbol{\beta}$. That is,

$$\begin{aligned} U(\boldsymbol{\beta}) = {} & \frac{1}{K^{3/2}}\sum_{l=1}^{n+1}\sum\sum\nolimits_{i<j}\hat{\omega}_{il}\hat{\omega}_{jl}\,\delta_{il}\delta_{jl}\,(\mathbf{x}_i - \mathbf{x}_j)\,\times \\ & \sum_{u=1}^{k_i}\sum_{v=1}^{k_j}\left[Y_i(t_u)\,e^{-\mathbf{x}_i^T\boldsymbol{\beta}} - Y_j(t_v)\,e^{-\mathbf{x}_j^T\boldsymbol{\beta}}\right]. \end{aligned}$$

This inference function $U(\boldsymbol{\beta})$ is unbiased, namely $\mathrm{E}\{U(\boldsymbol{\beta})\} = 0$. First note that under model (13.15), $\mathrm{E}\left[\exp\{-\mathbf{x}_i^T\boldsymbol{\beta}\}\,Y_i(t)\,|\,\mathbf{x}_i, R_i = t_l\right]$ is independent of index i. It follows that

$$\mathrm{E}[U(\boldsymbol{\beta})] = \frac{1}{K^{3/2}} \sum_{l=1}^{n+1} \sum\sum\nolimits_{i<j} \mathrm{E}\{\hat{\omega}_{il}\hat{\omega}_{jl}\, \delta_{il}\delta_{jl}\,(\mathbf{x}_i - \mathbf{x}_j) \times$$
$$\mathrm{E}\left[\sum_{u=1}^{k_i}\sum_{v=1}^{k_j}[Y_i(t_u)\, e^{-\mathbf{x}_i^T\boldsymbol{\beta}} - Y_j(t_v)\, e^{-\mathbf{x}_j^T\boldsymbol{\beta}}] \mid \text{all } \mathbf{x}_i, \text{ all } R_i\right]\},$$

which is equal to zero under the model (13.15). An estimate of $\boldsymbol{\beta}$, say $\hat{\boldsymbol{\beta}}$, is defined as the solution of equation $U(\boldsymbol{\beta}) = 0$. A main advantage of this estimating equation lies in the fact that the equation does not involve the underlying mean functions $\mu_0^{(l)}(t)$. It follows from the standard theory of estimating functions in Chapter 3 that $\hat{\boldsymbol{\beta}}$ is a consistent estimate of $\boldsymbol{\beta}$. Moreover, for large K, the distribution of $\sqrt{K}\,(\hat{\boldsymbol{\beta}} - \boldsymbol{\beta})$ can be approximated by a multivariate normal distribution with mean zero and observed Godambe Information covariance matrix given by

$$\mathbf{j}_o(\hat{\boldsymbol{\beta}}) = K\,\mathbf{S}_o^{-1}(\hat{\boldsymbol{\beta}})\,\hat{\mathbf{V}}_o(\hat{\boldsymbol{\beta}})\,\mathbf{S}_o^{-T}(\hat{\boldsymbol{\beta}}),$$

where $\mathbf{S}_o$ is the observed sensitivity matrix $\mathbf{S}_o(\boldsymbol{\beta}) = -\,\partial U(\boldsymbol{\beta})/\partial\boldsymbol{\beta}$, and $\mathbf{V}_o$ is the observed variability matrix given by

$$\hat{\mathbf{V}}_o(\boldsymbol{\beta}) = K^{-1}\sum_{i=1}^{K} \exp(-2\mathbf{x}_i^T\boldsymbol{\beta})\,\hat{\mathbf{a}}_i\,\hat{\mathbf{a}}_i^T.$$

Given $\hat{\boldsymbol{\beta}}$, a natural estimator of the baseline mean function $\mu_0^{(l)}(t)$ is given by

$$\hat{\mu}_0^{(l)}(t_k) = \frac{\sum_{i=1}^{K} \delta_{il}\, I(t_k \le t_{k_i})\, Y_i(t_k)\,/\,\exp(\mathbf{x}_i^T\hat{\boldsymbol{\beta}})}{\sum_{i=1}^{K} \delta_{il}\, I(t_k \le t_{k_i})} \tag{13.16}$$

at time t_k, and by the straight line between $\hat{\mu}_0^{(l)}(t_{k-1})$ and $\hat{\mu}_0^{(l)}(t_k)$ for $t_{k-1} < t \le t_k$, where $t_0 = 0$, $k = 1, \ldots, n$.

Example 13.7 (Analysis of Schizophrenia Trial Data).

This example continues the analysis of the schizophrenics data given in Example 13.1, where the dropout process has been found to be nonignorable. Now the above semiparametric pattern mixture model is applied to analyze the data and draw inference on the treatment effects.

The trial was designed to compare a standard anti-psychotic medication (ST) to a new one (NT), of which three doses (low, medium, and high) were administered in the trial. The variable of interest, BPRS, was assessed at week zero (randomization) and at weeks 1, 2, 3, 4, and 6 and ranges from 0 to 108 with higher scores indicating more severe symptoms. Thus, $n = 6$ and $t_1 = 0$, $t_2 = 1$, $t_3 = 2$, $t_4 = 3$, $t_5 = 4$, $t_6 = 6$. At each of these times, there were patient dropouts and the study protocol called for BPRS assessed at the time of dropout. The focus of this example is on the comparison of two treatment arms, ST (63 patients) and NT with medium dose (61 patients). Here only medium dose is considered in the analysis just for simplicity. During the study, patients who did not complete the trial were categorized into three dropout types, which are withdrawal due to lack of treatment effect, withdrawal due to adverse experience, and withdrawal due to other reasons.

Table 13.9. Estimates for probability of dropout at a given day, by treatment group.

Treatment	Day						
	0	1	2	3	4	6	∞ (Finish)
New ($\hat{\omega}_1$)	.016	.000	.037	.021	.022	.043	.861
Standard ($\hat{\omega}_0$)	.000	.000	.035	.057	.062	.067	.779

For the i-th patient, let $Y_i(t)$ denote his or her BPRS at week t and R_i the time to dropout due to lack of treatment effect. Other withdrawal times are considered as censoring times C_i. Define $T_i = 0$ if patient i received ST and 1 if patient i received NT. For convenience, let $Z_0 = \{i : x_i = 0\}$, the set of indices for patients receiving ST, and $Z_1 = \{i : x_i = 1\}$, the set of indices for patients receiving NT. Let $\omega_l^{(0)}$ and $\omega_l^{(1)}$ denote ω_{il} for patients in Z_0 and Z_1, respectively. Table 13.9 gives the estimates of $\omega_l^{(0)}$ and $\omega_l^{(1)}$, which were obtained by differencing the Kaplan-Meier estimates of the probabilities that patients remained on the study at a given week by treatment group.

The testing procedure gave $U = 4.8005$ and test statistic $W = 0.0007$, with a p-value of 0.98 according to the χ_1^2 distribution with 1 degree of freedom. This suggests that there is no significant difference globally between the standard and new treatments. The result is similar to those given in Hogan and Laird (1997), who tested the treatment difference by comparing the mean BPRS of the two arms.

In estimation of 1-dimensional β, note that with only two treatment arms, the estimate of β yields a closed form expression given by

$$\hat{\beta} = -\log\left(-\frac{\sum_{i\in Z_0} \hat{a}_i}{\sum_{i\in Z_1} \hat{a}_i}\right)$$

with the standard error equal to $\sqrt{K}\sigma(\hat{\beta})/|A(\hat{\beta})|$, where

$$A(\hat{\beta}) = -\frac{1}{\sqrt{K}} \sum_{i \in Z_0} \hat{a}_i,$$

$$\sigma^2(\hat{\beta}) = \frac{1}{K} \left\{ \sum_{i \in Z_0} \hat{a}_i^2 + e^{-2\hat{\beta}} \sum_{i \in Z_1} \hat{a}_i^2 \right\}.$$

A simple calculation leads to $\hat{\beta} = 0.0055$ with the corresponding standard error equal to 2.4162. This suggests again that for two patients taking ST and NT, respectively, their mean BPRS would not significantly differ if they stayed on the treatments for the same period of time.

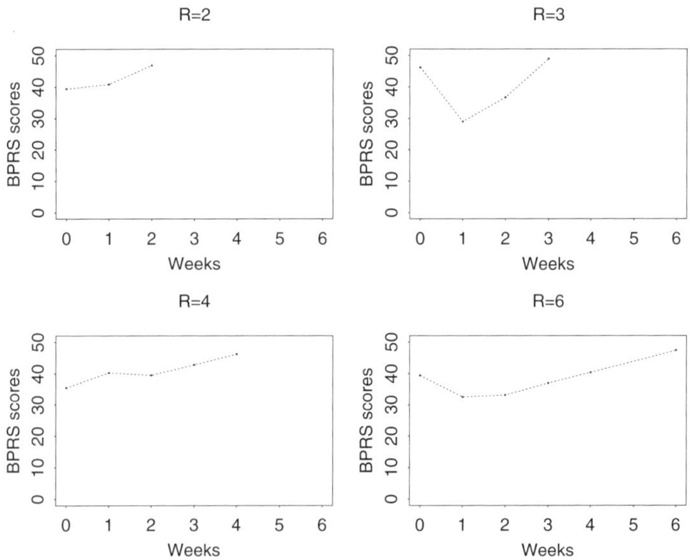

Fig. 13.4. The estimated baseline mean functions of BPRS scores by time of informative dropout.

To estimate the underlying baseline mean functions specified by model (13.15), the estimates given by (13.16) were obtained and are displayed in Figure 13.4 for each of four dropout patterns. These four baseline mean BPRS functions all indicate that subjects' symptoms on average got worse prior to their dropouts.

In contrast, Figure 13.5 shows the estimated underlying baseline mean function for subjects with $\delta_{in+1} = 1$, the completers and non-informative dropouts. It is easy to see that subjects' symptoms on average improved or stablized over the coure of the trial. In summary, Both Figures 13.4 and 13.5 clearly indicate an NMAR mechanism; that is, early dropout due to treatment effect is related to higher BPRS.

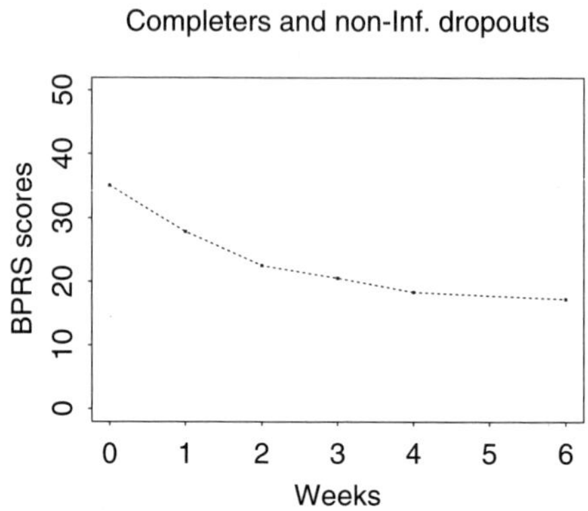

Fig. 13.5. The estimated baseline mean functions for completers and non-informative dropouts.

References

Abramowitz, M. and Stegun, I. A. (1970). *Handbook of Mathematical Functions*, Dover, New York.

Accardi, L., Cabrera, J. and Watson, G. S. (1987). Some stationary Markov processes in discrete time for unit vectors, *Metron* **45**, 115–133.

Aerts, M., Geys, H., Molenberghs, G. and Ryan, L. M. (2002). *Topics in Modelling of Clustered Data*, Chapman & Hall, London.

Albert, P. S. (1999). Tutorial in biostatistics: Longitudinal data analysis (repeated measures) in clinical trials, *Statistics in Medicine* **18**, 1707–1732.

Albert, P. S. and Waclawiw, M. A. (1998). A two-state markov chain for heterogeneous transitional data: a quasi-likelihood approach, *Statistics in Medicine* **17**, 1481–1493.

Anderson, J. A. and Pemberton, J. D. (1985). The grouped continuous model for multivariate ordered categorical variables and covariate adjustment, *Biometrics* **41**, 875–885.

Arnold, S. F. (1990). *Mathematical Statistics*, Prentice-Hall, New Jersey.

Artes, R., Paula, G. A. and Ranvaud, R. (2000). Analysis of circular longitudinal data based on generalized estimating equations, *Australian & New Zealand Journal of Statistics* **42**, 347–358.

Bahadur, R. R. (1961). A representation of the joint distribution of responses on n dichotomous items, *Studies in Item Analysis and Prediction, Stanford Mathematical Studies in the Social Sciences (H. Solomon, ed.)* **VI**, 158–168.

Barndorff-Nielsen, O. E. and Cox, D. R. (1989). *Asymptotic Techniques for Use in Statistics*, Chapman & Hall, London.

Barnhart, H. X. and Williamson, J. M. (1998). Goodness-of-fit tests for GEE modelling with binary responses, *Biometrics* **54**, 720–729.

Beal, S. L. and Sheiner, L. B. (1988). Heteroscedastic nonlinear regression, *Technometrics* **30**, 327–338.

Bishop, Y. M. M., Fienberg, S. E. and Holland, P. W. (1975). *Discrete Multivariate Analysis: Theory and Practice*, MIT Press, Cambridge.

Booth, J. G. and Hobert, J. P. (1999). Maximizing generalized linear mixed model likelihoods with an automated Monte Carlo EM algorithm, *Journal of the Royal Statistical Society, Series B* **61**, 265–285.

Borgioli, C., Martelli, M., Porri, F., D'Elia, A., Marchetti, G. M. and Scapini, F. (1999). Orientation in *talitrus saltator* (montagu): trends in intrapopulation variability related to environmental and intrinsic factors, *Journal of Experimental Marine Biology and Ecology* **238**, 29–47.

Box, G. E. P. and Jenkins, G. M. (1976). *Time Series Analysis: Forecasting and Control*, Holden-Day, San Francisco.

Breckling, J. (1989). *Analysis of Directional Time Series: Application to Wind Speed and Direction (Lecture Notes in Statistics 61)*, Springer-Verlag, Berlin.

Breslow, N. E. and Clayton, D. G. (1993). Approximate inference in generalized linear mixed models, *Journal of the American Statistical Association* **88**, 9–25.

Breslow, N. E. and Day, N. E. (1980). *Statistical Methods in Cancer Research*, IARC Scientific Publications No.32, Lyon.

Breslow, N. E. and Lin, X. (1995). Bias correction in generalised linear mixed models with a single component of dispersion, *Biometrika* **82**, 81–91.

Brockwell, P. J. and Davis, R. A. (1991). *Time Series: Theory and Methods*, 2nd edn, Springer, New York.

Brockwell, P. J. and Davis, R. A. (1996). *Introduction to Time Series and Forecasting*, Springer, New York.

Brooks, S. and Gelman, A. (1998). General methods for monitoring convergence of iterative simulations, *Journal of Computational and Graphical Statistics* **7**, 434–455.

Butler, S. M. and Louis, T. A. (1992). Random-effects models with non-parametric priors, *Statistics in Medicine* **11**, 1981–2000.

Carey, V., Zeger, S. L. and Diggle, P. J. (1993). Modelling multivariate binary data with alternating logistic regression, *Biometrika* **80**, 517–526.

Carlin, B. P. and Polson, N. G. (1992). Monte Carlo Bayesian methods for discrete regression models and categorical time series, *Bayesian Statistics* **4**, 577–586.

Carlin, B. P., Polson, N. G. and Stoffer, D. S. (1992). A Monte Carlo approach to nonnormal and nonlinear state-space modeling, *Journal of the American Statistical Association* **87**, 493–500.

Chaganty, N. R. (1997). An alternative approach to the analysis of longitudinal data via generalized estimating equations, *Journal of Statistical Planning and Inference* **63**, 39–54.

Chaganty, N. R. and Joe, H. (2004). Efficiency of generalized estimating equations for binary responses, *Journal of the Royal Statistical Society, Series B* **66**, 851–860.

Chamberlain, G. (1987). Asymptotic efficiency in estimation with conditional moment restrictions, *Journal of Econometrics* **34**, 305–334.

Chan, J. S. K., Kuk, A. Y. C., Bell, J. and McGilchrist, C. (1998). The analysis of methadone clinic data using marginal and conditional logistic models with mixture or random effects, *Australian & New Zealand Journal of Statistics* **40**, 1–10.

Chan, K. S. and Ledolter, J. (1995). Monte Carlo EM estimation for time series models involving observations, *Journal of the American Statistical Association* **90**, 242–252.

Chen, H. Y. and Little, R. J. A. (1995). A test of missing completely at random for generalised estimating equations with missing data, *Biometrika* **86**, 1–13.

Coles, S. (1998). Inference for circular distributions and processes, *Statistics and Computing* **8**, 105–113.

Cook, R. D. (1986). Assessment of local influence, *Journal of the Royal Statistical Society, Series B* **48**, 133–169.

Cook, R. J. and Ng, E. T. M. (1997). A logistic-bivariate normal model for overdispersed two-state markov processes, *Biometrics* **53**, 358–364.

Cook, R. J., Zeng, L. and Yi, G. Y. (2004). Marginal analysis of incomplete longitudinal binary data: a cautionary note on LOCF imputation, *Biometrics* **60**, 820–828.

Cox, D. R. (1970). *The Analysis of Binary Data*, Chapman & Hall, London.

Cox, D. R. (1981). Statistical analysis of time series, some recent developments, *Scandinavian Journal of Statistics* **8**, 93–115.

Crowder, M. (1986). On consistency and inconsistency of estimating equations, *Econometric Theory* **3**, 305–330.

Crowder, M. (1987). On linear and quadratic estimating function, *Biometrika* **74**, 591–597.

Crowder, M. (1995). On the use of a working correlation matrix in using generalized linear models for repeated measures, *Biometrika* **82**, 407–410.

Czado, C. and Song, P. X.-K. (2007). State space mixed models for longitudinal observations with binary and binomial responses, *Statistical Papers* **48**, to appear.

Davidian, M. and Giltinan, D. M. (1995). *Nonlinear Models for Repeated Measures*, Chapman & Hall, London.

Davis, C. S. (2002). *Statistical Methods for the Analysis of Repeated Measurements*, Springer, New York.

Davis, P. J. and Rabinowitz, P. (1984). *Methods of Numerical Integration*, Academic Press, New York.

de Jong, P. (1991). The diffuse Kalman filter, *Annals of Statistics* **2**, 1073–1083.

de Jong, P. and Shephard, N. (1995). The simulation smoother for time series models, *Biometrika* **82**, 339–350.

D'Elia, A., Borgioli, C. and Scapini, F. (2001). Orientation of sandhoppers under natural conditions in repeated trials: an analysis using longitudinal directional data, *Estuarine, Coastal and Shelf Science* **53**, 839–847.

Dempster, A. P., Laird, N. M. and Rubin, D. B. (1977). Maximum-likelihood from incomplete data via the EM algorithm (with discussion), *Journal of the Royal Statistical Society, Series B* **39**, 1–38.

Devroye, L. (1984). A simple algorithm for generating random variates with a log-concave density, *Computing* **33**, 247–257.

Diggle, P. J. (1989). Testing for random dropouts in repeated measurement data, *Biometrics* **45**, 1255–1258.

Diggle, P. J. (1990). *Time Series: An Biostatistical Introduction*, Clarendon Press, Oxford.

Diggle, P. J. and Kenward, M. G. (1994). Informative dropout in longitudinal data analysis (with discussion), *Applied Statistics* **43**, 49–93.

Diggle, P. J., Heagerty, P., Liang, K.-Y. and Zeger, S. L. (2002). *Analysis of Longitudinal Data*, 2nd edn, Oxford University Press, New York.

Drasgow, F. (1988). Polychoric and polyserial correlations, *in* L. Kotz and N. Johnson (eds), *Encyclopedia of Statistical Sciences*, Vol. 7, Wiley, New York, pp. 69–74.

Durbin, J. (1960). Estimation of parameters in time-series regression models, *Journal of the Royal Statistical Society, Series B* **22**, 139–153.

Durbin, J. and Koopman, S. J. (1997). Monte Carlo maximum likelihood estimation for non-Gaussian state space models, *Biometrika* **84**, 669–684.

Efron, B. and Tibshirani, R. J. (1993). *An Introduction to the Bootstrap*, Chapman & Hall, London.

Fahrmeir, L. (1992). Posterior mode estimation by extended Kalman filtering for multivariate dynamic generalized linear models, *Journal of the American Statistical Association* **87**, 501–509.

Fahrmeir, L. and Kaufmann, H. (1987). Regression models for nonstationary categorical time series, *Journal of time Series Analysis* **8**, 147–160.

Fahrmeir, L. and Tutz, G. (1994). *Multivariate Statistical Modelling Based on Generalized Linear Models*, Springer, Berlin.

Fan, J. and Gijbels, I. (1996). *Local Polynomial Modeling and Its Applications*, Chapman & Hall, London.

Ferguson, T. S. (1958). A method of generating best asymptotically normal estimates with application to the estimation of bacterial densities, *Annals of Mathematical Statistics* **29**, 1046–1062.

Fisher, N. I. (1993). *Statistical Analysis of Circular Data*, University Press, Cambridge.

Fisher, N. I. and Lee, A. J. (1992). Regression models for an angular response, *Biometrics* **48**, 665–677.

Fisher, N. I. and Lee, A. J. (1994). Time series analysis for circular data, *Journal of the Royal Statistical Society, Series B* **56**, 327–339.

Fisher, R. A. (1935). The fiducial argument in statistical inference, *Annals of Eugenics* **6**, 391–398.

Fitzmaurice, G. M. and Laird, N. M. (1995). Regression models for a bivariate discrete and continuous outcome with clustering, *Journal of the American Statistical Association* **90**, 845–852.

Fitzmaurice, G. M. and Laird, N. M. (2000). Generalized linear mixture models for handling nonignorable dropouts in longitudinal studies, *Biostatistics* **1**, 141–156.

Fitzmaurice, G. M., Laird, N. M. and Rotnitzky, A. (1993). Regression models for discrete longitudinal responses, *Statistical Science* **8**, 284–309.

Fitzmaurice, G. M., Laird, N. M. and Ware, J. H. (2004). *Applied Longitudinal Analysis*, Wiley, New York.

Fitzmaurice, G. M., Laird, N. M. and Zahner, E. P. (1996). Multivariate logistic models for incomplete binary responses, *Journal of the American Statistical Association* **91**, 99–108.

Follmann, D. A. and Wu, M. C. (1995). An approximate generalized linear model with random-effects for informative missing data, *Biometrics* **51**, 151–168.

Gelfand, A. E. and Smith, A. F. M. (1990). Sampling based approaches to calculating marginal densities, *Journal of the American Statistical Association* **85**, 398–409.

Gelfand, A. E., Sahu, S. K. and Carlin, B. P. (1995). Efficient parameterizations for normal linear mixed models, *Biometrika* **82**, 479–488.

Gelfand, A. E., Sahu, S. K. and Carlin, B. P. (1996). Efficient parametrizations for generalized linear mixed models, *Bayesian Statistics* **5**, 165–180.

Gelman, A. and Rubin, D. B. (1992). Inference from iterative simulation using multiple sequences (with discussion), *Statistical Science* **7**, 457–511.

Geman, S. and Geman, D. (1984). Stochastic relaxation, Gibbs distributions and the Bayesian restoration of images, *IEEE Transactions on Pattern Analysis and Machine Intelligence* **6**, 721–741.

Genz, A. (1992). Numerical computation of multivariate normal probabilities, *Journal of Computational and Graphical Statistics* **1**, 141–149.

Geweke, J. (1992). Evaluating the accuracy of sampling-based approaches to the calculation of posterior moments, *in* A. D. J.O. Berger, J.M. Bernardo and A. Smith (eds), *Bayesian Statistics 4*, Oxford University Press, Oxford, pp. 169–194.

Geyer, C. J. and Thompson, E. A. (1992). Constrained Monte Carlo maximum likelihood for dependent data, *Journal of the Royal Statistical Society, Series B* **54**, 657–699.

Gilks, W. R., Richardson, S. and Spiegelhalter, D. J. (1996). *Markov Chain Monte Carlo in Practice*, Chapman & Hall, New York.

Godambe, V. P. (1960). An optimum property of regular maximum likelihood estimation, *Annals of Mathematical Statistics* **31**, 1208–1212.

Godambe, V. P. (1976). conditional likelihood and unconditional optimum estimating equations, *Biometrika* **63**, 277–284.

Godambe, V. P. (1991). *Estimating Functions*, Oxford University Press, Oxford.

Godambe, V. P. and Thompson, M. E. (1974). Estimating equations in the presence of a nuisance parameter, *Annals of Statistics* **2**, 568–571.

Godambe, V. P. and Thompson, M. E. (1978). Some aspects of the theory of estimating equations, *Journal of Statistical Planning and Inference* **2**, 95–104.

Green, P. J. (1987). Penalized likelihood for general semi-parametric regression models, *International Statistical Review* **55**, 245–259.

Hall, A. R. (2005). *Generalized Method of Moments*, Oxford University Press, Oxford.

Hansen, L. (1982). Large sample properties of generalized methods of moments estimators, *Econometrica* **50**, 1029–1055.

Hardin, J. W. and Hilbe, J. M. (2003). *Generalized Estimating Equations*, Chapman & Hall, Boca Raton, FL.

Harris, B. (1988). Tetrachoric correlation coefficient, *in* L. Kotz and N. Johnson (eds), *Encyclopedia of Statistical Sciences*, Vol. 9, Wiley, New York, pp. 223–225.

Harvey, A. C. (1981). *Time Series Models*, Allan, Oxford.

Harvey, A. C. (1990). *Forecasting, Structural Time Series Models and the Kalman Filter*, Cambridge University Press, Cambridge.

Harville, D. A. (1977). Maximum likelihood approaches to variance component estimation and to related problems, *Journal of the American Statistical Association* **72**, 320–340.

Heagerty, P. J. and Zeger, S. L. (1996). Marginal regression models for clustered ordinal measurements, *Journal of the American Statistical Association* **91**, 1024–1036.

Heagerty, P. J. and Zeger, S. L. (1998). Lorelogram: a regression approach to exploring dependence in longitudinal categorical responses, *Journal of the American Statistical Association* **93**, 150–162.

Hedeker, D. and Mermelstein, R. J. (2000). Analysis of longitudinal substance use outcomes using ordinal random-effects regression models, *Journal of Addiction* **3**, 381–394.

Heidelberger, P. and Welch, P. (1983). Simulation run length control in the presence of an initial transient, *Operations Research* **31**, 1109–1144.

Heyde, C. C. (1997). *Quasi-likelihood and Its Application*, Springer-Verlag, New York.

Hogan, J. W. and Laird, N. M. (1997). Mixture models for the joint distribution of repeated measures and event times, *Statistics in Medicine* **16**, 239–258.

Horvitz, D. G. and Thompson, D. J. (1952). A generalization of sampling without replacement from a finite universe, *Journal of the American Statistical Association* **47**, 663–685.

Ibrahim, J. G., Chen, M.-H. and Lipsitz, S. R. (2001). Missing responses in generalized linear mixed models when the missing data mechanism is nonignorable, *Biometrika* **88**, 551–564.

Ishwaran, H. and Takahara, G. (2002). Independent and identically distributed Monte Carlo algorithms for semiparametric linear mixed models, *Journal of the American Statistical Association* **97**, 1154–1166.

Joe, H. (1995). Approximations to multivariate normal rectangle probabilities based on conditional expectations, *Journal of the American Statistical Association* **90**, 957–964.

Joe, H. (1997). *Multivariate Models and Dependence Concepts*, Chapman & Hall, London.

Jones, B. and Kenward, M. G. (1989). *Design and Analysis of Cross-Over Trials*, Chapman & Hall, London.

Jørgensen, B. (1987). Exponential dispersion models (with discussion), *Journal of the Royal Statistical Society, Series B* **49**, 127–162.

Jørgensen, B. (1997). *The Theory of Dispersion Models*, Chapman & Hall, London.

Jørgensen, B. and Knudsen, S. J. (2004). Parameter orthogonality and bias adjustment for estimating functions, *Scandinavian Journal of Statistics* **31**, 93–114.

Jørgensen, B. and Song, P. X.-K. (1998a). Stationary time series models with exponential dispersion model margins, *Journal of Applied Probability* **35**, 78–92.

Jørgensen, B. and Song, P. X.-K. (1998b). Stationary state space models for longitudinal data, *Research Report #9*, Department of Statistics, University of Southern Denmark.

Jørgensen, B., Lundbye-Christensen, S., Song, P. X.-K. and Sun, L. (1996a). State-space models for multivariate longitudinal data of mixed types, *Canadian Journal of Statistics* **24**, 385–402.

Jørgensen, B., Lundbye-Christensen, S., Song, P. X.-K. and Sun, L. (1996b). A longitudinal study of emergency room visits and air pollution for Prince George, British Columbia, *Statistics in Medicine* **15**, 823–836.

Jørgensen, B., Lundbye-Christensen, S., Song, P. X.-K. and Sun, L. (1999). A state-space models for multivariate longitudinal count data, *Biometrika* **86**, 169–181.

Jørgensen, B., Martinez, J. R. and Tsao, M. (1994). Asymptotic behaviour of the variance function, *Scandinavian Journal of Statistics* **21**, 223–243.

Kalbfleisch, J. D. and Lawless, J. F. (1985). The analysis of panel data under a markov assumption, *Journal of the American Statistical Association* **80**, 863–873.

Kalman, R. E. (1960). A new approach to linear filtering and prediction problems, *Journal of Basic Engineering* **82**, 34–45.

Kalman, R. E. and Bucy, R. S. (1961). New results in linear filtering and prediction theory, *Journal of Basic Engineering* **83**, 95–108.

Karim, M. R. and Zeger, L. R. (1988). GEE: a SAS macro for longitudinal data analysis, *Technical Report #674*, Department of Biostatistics, The Johns Hopkins University, Baltimore.

Kass, R. and Raftery, A. (1995). Bayes factors and model uncertainty, *Journal of the American Statistical Association* **90**, 773–795.

Kendall, M. G. and Gibbons, J. D. (1990). *Rank Correlation Methods*, Oxford University Press, New York.

Kimball, B. F. (1946). Sufficient statistical estimation functions for the parameters of the distribution of maximum values, *Annals of Mathematical Statistics* **17**, 299–309.

Kitagawa, G. (1987). Non-Gaussian state-space modelling of non stationary time series (with discussion), *Journal of the American Statistical Association* **82**, 1032–1063.

Klein, J. P. and Moeschberger, M. L. (1997). *Survival Analysis: Techniques for Censored and Truncated Data*, Springer, New York.

Knight, K., Leroux, B. G., Millar, J. and Petkau, A. (1989). Air pollution and human health: A study based on emergency room visits data from Prince George, British Columbia, *Technical Report #136*, Department of Statistics, University of British Columbia.

Kosorok, M. R. and Chao, W. H. (1996). The analysis of longitudinal ordinal response data in continuous time, *Journal of the American Statistical Association* **91**, 807–817.

Laird, N. M. and Ware, J. H. (1982). Random-effects models for longitudinal data, *Biometrics* **38**, 963–974.

Lambert, M. L., Lamble, S. J. and Girard, R. W. (1987). *1986 Annual Air Quality Report for Prince George*, Waste Management Branch, Ministry of Environment & Parks of British Columbia, Prince George, British Columbia.

Lange, K. (1999). *Numerical Analysis for Statisticians*, Springer-Verlag, New York.

Lee, M. J. (1996). *Methods of Moments and Semiparametric Econometrics for Limited Dependent Variable Models*, Springer, New York.

Lee, Y. and Nelder, J. A. (1996). Hierarchical generalized linear models (with discussion), *Journal of the Royal Statistical Society, Series B* **58**, 619–678.

Lee, Y. and Nelder, J. A. (2004). Conditional and marginal models: Another view, *Statistical Science* **19**, 219–238.

Lewis, P. A. W., McKenzie, E. and Hugus, D. K. (1989). Gamma processes, *Communication in Statistics–Stochastic Models* **5(1)**, 1–30.

Liang, K.-Y. and Self, S. G. (1996). On the asymptotic behaviour of the pseudo-likelihood ratio test statistic, *Journal of the Royal Statistical Society, Series B* **58**, 785–796.

Liang, K.-Y. and Zeger, S. L. (1986). Longitudinal data analysis using generalized linear models, *Biometrika* **73**, 13–22.

Liao, J. G. and Qaqish, B. F. (2005). Discussion of "maximization by parts in likelihood inference" by Song, Fan, and Kalbfleisch, *Journal of the American Statistical Association* **100**, 1160–1161.

Lin, X. and Breslow, N. E. (1996). Bias correction in generalized linear mixed models with multiple components of dispersion, *Journal of the American Statistical Association* **91**, 1007–1016.

Lindsay, B. G. (1988). Composite likelihood methods, *Contemporary Mathematics* **80**, 221–240.

Lindsay, B. G. and Qu, A. (2003). Inference functions and quadratic score tests, *Statistical Science* **18**, 394–410.

Lindsey, J. K. (1999). *Models for Repeated Measurements*, 2nd edn, Oxford University Press, Oxford.

Lindstrom, M. J. and Bates, D. M. (1988). Newton-Raphson and EM-algorithms for linear mixed-effects models for repeated-measures data, *Journal of the American Statistical Association* **83**, 1014–1022.

Lindstrom, M. J. and Bates, D. M. (1990). Nonlinear mixed effects models for repeated measures data, *Biometrics* **46**, 673–687.

Lipsitz, S. R. and Ibrahim, J. G. (2000). Estimation with correlated censored survival data with missing covariates, *Biostatistics* **1**, 315–327.

Little, R. J. A. (1993). Pattern-mixture models for multivariate incomplete data, *Journal of the American Statistical Association* **88**, 125–134.

Little, R. J. A. (1994). A class of pattern-mixture models for normal incomplete data, *Biometrika* **81**, 471–483.

Little, R. J. A. and Rubin, D. B. (2002). *Statistical Analysis With Missing Data*, 2nd edn, Wiley, New York.

Louis, T. A. (1982). Finding the observed information matrix when using the EM algorithm, *Journal of the Royal Statistical Society, Series B* **44**, 226–233.

Mardia, K. V. (1972). *Statistics of Directional Data*, Academic Press, London.

McCullagh, P. (1983). Quasi-likelihood functions, *Annals of Statistics* **11**, 59–67.

McCullagh, P. and Nelder, J. A. (1989). *Generalized Linear Models*, 2nd edn, Chapman & Hall, London.

McCulloch, C. E. (1997). Maximum likelihood algorithms for generalized linear mixed models, *Journal of the American Statistical Association* **92**, 162–170.

McCulloch, C. E. and Searle, S. R. (2001). *Generalized, Linear, and Mixed Models*, Wiley & Sons, New York.

McGilchrist, C. A. (1994). Estimation in generalized mixed models, *Journal of the Royal Statistical Society, Series B* **56**, 61–69.

McLeish, D. L. and Small, C. G. (1988). *The Theory and Applications of statistical Inference Functions*, Lecture Notes in Statistics 44, Springer-Verlag, New York.

Meyers, S. M., Ambler, J. S., Tan, M., Werner, J. C. and Huang, S. S. (1992). Variation of perfluorpropane disappearance after vitrectomy, *Retina* **12**, 359–363.

Minini, P. and Chavance, M. (2004). Sensitivity analysis of longitudinal binary data with non-monotone missing values, *Biostatics* **5**, 531–544.

Molenberghs, G., Kenward, M. G. and Lesaffre, E. (1997). The analysis of longitudinal ordinal data with nonrandom drop-out, *Biometrika* **84**, 33–44.

Muenz, L. R. and Rubinstein, L. V. (1985). Markov models for covariate dependence of binary sequence, *Biometrics* **43**, 863–871.

Neal, R. (1998). Suppressing random walks in Markov chain Monte Carlo using ordered over-relaxation, *in* M. Jordan (ed.), *Learning in Graphical Models*, Kluwer Academic Publishers, Dordrecht, pp. 205–230.

Neuhaus, J. M. and Kalbfleisch, J. D. (1998). Between- and within-cluster covariate effects in the analysis of clustered data, *Biometrics* **54**, 638–645.

Neuhaus, J. M., Hauck, W. W. and Kalbfleisch, J. D. (1992). The effects of mixture distribution misspecification when fitting mixed-effects logistic models, *Biometrika* **1992**, 755–762.

Neuhaus, J. M., Kalbfleisch, J. D. and Hauck, W. W. (1991). A comparison of cluster-specific and population-averaged approaches for analyzing correlated binary data, *International Statistics Review* **59**, 25–35.

Newey, W. K. and McFadden, D. (1994). Large sample estimation and hypothesis testing, *Handbook of Econometrics* **4**.

Neyman, J. (1949). Contribution to the theory of χ^2 test, *in* J. Neyman (ed.), *Berkeley Symposium in Mathematical Statistics and Probability*, University of California Press, Berkeley, CA, pp. 239–273.

Olkin, I. and Pratt, J. W. (1958). A multivariate Tchebycheff inequality, *Annals of Mathematical Statistics* **29**, 226–234.

Olsson, U. (1979). Maximum likelihood estimation of the polychoric correlation coefficient, *Psychometrika* **44**, 443–460.

Paik, M. C. (1992). Parametric variance function estimation for nonnormal repeated measurement data, *Biometrics* **48**, 19–30.

Paik, M. C. (1997). The generalized estimating equation approach when data are not missing completely at random, *Journal of the American Statistical Association* **92**, 1320–1329.

Pan, W. (2001). Model selection in estimating equations, *Biometrics* **57**, 529–534.

Pan, W. (2002). Goodness-of-fit tests for GEE with correlated binary data, *Scandinavian Journal of Statistics* **29**, 101–110.

Penrose, K., Nelson, A. and Fisher, A. (1985). Generalized body composition prediction equation for men using simple measurement techniques (abstract), *Medicine and Science in Sports and Exercise* **17**, 189.

Petkau, A. J. and White, R. A. (2003). Statistical approaches to assessing the effects of neutralizing antibodies: IFNbeta-1b in the pivotal trial of relapsing-remitting multiple sclerosis, *Neurology* **61(9 Suppl 5)**, 35–37.

Petkau, A. J., White, R. A., G. C. Ebers, A. T. R., Sibley, W. A., Lublin, F. D. and Paty, D. W. (2004). Longitudinal analysis of the effects of neutralizing antibodies on interferon beta-1b in relapsing-remitting multiple sclerosis, *Multiple Sclerosis* **10**, 126–138.

Pinheiro, J. C., Liu, C. and Wu, Y. N. (2001). Efficient algorithms for robust estimation in linear mixed-effects models using the multivariate t distribution, *Journal of Computational and Graphical Statistics* **10**, 249–276.

Preisser, J. S. and Qaqish, B. F. (1996). Deletion diagnostics for generalized estimating equations, *Biometrika* **83**, 551–562.

Prentice, R. L. (1988). Correlated binary regression with covariates specific to each binary observation, *Biometrics* **44**, 1033–1048.

Prentice, R. L. and Zhao, L. P. (1991). Estimating equations for parameters in means and covariances of multivariate discrete and continuous responses, *Biometrics* **47**, 825–839.

Qiu, Z., Song, P. X.-K. and Tan, M. (2002). Bayesian hierarchical analysis of multilevel repeated ordinal data using winbugs, *Journal of Biopharmaceutical Statistics* **12**, 121–135.

Qu, A. and Song, P. X.-K. (2002). Testing ignorable missingness in estimating equation, *Biometrika* **89**, 841–850.

Qu, A. and Song, P. X.-K. (2004). Assessing robustness of generalised estimating equations and quadratic inference functions, *Biometrika* **91**, 447–459.

Qu, A., Lindsay, B. G. and Li, B. (2000). Improving generalized estimating equations using quadratic inference functions, *Biometrics* **87**, 823–836.

Qu, Y. and Tan, M. (1998). Analysis of clustered ordinal data with subclusters via a Bayesian hierarchical model, *Communications in Statistics: Theory and Methods* **27**, 1461–1475.

Rao, C. R. (1973). *Linear Statistical Inference and Its Applications*, 2nd edn, Wiley, New York.

Raudenbush, S. W., Yang, M.-L. and Yosef, M. (2000). Maximum likelihood for generalized linear models with nested random effects via high-order, multivariate Laplace approximation, *Journal of Computational and Graphical Statistics* **9**, 141–157.

Rivest, L.-P. (1997). A decentred predictor for circular-circular regression, *Biometrika* **84**, 717–726.

Robert, C. P. (1995). Simulation of truncated normal variables, *Statistics and Computing* **5**, 121–125.

Robins, J. M., Rotnitzky, A. and Zhao, L. P. (1994). Estimation of regression coefficients when some regressors are not always observed, *Journal of the American Statistical Association* **89**, 846–866.

Robins, J. M., Rotnitzky, A. and Zhao, L. P. (1995). Analysis of semiparametric regression models for repeated outcomes under the presence of missing data, *Journal of the American Statistical Association* **90**, 106–121.

Rotnitzky, A. and Jewell, N. (1990). Hypothesis testing of regression parameters in semiparametric generalized linear models for cluster correlated data, *Biometrika* **90**, 485–497.

Rubin, D. B. (1976). Inference and missing data, *Biometrika* **63**, 581–592.

Rubin, D. B. (1987). *Multiple Imputation for Nonresponse in Surveys*, Wiley, New York.

Rubin, D. B., Stern, H. S. and Vehovar, V. (1994). Handling "don't know" survey responses: the case of the slovenian plebiscite, *Journal of the American Statistical Association* **90**, 822–828.

Ruppert, D. (2005). Discussion of "maximization by parts in likelihood inference" by Song, Fan, and Kalbfleisch, *Journal of the American Statistical Association* **100**, 1161–1163.

Schafer, J. L. (1997). *Analysis of Incomplete Multivariate Data*, Chapman & Hall, London.

Schall, R. (1999). Estimation in generalized linear models with random effects, *Biometrika* **78**, 719–727.

Shih, W. J. and Quan, H. (1997). Testing for treatment differences with dropouts present in clinical trials–a composite approach, *Statistics in Medicine* **16**, 1225–1239.

Sklar, A. (1959). Fonctions de répartition à n dimensions et leurs marges, *Publications de L'Institut de Statistiques de l'Universite de Paris* **8**, 229–231.

Sommer, A., Katz, J. and Tarwotjo, I. (1984). Increased risk of respiratory infection and diarrhea in children with pre-existing mild vitamin A deficiency, *American Journal of Clinical Nutrition* **40**, 1090–1095.

Song, P. X.-K. (2000a). Multivariate dispersion models generated from Gaussian copula, *Scandinavian Journal of Statistics* **27**, 305–320.

Song, P. X.-K. (2000b). Monte Carlo Kalman filter and smoothing for multivariate discrete state space models, *Canadian Journal of Statistics* **28**, 641–652.

Song, P. X.-K. and Tan, M. (2000). Marginal models for longitudinal continuous proportional data, *Biometrics* **56**, 496–502.

Song, P. X.-K., Fan, Y. and Kalbfleisch, J. D. (2005). Maximization by parts in likelihood inference (with discussion), *Journal of the American Statistical Association* **100**, 1145–1167.

Song, P. X.-K., Qiu, Z. and Tan, M. (2004). Modelling heterogeneous dispersion in marginal simplex models for longitudinal continuous proportional data, *Biometrical Journal* **46**, 540–553.

Song, P. X.-K., Zhang, P. and Qu, A. (2007). Maximum likelihood inference in robust linear mixed-effects models using multivariate t distributions, *Statistica Sinica* **17**, to appear.

Spiegelhalter, D. J., Best, N. G., Carlin, B. P. and van der Linde, A. (2002). Bayesian measures of model complexity and fit (with discussion), *Journal of the Royal Statistical Society, Series B* **64**, 583–639.

Stiratelli, R., Laird, N. M. and Ware, J. H. (1984). Random-effects models for serial observations with binary response, *Biometrics* **40**, 961–971.

Stoffer, D. S., Scher, M. S., Richardson, G. A., Day, N. L. and Coble, P. A. (1988). A Walsh-Fourier analysis of the effects of moderate maternal alcohol consumption on neonatal sleep-state cycling, *Journal of the American Statistical Association* **83**, 954–963.

Sun, D., Tsutakawa, R. K. and He, Z. (2001). Propriety of posteriors with improper priors in hierarchical linear mixed models, *Statistica Sinica* **11**, 77–95.

Sun, J. and Song, P. X.-K. (2001). Statistical analysis of repeated measurements with informative censoring times, *Statistics in Medicine* **20**, 63–73.

Sutradhar, B. C. and Dass, K. (1999). On the efficiency of regression estimators in generalized linear models for longitudinal data, *Biometrika* **86**, 459–465.

Tanner, M. A. and Wong, W. H. (1987). The calculation of posterior distributions by data augmentation, *Journal of the American Statistical Association* **82**, 528–540.

Thall, P. F. and Vail, S. C. (1990). Some covariance models for longitudinal count data with overdispersion, *Biometrics* **46**, 657–671.

van der Vaart, A. W. and Wellner, J. A. (1996). *Weak Convergence and Empirical Processes*, Springer-Verlag, New York.

Verbeke, G. and Lesaffre, E. (1997). The effect of misspecifying the random effects distribution in linear mixed models for longitudinal data, *Journal of Computational Statistics and Data Analysis* **23**, 541–556.

Verbeke, G. and Molenberghs, G. (2000). *Linear Mixed Models for Longitudinal Data*, Springer-Verlag, New York.

Vonesh, E. F. and Carter, R. L. (1987). Efficient inference for random-coefficient growth curve models with unbalanced data, *Biometrics* **43**, 617–628.

Wang, Y.-G. and Carey, V. J. (2003). Working correlation structure misspecification, estimation and covariance design: Implications for the GEE performance, *Biometrika* **90**, 29–41.

Wedderburn, R. W. M. (1974). Quasi-likelihood functions, generalized linear models and the Gauss-Newton method, *Biometrika* **61**, 439–447.

Wei, G. C. G. and Tanner, M. A. (1990). A Monte Carlo implementation of the EM algorithm and the poor man's data augmentation algorithm, *Journal of the American Statistical Association* **85**, 699–704.

Weiss, R. E. and Lazaro, C. G. (1992). Residual plots for repeated measures, *Statistics in Medicine* **11**, 115–124.

West, M. and Harrison, P. J. (1997). *Bayesian Forecasting and Dynamic Models*, 2nd edn, Springer-Verlag, New York.

Wu, H. and Zhang, J.-T. (2006). *Nonparametric Regression Methods for Longitudinal Data Analysis: Mixed-Effects Modeling Approaches*, Wiley, New York.

Wu, M. C. and Carroll, R. J. (1988). Estimation and comparison of changes in the presence of informative right censoring: conditional linear model, *Biometrics* **44**, 175–188.

Xing, B. (2004). *Best Quadrature Formulas, Mixture of Normal Approximation and State Space Models*, PhD thesis, Department of Mathematics and Statistics, York University, Toronto, Ontario.

Yi, G. Y. and Cook, R. J. (2002). Marginal methods for incomplete longitudinal data arising in clusters, *Journal of the American Statistical Association* **97**, 1071–1080.

Zeger, S. L. (1988). A regression model for time series of counts, *Biometrika* **75**, 621–629.

Zeger, S. L. and Karim, R. (1991). Generalized linear models with random effects: a Gibbs sampling approach, *Journal of the American Statistical Association* **86**, 79–86.

Zeger, S. L. and Qaqish, B. (1988). Markov regression models for time series: A quasi-likelihood approach, *Biometrics* **44**, 1019–1031.

Zeger, S. L., Liang, K.-Y. and Albert, P. S. (1988). Models for longitudinal data: a generalized estimating equation approach, *Biometrics* **44**, 1049–1060.

Zhang, D. and Davidian, M. (2001). Linear mixed models with flexible distributions of random effects for longitudinal data, *Biometrics* **57**, 795–802.

Zhang, P. (2006). *Contributions to Mixed-Effects Models for Longitudinal Data*, PhD thesis, Department of Statistics and Actuarial Science, University of Waterloo, Waterloo, Ontario.

Zhao, L. P. and Prentice, R. L. (1990). Correlated binary regression using a generalized quadratic model, *Biometrika* **77**, 642–648.

Ziegler, A., Kastner, C. and Blettner, M. (1998). The generalized estimating equations: An annotated bibliography, *Biometrical Journal* **40**, 115–139.

Zorn, C. J. W. (2001). Generalized estimating equation models for correlated data: A review with applications, *American Journal of Political Science* **45**, 470–490.

Index

Springer Series in Statistics (continued from p. ii)

Küchler/Sørensen: Exponential Families of Stochastic Processes
Kutoyants: Statistical Inference for Ergodic Diffusion Processes
Lahiri: Resampling Methods for Dependent Data
Lavallée: Indirect Sampling
Le Cam: Asymptotic Methods in Statistical Decision Theory
Le Cam/Yang: Asymptotics in Statistics: Some Basic Concepts, 2nd edition
Le/Zidek: Statistical Analysis of Environmental Space-Time Processes
Liu: Monte Carlo Strategies in Scientific Computing
Manski: Partial Identification of Probability Distributions
Mielke/Berry: Permutation Methods: A Distance Function Approach, 2nd edition
Molenberghs/Verbeke: Models for Discrete Longitudinal Data
Mukerjee/Wu: A Modern Theory of Factorial Designs
Nelsen: An Introduction to Copulas, 2nd edition
Pan/Fang: Growth Curve Models and Statistical Diagnostics
Politis/Romano/Wolf: Subsampling
Ramsay/Silverman: Applied Functional Data Analysis: Methods and Case Studies
Ramsay/Silverman: Functional Data Analysis, 2nd edition
Reinsel: Elements of Multivariate Time Series Analysis, 2nd edition
Rosenbaum: Observational Studies, 2nd edition
Rosenblatt: Gaussian and Non-Gaussian Linear Time Series and Random Fields
Särndal/Swensson/Wretman: Model Assisted Survey Sampling
Santner/Williams/Notz: The Design and Analysis of Computer Experiments
Schervish: Theory of Statistics
Shao/Tu: The Jackknife and Bootstrap
Simonoff: Smoothing Methods in Statistics
Song: Correlated Data Analysis: Modeling, Analytics, and Applications
Sprott: Statistical Inference in Science
Stein: Interpolation of Spatial Data: Some Theory for Kriging
Taniguchi/Kakizawa: Asymptotic Theory for Statistical Inference for Time Series
Tanner: Tools for Statistical Inference: Methods for the Exploration of Posterior Distributions and Likelihood Functions, 3rd edition
Tillé: Sampling Algorithms
Tsaitis: Semiparametric Theory and Missing Data
van der Laan/Robins: Unified Methods for Censored Longitudinal Data and Causality
van der Vaart/Wellner: Weak Convergence and Empirical Processes: With Applications to Statistics
Verbeke/Molenberghs: Linear Mixed Models for Longitudinal Data
Weerahandi: Exact Statistical Methods for Data Analysis

Printed in the United States of America.